BORDERLESS EDUCATION AS A CHALLENGE IN THE 5.0 SOCIETY

PROCEEDINGS OF THE 3RD INTERNATIONAL CONFERENCE ON EDUCATIONAL SCIENES (ICES 2019), 7 NOVEMBER 2019, BANDUNG, INDONESIA

Borderless Education as a Challenge in the 5.0 Society

Editors

Ade Gafar Abdullah, Vina Adriany & Cep Ubad Abdullah

Universitas Pendidikan Indonesia, Indonesia

CRC Press is an imprint of the
Taylor & Francis Group, an **informa** business

A BALKEMA BOOK

CRC Press/Balkema is an imprint of the Taylor & Francis Group, an informa business

Typeset by Integra Software Services Pvt. Ltd., Pondicherry, India

Library of Congress Cataloging-in-Publication Data

Applied for

Published by: CRC Press/Balkema
Schipholweg 107C, 2316XC Leiden, The Netherlands
e-mail: Pub.NL@taylorandfrancis.com
www.routledge.com – www.taylorandfrancis.com

ISBN: 978-0-367-61960-2 (Hbk)
ISBN: 978-1-003-10727-9 (eBook)
DOI: 10.1201/9781003107279
https://doi.org/10.1201/9781003107279

Borderless Education as a Challenge in the 5.0 Society – Abdullah, Adriany & Abdullah (eds)
© 2021 Taylor & Francis Group, London, ISBN 978-0-367-61960-2

Table of contents

Borderless Education as a Challenge in the 5.0 Society – Abdullah, Adriany & Abdullah (eds)
© 2021 Taylor & Francis Group, London, ISBN 978-0-367-61960-2

Preface

For the past 10 years, the world has been undergoing transformative changes in almost all aspects of life. Education is also no exception when it comes to the changes in our society. One of the most significant alters that converts human interaction and the way education runs is due to the advancement of technology.

These conversions have resulted in the emergence of two dominant discourses in education. The first discourse is revolution 4.0, and the second discourse is society 5.0. The two discourses are related with one another, both emphasize the vicissitudes resulted from the technology progression.

This conference aims to problematize the two discourses by inviting all the authors to redefine what we mean by education in the current state, and to ask in critical ways to what extent can technology facilitate teaching practices, and what are missing by the extensive use of technology? The conference also becomes a venue that allows us to whether care and education can be completely replaced by technology.

However, the conference does not aim to utterly diminish the advantage the use of technology. We also recognize that the technology does offer new ways of doing education and somehow it also potentially creates more democratic and equal atmosphere. Hence, some of the papers in this conference also focus on the benefits of technology in education as well as offering strategies on how to incorporate it into teaching practices.

Finally, it is hoped that the conference, as reflected in the variety of papers would allow educators, teachers, academics, researchers, as well as practitioners to continue being engaged in the process of redefining education. Education however is situated within specific socio-cultural-historical context and hence, we need to be constantly involved in the process of reconceptualizing it.

Editorial Team

Borderless Education as a Challenge in the 5.0 Society – Abdullah, Adriany & Abdullah (eds)
© 2021 Taylor & Francis Group, London, ISBN 978-0-367-61960-2

Scientific committee

Advisor

1. Prof. Dr. H. R. Asep Kadarohman, M.Si. (Rector of Universitas Pendidikan Indonesia)

Honorary Chairs

1. Dr. H. Edi Suryadi, M.Si. (Vice Rector for Finance, Resources and General Affairs)
2. Prof Dr. H. Didi Sukyadi, M.A. (Vice Rector for Research, Partnership and Business)
3. Dr. Agus Taussfiq, M.Pd. (Dean of Education Sciences Faculty UPI)

Steering Committee

1. Dr. Rudi Susilana, M.Si. (Vice Dean on Academic Affair of Faculty of Educational Sciences)
2. Dr. Aceng Muhtaram Mirfani, M.Pd. (Vice Dean on Human Resources and Finance of Faculty of Educational Sciences)
3. Dr. Sardin, M.Pd. (Vice Dean on Student Affarir of Faculty of Educational Sciences)

Reviewers Team

1. Prof. Dr. Tatat Hartati, M.Pd
2. Vina Adriyani, Ph.D
3. Dr. Henny Djohaeni, M.Si
4. Dr. Hani Yulindrasari, M. Gendst
5. Dr. Yeni Rachmawati, M.Pd.

Organizing committee

Conference Chair
Dr. Riche Cynthia J., M.Si.

Secretary
Dr. Iip Saripah, M.Pd

Finance Chair
Rum Astuti, S.E.

Finance Co-Chair
Kardi Sunarya, S.Pd.

Event/Protocol Section Chair
Dr. Ipah Saripah, M.Pd

Session Chair
1. Dr. Laksmi Dewi, M.Pd
2. Nadia Hanoum, M.Pd

Moderator
Dr. Herlina, M.Psi

Publication and Documentation Chairs
Gema Rullyana, M.Ikom

Secretariat and Website Chairs
1. Angga Hadiapurwa, M.Ikom
2. Ardiansyah, M.IKom
3. Ramdani, S.Pd.

Refreshments and Accommodation Chairs
Tim Tendik Keuangan dan Akademik FIP UPI

Public Relations and Exhibition Chairs
Tim Tendik Keuangan dan Akademik FIP UPI

Acknowledgements

The committee would like to express gratitude to all who have been involved in the conference. Our highest appreciation goes to the Rector of Universitas Pendidikan Indonesia for his constant support. The same thankfulness also goes to the Vice Rector of Research, Partnership and Business for his insight that helps the committee to execute the conference.

The committee would also like to thank to the members, reviewers, as well as publication team who have collaborated together to ensure the production of both the conference and the proceeding. Our special thanks is also dedicated to the team from CRC, Routledge who provide a space for the research from our participants to be published and disseminated further.

Finally, our gratitude also goes to all participants who have made our conference successful.

The Conference Committee

Characteristics of millennial generation and challenges of educational institutions in the industry 4.0 era

Borderless Education as a Challenge in the 5.0 Society – Abdullah, Adriany & Abdullah (eds)
 ISBN 978-0-367-61960-2

The readiness of accounting education students as millennial generation enters the disruption Era

Y. Septiana
Universitas Negeri Yogyakarta, Yogyakarta, Indonesia

ABSTRACT: This survey study aims to determine the level of readiness of Accounting Education students as the millennial generation enters the disruption era. The population of this study is the 289 students of the Accounting Education of Yogyakarta State University from 2016 to 2019, of whom 158 students were taken as research samples using proportionate random sampling. Data collection techniques in this study were questionnaires with data analysis using factor analysis and quantitative descriptive statistical techniques. This factor analysis produces factors that can be used to measure the level of readiness of Accounting Education students as the millennial generation is entering the disruption era, including Disruptive Mindset factors, Connected factors, Creative factors, Confidence factors, Readiness factors, and Spiritual factors. The analysis shows that the level of readiness of Accounting Education students in entering the era of major disruption is in the middle category, which means that students have sufficient abilities and indicates that several aspects require attention to achieve readiness.

1 INTRODUCTION

Technology, especially information and communication technology, is currently developing rapidly. The development of information and communication technology is helping humans to communicate long-distance as easily as short-distance. Everyone can provide news and information to other parts of the world. Today, people are more familiar with the big-data era or the disruption era. The disruption era is a time in which change occurs unexpectedly and fundamentally in almost all aspects of life. The disruption era provides flexibility for the development of increasingly sophisticated information and communication technology.

In 2020, this world is entering the era of the industrial revolution 4.0. Some of the characteristics of the industrial revolution 4.0 are digitalization, the Internet of Things, the Internet of People, big data, iCloud data, and artificial intelligence. Internet of Things (IoT) is the concept of objects that can transfer data through networks via human to human or human to computer devices. Internet of People (IoP) is an open-source and decentralized technology with open social graphics, which consists of interconnected peer-to-peer networks that consist of the profiles, connections, and identity information of people. Furthermore, some of the characteristics of the disruption era are big volume, variety, velocity, veracity, and transitoriness (Subiyantoro & Nugroho 2018).

At this time, humans use information technology facilities. The current generation, also often referred to as the millennial generation, includes young people aged around 17–38 years (Wahyuningsih 2019). They are social media users, for personal, group, economic, existence, and information-seeking activities (Ainiyah 2018). This generation is synonymous with technology, especially the Internet, social media and entertainment (Sabani 2018, Wahyuningsih 2019).

The hallmark of the millennial generation is a high level of interest in new things. Therefore, the millennial generation will easily adapt to these new things. They are often confident in their academic ability, high social spirit, and technological literacy, but they are very interested in personal achievement and recognition. The self-image of the millennial student is predicated on successful competition against others. Millennial education is not about learning but rather about personal achievement and securing credentials for employment (Millenbah et al. 2011).

The millennial generation arose during the internet explosion (Sharon 2015), meaning that this generation is closely related to the lifestyle of internet users and social media. The lifestyle of the millennial generation revolves around social media for existence, attention, opinions, growing image, communication and socialization, a place for achievement, adding insight, and sharing feelings and emotions (Ainiyah 2018). Besides, social media offers so many interesting features (Ainiyah 2018).

Many fields that experienced changes in this disruption era including the field of education (Afrianto 2018). The industrial revolution 4.0 has a positive impact on the world of education by providing opportunities to conduct research and innovative learning practices. Teachers also need to make changes to curriculum content where the curriculum will prepare students with 21st-century skills and choose the right learning model for the millennial generation. Teachers implement the blended learning model, accompanied by social media. Instructional media is used as a learning tool (Nicholas 2008). This disruption era provides an opportunity for teachers to give assignments flexibly, via a website (Hidayat & Lukitaningsih 2019). Therefore learning can be ICT-based, according to the results of research conducted by Rahamat et al. (2017), who explain that students are technologically, economically, and competently prepared for the use of ICT in learning, so it is necessary to develop an understanding of the issues associated with social and educational uses of ICTs in higher education (Bullen et al. 2011).

Teachers must also have a 21st-century learning mindset, namely digital literacy, continuous learning of new things, and be able to use the opportunities provided by the IR 4.0. This will encourage teachers to be productive and improve student learning outcomes (Afrianto 2018). Besides, skills and expertise are important for becoming an individual who can better adapt. Education will have an impact on the many challenges of workers and recruitment (Octavianus 2019).

In the recruitment process, students are expected to have the ability and skill to be qualified in their fields, because the machine has been able to replace worker resources. Therefore, students are expected to be ready to enter the disruption era with expertise in their field of study balanced with IT skills.

Accounting education students focus on studying in the field of accounting education and are expected to become accounting teachers, able to keep abreast of the times. Education in the disruption era requires students to be more active in using technology in learning, teaching using innovative learning models, and distance learning. Therefore, accounting education students are expected to display readiness upon entering the disruption era. The purpose of this analysis is to determine the readiness of accounting education students because students who are the current/millennial generation, and also prospective educators, certainly have an important role in education.

2 RESEARCH METHODS

This research is a survey study. The approach chosen for this research is a quantitative approach to facilitate the analysis of the readiness of accounting education students as the millennial generation entering the disruption era. The population in the study is 289 accounting education students from 2016 to 2019, with a research sample of 158 students (35 students of 2016, 37 of 2017, 51 of 2018, and 35 of 2019). Sampling was done by proportionate random sampling technique. Data collection techniques in this study were questionnaires with data analysis using factor analysis. The questionnaire was used to reveal the readiness of accounting education students as the millennial generation entering the disruption era. This research adapted the questionnaire from research conducted by Hidayah (2018), with answer choices in the form of four Likert scales. The analysis technique used in this study is quantitative descriptive statistical techniques. Research data will be presented in tables to make it easier to read.

Table 1. Categorization of readiness scores.

Score	Category	Criteria
$1 < X < 1.75$	Low	Shows that some aspects really need urgent attention to achieve readiness
$1.75 < X \leq 3.25$	Medium	Demonstrates that students have adequate abilities, but are not mature enough in their readiness and several aspects require attention to achieve readiness
$X \geq 3.25$	High	Demonstrates that students have adequate and mature abilities in their readiness to entering disruption era

Categorizing the readiness level of students uses scores as listed in Table 1, where the category of readiness level refers to research by Hidayah (2018).

3 RESULTS AND DISCUSSION

Factor analysis was used to determine what influences the readiness of accounting education students as the millennial generation enters the disruption era. The analysis shows the KMO and Bartlett's test value is 0.831, meaning more than 0.05, so the data can be continued for the next factor analysis. From the 20 statements analyzed, the MSA number criterion is above 0.5, which means that all statements can be used as further predictions. Then all statement items can be explained by the factors formed with the provision that the greater the communalities, the closer the statement related to the factors formed.

Based on the results of the analysis, Total Variance Explained shows that the 20 statements analyzed can be grouped into six factors. This can be seen from the total initial eigenvalues that indicate a number greater than 1, so there are six factors formed. Determination of the variables that enter each factor is done by comparing the magnitude of the correlation in each line. While the numbers on the results of the Rotated Component Matrix analysis are factor loadings, or the magnitude of the correlation between a variable with a factor 1, 2, 3, 4, 5 or 6. The formation of factors in this study followed the theory that there are six factors determining student readiness in entering the disruption era. Factor 1 consists of items 1–4, factor 2 consists of items 5–7, factor 3 consists of items 8–10, factor 4 consists of 11–13, factor 5 consists of items 14 and 15, and factor 6 consists of factors 16–20. The results of variable interpretation are in Table 2.

3.1 *Factor 1 is disruptive mindset*

The first factor is called Disruptive Mindset. This factor can explain the diversity of variance of 35.885%. The variable that most influences the Disruptive Mindset factor is "sensitive to the surrounding environment." This sensitivity aspect is important because it indicates the advancement of information technology is offset by still having a mindset sensitive to the environment.

3.2 *Factor 2 is connected*

The second factor is called Connected. This factor can explain the diversity of variance of 10.465%. The variable that most influences the Connected factor is the person who is good at socializing. In this case, accounting education students as the millennial generation have good social skills, particularly communication and adaptation skills both orally and in writing; skills in collaboration, cooperation, associations and assimilation; tolerance, respect for the rights of others, sensitivity and social care; self-control; literacy skills in technology, social and human sciences; narrative skills and the ability to argue logically and to share experiences with others. Thus, lecturers can use learning models that can improve students' social skills

Table 2. The result of variable interpretation.

Variable	Factor	Eigenvalues	Loading factor	% Variance	Cumulative %
A1	Disruptive Mindset	7.177	0.844	35.885	35.885
A2			0.736		
A3			0.525		
A4			0.648		
B1	Connected	2.093	0.838	10.465	46.350
B2			0.853		
B3			0.648		
C1	Creative	1.544	0.900	7.718	54.068
C2			0.785		
C3			0.849		
D1	Confidence	1.448	0.877	7.241	61.308
D2			0.878		
D3			0.721		
E1	Readiness	1.214	0.843	6.072	67.381
E2			0.776		
F1	Spiritual	1.058	0.763	5.289	72.670
F2			0.630		
F3			0.765		
F4			0.671		
F5			0.747		

(Hidayat & Lukitaningsih 2019). The ability to socialize is important for students to find new friends, new experiences, and new collaborations.

3.3 *Factor 3 is creative*

The third factor is called Creative. This factor can explain the diversity variance of 7.718%. The variable that most influences the Creative factor is the person who is used to thinking outside the box, rich in ideas and new ideas. Thinking outside the box means someone will produce something novel, which changes from before. Someone who thinks outside the box tends not to doubt the existence of change and innovation. Accounting education students can think creatively, a factor that influences the readiness of the millennial generation to entering the disruption era.

3.4 *Factor 4 is confidence*

The fourth factor is called Confidence. This factor can explain the variance of 7.241%. The variable that most influences the Confidence factor is the courage to express an opinion and not hesitate to debate in public, part of a person's ability to communicate well. Communication skills are an important aspect for humans because by communicating we can build relationships and cooperation. Communication skills are essential for accounting education students to gain employment and experience. As prospective accounting teachers, students are required to be able to communicate and deliver material to students smoothly. Therefore, this confidence factor is very important for accounting education students to entering the disruption era.

3.5 *Factor 5 is readiness*

The fifth factor is called Readiness. This factor can explain the diversity of variance of 6.072%. The variable that most influences the Readiness factor is being ready to enter the disruption era with currently relevant capabilities. The results of this study are in line with the results of research conducted by Rahamat et al. (2017), showing that students have a readiness and a positive perception of the use of mobile technology when learning, indicating students of the millennial generation are ready to enter this era of digital disruption supported by qualified knowledge and skills. Because of this, this factor is important for the millennial generation to entering the disruption era.

3.6 *Factor 6 is spiritual*

The sixth factor is called Spiritual. This factor can explain the diversity of variance of 5.289%. The variable that most influences the spiritual factor is the F3 variable, which is

Table 3. Average score of accounting education student readiness levels based on variables.

Var.	Statement	Mean	Std. Dev.	Category	< 2 (%)	(%)	> 3 (%)
A1	I am a person who is sensitive to the surrounding environment.	3.20	0.56	Medium	0.00	72.78	27.22
A2	I am a person who does things on time.	3.11	0.66	Medium	1.27	72.15	26.58
A3	I am a person who immediately follows up on a matter, not postponing it.	2.91	0.63	Medium	0.63	84.81	14.56
A4	I am the one who always gives a way out or a solution.	3.01	0.59	Medium	0.00	82.28	17.72
B1	I am a person who is good at socializing.	2.98	0.73	Medium	1.27	74.68	24.05
B2	I am a person who actively surfs social media and the Internet.	3.04	0.78	Medium	2.53	67.09	30.38
B3	I always use internet assistance to complete my assignments.	3.33	0.68	High	1.27	55.06	43.67
C1	I am a person who used to think outside the box, rich in ideas and ideas.	2.64	0.66	Medium	0.63	89.87	9.49
C2	I am a person who is able to communicate my ideas brilliantly.	2.63	0.68	Medium	1.27	88.61	10.13
C3	I am a creative person and full of innovation	2.65	0.66	Medium	0.63	89.87	9.49
D1	I am a confident person.	2.90	0.77	Medium	2.53	75.32	22.15
D2	I dare to express my opinion and not hesitate to debate in public.	2.75	0.84	Medium	5.06	74.68	20.25
D3	I am a person who finds it easy to interact with new and communicative people.	2.82	0.78	Medium	3.80	77.22	18.99
E1	We are facing a new era, the disruption era. Disruption has the potential to replace old physical technology with digital technology that produces something completely new and more efficient. As a millennial generation, I am ready to enter the disruption era with the capabilities that I have today.	3.09	0.56	Medium	0.00	79.11	20.89
E2	I am ready to compete with my opponents so that they can survive in the era of disruption.	3.15	0.54	Medium	0.63	76.58	22.78
F1	I always read my religious books and try to understand them.	3.11	0.68	Medium	1.27	70.89	27.85
F2	I am a person who likes to share (generous).	3.13	0.60	Medium	1.27	74.68	24.05
F3	I am a person who always prays before and after doing something.	3.28	0.60	High	0.00	64.56	35.44
F4	I am a person who is always grateful for any circumstance.	3.28	0.60	High	0.00	63.92	36.08
F5	I am a devout person in carrying out worship.	3.15	0.57	Medium	0.00	75.32	24.68
	Total average	3.01	0.66	Medium			

Table 4. Average score of student readiness levels based on factors.

Variable	Factor	Mean	Mean Cumulative	Category
A1	Disruptive Mindset	3.20	3.06	Medium
A2		3.11		
A3		2.91		
A4		3.01		
B1	Connected	2.98	3.12	Medium
B2		3.04		
B3		3.33		
C1	Creative	2.64	2.64	Medium
C2		2.63		
C3		2.65		
D1	Confidence	2.90	2.82	Medium
D2		2.75		
D3		2.82		
E1	Readiness	3.09	3.12	Medium
E2		3.15		
F1	Spiritual	3.11	3.19	Medium
F2		3.13		
F3		3.28		
F4		3.28		
F5		3.15		

always praying before and after doing something. The activity of always praying before and after doing something is an aspect of the spirituality of humans. This spirituality is the relationship between humans and God. Humans go through life to achieve a life goal based on the beliefs they hold. When entering this era of disruption it is important to prioritize aspects of spirituality in life because, with advances in increasingly sophisticated information technology, humans still have a strong spiritual grip so that these advances are balanced with spiritual aspects and good attitudes. Accounting education students have good spiritual aspects because students are agents of change, so they need to have a good spiritual foundation. Then, accounting education students who become accounting teachers in the future can teach the material well and can, in turn, teach their students about politeness and spirituality.

Table 3 shows that overall, the readiness score of accounting education students as the millennial generation enters the disruption era is in the medium category, which is 3.01. This means that, generally, accounting education students have adequate abilities but are not mature enough in their readiness and some aspects require attention to achieve readiness, as shown in Table 4.

Table 4 shows that all factors – namely Disruptive Mindset Factors, Connected Factors, Creative Factors, Confident Factors, Readiness Factors, and Spiritual Factors – fall into the middle (medium) category. This means that Accounting Education Students in entering this era of disruption have adequate abilities, but are not yet mature enough in their readiness. Thus these aspects need to be improved.

4 CONCLUSION

The readiness of Accounting Education students as the millennial generation enters the disruption era consists of six factors: Disruptive Mindset factors, Connected factors, Creative factors, Confidence factors, Readiness factors, and Spiritual factors. The analysis shows that at this time, the Accounting Education students' level of readiness is in the middle category, which means that students have sufficient abilities, but are not mature enough in their readiness and that several aspects require attention to achieve readiness. Meanwhile, three items are at a high level of category readiness.

REFERENCES

Afrianto. 2018. Being a professional teacher in the era of industrial revolution 4.0: Opportunities, challenges and strategies for innovative classroom practices Afrianto Faculty of Teachers Training and Education (FKIP), Universita. *English Language Teaching and Research* 2(1): 1–13.

Ainiyah, N. 2018. Remaja Millenial dan Media Sosial: Media Sosial Sebagai Media Informasi Pendidikan Bagi Remaja Millenial. *Jurnal Pendidikan Islam Indonesia* 2(2): 221–236.

Bullen, M., Morgan, T., & Qayyum, A. 2011. Digital learners in higher education: Generation is Not the issue. Apprenants numériques en enseignement supérieur: la génération n'est pas en cause. *Canadian Journal of Learning and Technology* 37(1): 1–24.

Hidayah, N. 2018. Analisis Kesiapan Mahasiswa Sebagai Millennials Generation Dalam Memasuki Era Disruption (Studi Kasus: Mahasiswa UGM, UNY, UIN SUKA, UII, UMY, dan UAD). 10(2): 1–15.

Hidayat, H., & Lukitaningsih, L. 2019. Strengthening social skills of social studies teacher in the era of disruption. In *1st International Conference on Education Social Sciences and Humanities (ICESSHum 2019)*. Atlantis Press: 419–424.

Millenbah, K. F., Wolter, B. H., & Taylor, W. W. 2011. Education in the era of the millennials and implications for future fisheries professionals and conservation. *Fisheries* 36(6): 300–304.

Nicholas, A. J. 2008. Preferred learning methods of the Millennial Generation. *The International Journal of Learning: Annual Review* 15(6): 27–34.

Octavianus, S. 2019. The cultivation of Indonesia's education financing policy in disruption era. *International Journal of Advances in Social and Economics* 1(1): 16.

Rahamat, R. B., Shah, P. M., Din, R. B., & Abd Aziz, J. B. 2017. Students' readiness and perceptions towards using mobile technologies for learning the English language literature component. *The English Teacher* 16: 69–84.

Sabani, N. 2018. Generasi Millenial Dan Absurditas Debat Kusir Virtual. *Informasi* 48(1): 95.

Sharon, A. 2015. Understanding the Millennial Generation. *Journal of Financial Service Professionals* 69(6): 11–14.

Subiyantoro, S., & Nugroho, A. 2018. Android-based instructional media development procedure to enhance teaching and learning in the age of disruption 4.0. In *International Conference on Applied Science and Engineering (ICASE 2018)*. Atlantis Press.

Wahyuningsih, E. 2019. Model of entrepreneurship spirit in Millennial Generation. *Russian Journal of Agricultural and Socio-Economic Sciences* 85(1): 3–14.

Borderless Education as a Challenge in the 5.0 Society – Abdullah, Adriany & Abdullah (eds)
 ISBN 978-0-367-61960-2

Developing digital flipped classroom learning design

R. Mariyana, B. Zaman & R. Rudiyanto
Universitas Pendidikan Indonesia, Bandung, Indonesia

ABSTRACT: This study aims to develop and improve the quality of learning by developing flipped classroom learning design that is easy to understand, easy to use by following the flow and digital-based media so that learners can quickly understand the learning material and helping to understand the learning. To describe the result of research from the improvement of learning design of flipped classroom based on digital media, this research utilised a descriptive method, with a purposive sampling technique that studied active students in Early Childhood Teacher Education Program (henceforth-ECTE Program). Data from students' final examination scores, questionnaires, interviews were analyzed using quantitative and qualitative approach. The final products were a flipped classroom software along with its digital-based classroom design. As the participants perceived the software as 'quick to understand' and 'easy to use', it is hoped that the software can be widely used not only in ECTE programs but also in other.

1 INTRODUCTION

Digital technologies have become an integral part of the teaching and learning processes. This process enables to incorporate Information and Communication Technology (ICT) in a variety of teaching and learning methods in higher education (Johnson et al. 2014). Flipped Classroom is a reversal of traditional learning procedures, where what is usually done in class in traditional learning is carried out at home in a flipped classroom, and which is usually carried out at home as homework or lecture assignments in traditional learning become carried out in class in flipped classroom. Utilization of information and communication technology has helped to improve the quality of education, and make the teaching-learning process more effective, efficient and enjoyable, but there are still many uses of technology in education that have not fully answered the learning needs of this 4.0 era. Learning is still traditional, not learning centered on students, where the objectives to be achieved are not just learning outcomes, but the learning process experienced by students, so that the learning process is meaningful and student learning outcomes can be achieved optimally.

This study aims to develop digital-based flipped classroom learning designs. The resulting product is a digital-based flipped classroom learning design that can be used in learning.

1.1 *Learning flipped classroom*

In 1993, Alison King published "From Sage on the Stage to Guides on the Side", which focused on the importance of using class time for the construction of meaning rather than mere information transmission. Although it does not directly illustrate the concept of "flipping" a classroom, King's (1993) work is often referred to as an impetus for inversion to allow educational space for active learning.

Lage et al. (2000) published a paper titled "Reverse Classes: A Gateway to Creating an Inclusive Learning Environment", which discusses their research on inverted classrooms at the college level. In their research focusing on two college economics courses, Lage et al. (2000) emphasized that one can utilize class time available from class inversion (transferring information presentations through lectures outside the classroom to media such as computers or VCRs) to meet the needs students with various learning styles. University of Wisconsin-Madison distributes software to replace lectures in large lecture-based computer science courses with coordinated lecturer video streaming and slides. In the late 1990s, J. Wesley Baker experimented with the same ideas at Cedarville University. He presented a paper discussing what he called "class flip" at an educational conference in 2000 in what might be the first published mention of the word "flip" related to the learning model.

Flipped Classroom is a learning model that "flips" traditional methods, where material is usually given in class and students work on assignments at home. The concept of Flipped Classroom includes active learning, student involvement, and podcasting. In a flipped classroom, material is given first that students must learn in their homes. Instead, class sessions are used for group discussions and assignments. Here, the teacher acts as a coach or adviser. Flipped classroom is a learning model and type of blended learning that reverses the traditional learning environment by providing learning content, often online, outside the classroom. Transferring activities, including those that might traditionally be considered homework, into the classroom. In reverse classes, students watch online lectures, collaborate on online discussions, or conduct research at home while engaging in concepts in class with the guidance of a mentor.

The application of the flipped classroom model has many advantages over the traditional learning model. The availability of material in the form of video gives students the freedom to stop or repeat the material at any time in parts they do not understand. In addition, the use of classroom learning sessions for projects or group assignments makes it easy for students to interact and learn from each other.

2 RESEARCH METHODS

The research method uses quasi-experiments. The study was conducted on two groups of students, the experimental group used a digital based learning collaboration application and flipped classroom, while the control group used a digital-based classical system. The design used was a posttest pretest design using a control group without random assignments. Research subjects were taken in the form of class groups without random assignments. Probability sampling technique or random sampling is a sampling technique that is done by providing opportunities or opportunities for all members of the population to be sampled. Thus the sample obtained is expected to be a representative sample.

3 RESULTS AND DISCUSSION

3.1 *Scores that use classroom digital learning strategies*

Based on data collected from respondents of 20 students, scores were obtained) students who used the flipped digital classroom learning strategy obtained the highest score of 82; lowest score of 65; average score of 69.33, median 72.5, and mode 66; and standard deviation 4.81.

3.2 *Scores using the classroom flipped digital learning strategy*

Based on data collected from respondents of 20 students, scores were obtained) students who used the flipped digital classroom learning strategy obtained the highest score of 82; lowest score of 65; average score of 69.33, median 72.5, and mode 66; and standard deviation 4.81.

Then the two-average test is performed. Two-average similarity test is used to test the similarity between the two averages, in this case between the digital classroom score and flipped digital classroom.

$$H_o : \mu_e = \mu_k$$

$$H_i : \mu_e \neq \mu_k$$

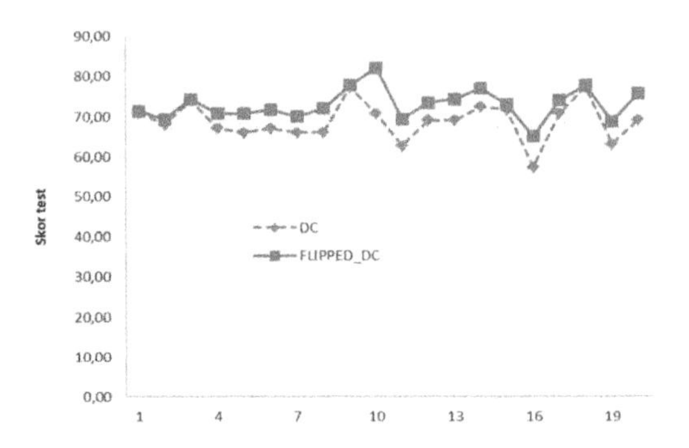

Figure 1. Graph of test score values.

For posttest data results, From the calculation results obtained t-test = 2.960 and t-table = t (1-α/2) (20) = t (1-0.95) (20) = 2.093. Because t-count = 2.960 is at an interval of -2.093 <2.960 and 2.960> 2.093, H_o is rejected, so the experimental class value data (flipped digital classroom) is different from digital classroom. The average value of flipped digital classroom is greater than the average value of digital classroom 72.88> 68.75.

For more details can be seen in the following graph (see Figure 1).

Instructors implementing a flipped classroom use various methods for preparing the online content. Strayer (2007) made useful observations and suggestions for instructors who consider using the flipped classroom model. Active learning pedagogies continue to evolve, and new methods of delivering course material are being developed. Assimilating active learning can be as simple as integrating in-class activities alongside traditional lectures. Yet educators in elementary through post-secondary education are finding innovative ways to restructure the classroom (Strayer 2007) in order to focus attention on the learner (Bergmann & Sams 2012). Instructors adopting the flipped classroom model assign the class lecture or instructional content as homework. In preparation for class, students are required to view the lecture. According to Tucker (2012), students utilize the time in class to work through problems, advance concepts and engage in collaborative learning. With internet access widely available on most college and university campuses, students may view web-based instruction on their own time, at their own pace. This provides opportunities to utilize the classroom for the application of information addressed in the online lecture. Because students have viewed the lecture prior to class, contact hours can be devoted to problem-solving, skill development, and gaining a deeper understanding of the subject matter (Bergmann & Sams 2012). The teacher is able to provide students with a wide range of learner-centered opportunities in class for greater teacher-to-student. mentoring and peer-to-peer collaboration, increasing the possibility to engage Millennial students (Prensky 2010).

4 CONCLUSION

In traditional flipped classroom (FC), learning of new content mostly occurs through watching videos and transferring information from instructor to students utilizing technological tools, (Blau & Shamir-Inbal 2017). The research analyzed learning experiences and their interpretations by the students. In contrast to traditional FC model, the findings revealed active learning of students in both in- and out-of-class settings that took place before, during, and after the lesson. The instructor promoted extensive independent learning, learning regulation, continuous dialogue and collaborative interactions among peers. The re-designed model highlights co-creation of the course content and of digital learning outcomes by students, self-regulation and teamwork coregulation, which are rare in higher education.

REFERENCES

Bergmann, J., & Sams, A. 2012. *Flip your classroom: reach every student in every class every day*. Washington, DC: International Society for Technology in Education.

Blau, I. & Shamir-Inbal, T. 2017. Re-designed flipped learning model in an academic course: The role of co-creation and co-regulation. *Computers & Education* 115: 69–81.

Johnson, G. M. & Davies, S. M. 2014. Self-regulated learning in digital environments: Theory, research, praxis. *British Journal of Research* 1(2): 1–14.

King, A. 1993. From sage on the stage to guide on the side. *College Teaching* 41(1): 30–35.

Lage, M. J., Platt, G.J., & Treglia, M. 2000. Inverting the classroom: A gateway to creating an inclusive learning environment. *The Journal of Economic Education* 31: 30–43.

Prensky, M. 2010. *Partnering: Teaching digital natives Partnering for real learning*: 9-29. Thousand Oaks, CA: Corwin Press.

Strayer, J. 2007. *The effects of the classroom flip on the learning environment: A comparison of learning activity in a traditional classroom and a flip classroom that used an intelligent tutoring system* (Doctoral dissertation, The Ohio State University).

Tucker, B.2012. The Flipped Classroom. *Education Next* 12(1).

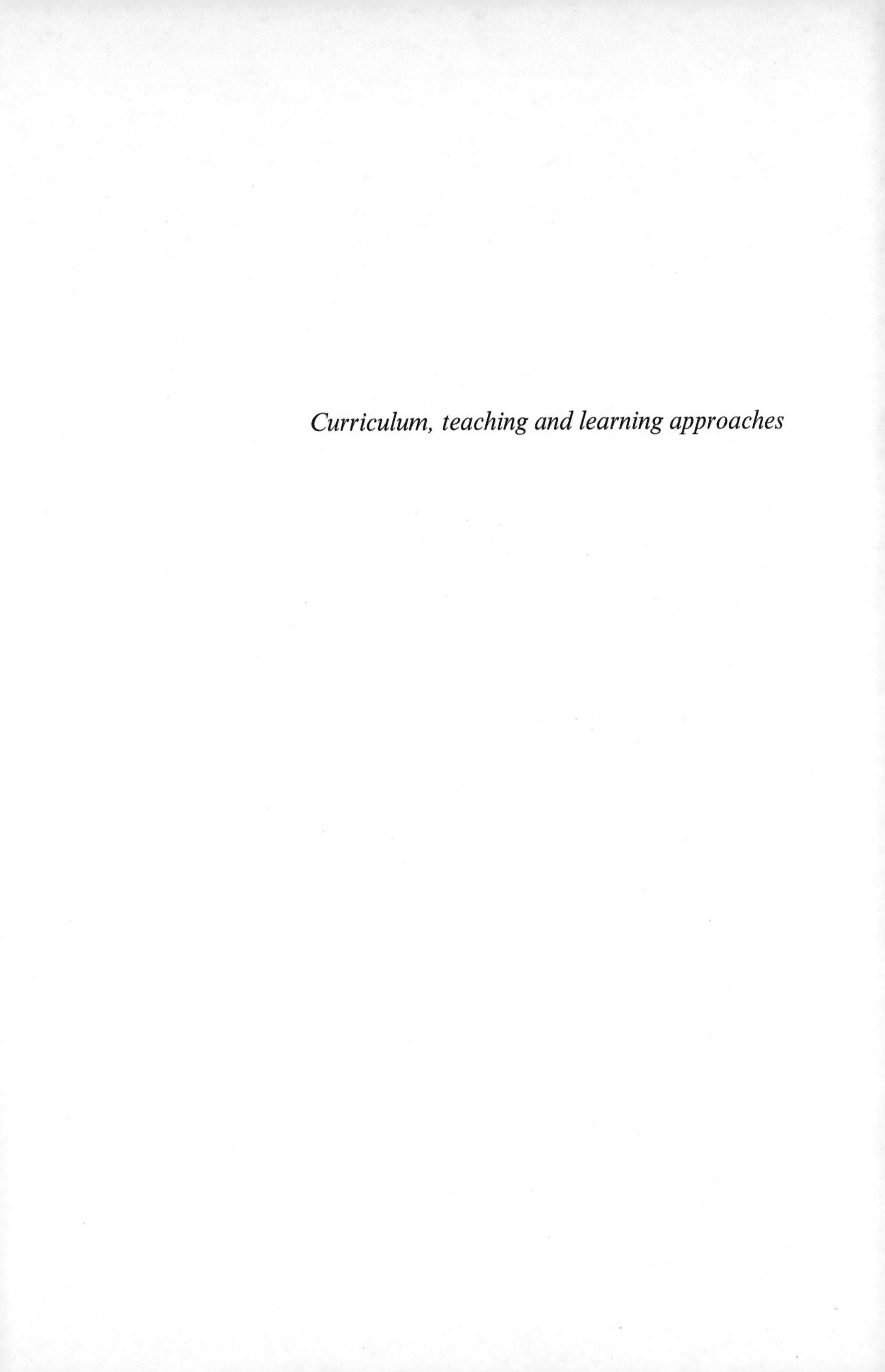

Curriculum, teaching and learning approaches

Borderless Education as a Challenge in the 5.0 Society – Abdullah, Adriany & Abdullah (eds)
 ISBN 978-0-367-61960-2

The evaluation of the application of scientific approach in the 2013 curriculum implementation

I. Fauzziyah & T. Ruhimat
Universitas Pendidikan Indonesia, Bandung, Indonesia

ABSTRACT: This study aims to measure the compatibility of the application of scientific approach in the 2013 Curriculum implementation at SMPN 1 Ciawi Bogor which is evaluated from the aspects of planning, implementation, and the learning process evaluation. The method used is a descriptive evaluation method with the Countenance Stake evaluation model. The instrument used is a questionnaire. Data analysis uses quantitative descriptive statistics in the form of a percentage. The results shows; (1) the application of scientific approach of the 2013 Curriculum implementation at SMPN 1 Ciawi Bogor in the aspect of planning is in the excellent category (82%), (2) the application of scientific approach of the 2013 Curriculum implementation at SMPN 1 Ciawi Bogor in the aspect of implementation is in the excellent category (88%), and (3) the application of scientific approach of the 2013 Curriculum implementation at SMPN 1 Ciawi Bogor in the aspect of assessment is in excellent category (82%).

1 INTRODUCTION

A curriculum is a set of plans and settings regarding objectives, content, and learning materials as well as the methods used to guide the process of learning in order to achieve certain goals (Undang-Undang Nomor 20 2003). Referring to this definition, it can be interpreted that the curriculum functions as a guideline in the implementation of education at all levels and types. In Indonesia, the curriculum has undergone 11 changes, starting from the pre-independence era with very simple form of curriculum to the post-independence era which is continuously refined, namely in 1947, 1952, 1964, 1968, 1975, 1984, 1994, 2004, 2006, and 2013 (Machali 2014). These changes are based on existing developments, both in terms of increasingly sophisticated technology, the development of students, and the demands of the standards that are needed to be achieved (Kurniaman & Noviana 2017).

The 2013 curriculum is a development of the previous curriculum, namely the 2006 curriculum or better known as the Education Unit Level Curriculum (*Kurikulum Tingkat Satuan Pendidikan* or KTSP). In general, there are several differences between the 2006 Curriculum and the 2013 Curriculum starting from the elementary, secondary, up to the high and vocational levels. *Firstly,* the learning process supports three domains of competence, namely attitudes, knowledge and skills competencies. *Secondly,* the usage of a scientific approach in the learning process on all subjects. *Thirdly,* competency-based assessment, namely the shift of assessment through tests (measuring the knowledge competencies based on results only) towards authentic assessments (measuring the competencies of attitudes, knowledge, and skills based on process and results). *Fourthly,* at the elementary level, learning activities use integrative thematic for all classes (Musfiqon & Nurdyansyah 2015).

The implementation of the 2013 Curriculum began in the 2013/2014 school year at the level of elementary education (elementary school/equivalent) to secondary education (junior/senior/vocational high school/equivalent). One distinctive feature of the 2013 curriculum is the application of scientific approach into the learning process, as explained in the Regulation of Ministry of Education and Culture No. 103 of 2014 concerning Learning Process in Elementary and Secondary Education that learning in the 2013 Curriculum uses a scientific approach or a scientific process-based approach (Iryani & Radiusni 2017, Kemdikbud 2014).

A scientific approach or scientific process-based approach is an organization of learning experiences in a logical sequence including learning to observe, ask questions, gather information or attempt, reason or associate and communicate (the Regulation of Ministry of Education and Culture No. 103 of 2014 concerning Learning in Primary and Secondary Education). Referring to the explanation, the scientific approach covers five core activities, namely observing, asking questions, gathering information, associating, and communicating (better known as 5M: *mengamati, menanya, mengumpulkan informasi, mengasosiasi, dan mengomunikasikan*). The five core activities can be applied sequentially or non-sequentially and not all of the activities must be applied in each meeting, but it can be adjusted to the teaching materials and the time allocation.

The 2013 Curriculum is implemented at schools that are appointed by the government or schools that are ready to implement the new curriculum, starting from elementary to secondary education. (Iryani & Radiusni 2017). At the beginning of its implementation, the 2013 Curriculum raised a lot of criticism and protest because it was considered to cause more problems. According to Ahmad Syawran (Iryani & Radiusni 2017), the problems that occur in the 2013 Curriculum implementation are including the contents and packaging of the curriculum, teacher readiness, and the emergence of multiple interpretations in its implementation. The problems that occur are considered natural, but the best solution must be immediately sought in order to achieve the goals expected from the curriculum itself (Iryani & Radiusni 2017).

One of the schools that has applied the scientific approach into the learning process since the 2013/2014 school year is SMPN 1 Ciawi Bogor. At the beginning of its implementation, the teacher of SMPN 1 Ciawi experienced some difficulties at each stage of the implementation. At the planning stage, teachers had difficulties in preparing the assessment tools. At the implementation stage, the teachers were not used to student-centered learning. While in the assessment stage, teachers were still confused by the authentic assessment.

Based on the existing problems, it can be said that the application of scientific approach in the 2013 Curriculum implementation at SMPN 1 Ciawi Bogor was still not yet optimal. Therefore, it is needed to conduct an evaluation in order to see the compatibility of the application of scientific approach in the 2013 Curriculum implementation at SMPN 1 Ciawi Bogor.

This research discusses; (1) how is the compatibility of the application of scientific approach in the 2013 Curriculum implementation at SMPN 1 Ciawi Bogor on the aspect of planning? (2) how is the compatibility of the application of scientific approach in the 2013 Curriculum implementation at SMPN 1 Ciawi Bogor on the aspect of implementation? (3) how is the compatibility of the application of scientific approach in the 2013 Curriculum implementation at SMPN 1 Ciawi Bogor on the aspect of evaluation?

2 RESEARCH METHODS

This research was conducted at SMPN 1 Ciawi Bogor with the sample of 20 randomly selected teachers. The method used is a descriptive method with evaluation study type using the Countenance Stake evaluation model. The Countenance Stake Model divides evaluation activities into two aspects, namely description and judgment on the planning or context stage (antecedents), process (transaction), and outcomes (Iryani & Radiusni).

The results of the study were compared with the standards referred to the Regulation of Ministry of Education and Culture No. 81A of 2013 concerning the curriculum implementation, the Regulation of Ministry of Education and Culture No. 65 of 2013 concerning the Process Standards of Elementary and Secondary Education, and the Regulation of Ministry of Education and Culture No. 22 of 2016 concerning the Process Standards of the Elementary and Secondary Education.

The instrument used is a questionnaire. The scale used in the questionnaire is the Likert scale. The statement given is positive statements divided into 5 ranges, with the following

Table 1. Criteria interpretation (Arikunto 2014).

Percentage	Criteria
81 – 100%	Excellent
61 – 80%	Good
41 – 60%	Sufficient
21 – 40%	Deficient
< 21%	Very Deficient

information and weights: SL = Always (5); SR = Frequent (4); KD = Sometimes (3); JR = Rarely (2); and TP = Never (1). Data analysis uses quantitative descriptive statistics in the form of a percentage with the following formula.

$$\text{Percentage} = \frac{\text{mean score}}{\text{ideal score}} \text{ x } 100\% \tag{1}$$

After calculating the percentage value (%), the following criteria interpretation will be done (Table 1).

3 RESULTS AND DISCUSSION

3.1 *Planning for learning*

The evaluation result of the application of scientific approach in the 2013 Curriculum Implementation on the aspect of planning is showed in the following results (Table 2).

Based on the data in the table, it can be seen that the syllabus study is 90% in accordance with the Regulation of Ministry of Education and Culture No. 81A of 2013 concerning the Curriculum Implementation and Regulation of the Ministry of Education and Culture No. 22 of 2016 concerning the Standards Process of Elementary and Secondary Education, the design of Lesson Plan (RPP) is 76% in accordance with The Regulation of Ministry of Education and Culture No. 81A of 2013 concerning the Curriculum Implementation, and Regulation of the Ministry of Education and Culture No. 22 of 2016 concerning the Standards Process of Elementary and Secondary Education, media and teaching material selection is 83% in accordance with the Regulation of

Table 2. The evaluation result of the application of scientific approach in the 2013 curriculum implementation on the aspect of planning.

		Description Matrix			Judgement Matrix
Step	Aspect	Intensity	Observation (%)	Standard (%)	Judgement
Planning (antecedents)	Syllabus study	In accordance with the Regulation of Ministry of Education and Culture No. 81A of 2013 and the Regulation of Ministry of Education and Culture No. 22 of 2016	90	100	Available
	Lesson Plan (RPP) Design		76		
	Selection of media and teaching materials		83		
	Making evaluation instruments		81		

Ministry of Education and Culture No. 81A of 2013 concerning the Curriculum Implementation and the Regulation of Ministry of Education and Culture No. 22 of 2016 concerning the Standards Process of Elementary and Secondary Education, and last but not least, the evaluation instruments is 81% in accordance with the Regulation of Ministry of Education and Culture No. 81A of 2013 concerning the Curriculum Implementation and the Regulation of Ministry of Education and Culture No. 22 of 2016 concerning the Standards Process of Elementary and Secondary Education (Kemdikbud 2013a,b, & Kemdikbud 2016).

Thus, the overall calculation is conducted and it is obtained that the evaluation results on the aspects of planning for learning with scientific approach is 82% which means that it is in the excellent category.

3.2 *Learning implementation*

Following is the evaluation results of the application of scientific approach in the 2013 Curriculum implementation on the aspect of implementation (Table 3).

Based on the table it can be concluded that the learning process is 89% in accordance with the Regulation of Ministry of Education and Culture No. 81A of 2013 concerning the Curriculum Implementation and the scientific learning is 87% is accordance with the Regulation of Ministry of Education and Culture No. 65 of 2013 concerning the Standards Process of Elementary and Secondary Education and the Regulation of Ministry of Education and Culture No. 22 of 2016 concerning the Standards Process of Elementary and Secondary Education.

Thus, the overall calculation is conducted and it is obtained that the evaluation results on the aspects of learning implementation with scientific approach is 88% which means it is within the excellent category.

3.3 *Learning evaluation*

The evaluation result of the application of scientific approach in the 2013 Curriculum implementation on the aspect of evaluation shows the following results (Table 4).

Based on the table, it can be seen that the assessment on the attitude aspect is 80% in accordance with the Regulation of Ministry of Education and Culture No. 81A of 2013 concerning the Curriculum Implementation, and the assessment of the knowledge aspect is 87% in accordance with the Regulation of Ministry of Education and Culture No. 81A of 2013 concerning the Curriculum Implementation, and assessment on the skill aspect is 81% in accordance with The Regulation of Ministry of Education and Culture No. 81A of 2013 concerning Curriculum Implementation.

Thus, the overall calculation is conducted and it is obtained that the evaluation results on the aspects of learning implementation with scientific approach is 82% which means it is within the excellent category.

Table 3. The evaluation results of the application of scientific approach in the 2013 curriculum implementation on the aspect of implementation.

		Description Matrix			Judgement Matrix
Step	Aspect	Intens	Observation (%)	Standard (%)	Judgement
Process (transaction)	Learning process	In accordance with the Regulation of Ministry of Education and Culture No. 81A of 2013 and the	89	100	Available
	Scientific approach	Regulation of Ministry of Education and Culture No. 22 of 2016	87		

Table 4. The evaluation result of the application of scientific approach in the 2013 curriculum implementation on the aspect of evaluation.

		Description Matrix			Judgement Matrix
Step	Aspect	Intens	Observation (%)	Standard (%)	Judgement
Evaluation (outcomes)	Attitude assessment	In accordance with the Regulation of Ministry of Education and Culture No. 81A of 2013	80	100	Available
	Knowledge assessment		87		
	Skill assessment		81		

4 CONCLUSION

The results of this study conclude several things, namely; (1) lesson planning with scientific approach in the 2013 Curriculum implementation at SMPN 1 Ciawi Bogor is in the excellent category (82%) and in accordance with predetermined criteria, namely the Regulation of Ministry of Education and Culture No. 81A of 2013 concerning the Curriculum Implementation and the Regulation of Ministry of Education and Culture No. 22 of 2016 concerning the Standards Process of Elementary and Secondary Education, (2) the implementation of learning with scientific approach in the 2013 Curriculum implementation at SMPN 1 Ciawi is in the excellent category (80%) and in accordance with predetermined criteria, namely the Regulation of Ministry of Education and Culture No. 81A of 2013 concerning the Curriculum Implementation, the Regulation of Ministry of Education and Culture No. 103 of 2014 concerning Learning Process in Elementary and Secondary Education, and the Regulation of Ministry of Education and Culture No. 65 of 2013 concerning the Standards Process of Elementary and Secondary Education and the Regulation of Ministry of Education and Culture No. 22 of 2016 concerning the Standards Process of Elementary and Secondary Education, and (3) learning assessment using scientific approach in the 2013 Curriculum implementation at SMPN 1 Ciawi Bogor is in the excellent category (82%) and in accordance with predetermined criteria, namely the Regulation of Ministry of Education and Culture No. 81A of 2013 concerning Curriculum Implementation.

5 RECOMMENDATION

Based on the conclusions that have been presented previously, there are a number of recommendations related to the results of the evaluation study about the compatibility of the implementation of scientific learning in the 2013 Curriculum at SMP Negeri 1 Ciawi, Bogor Regency, as follows.

- School is desired to continue facilitating teachers in the Teacher Working Group (KKG) activities, Subject Teachers' Discussion (MGMP), as well as trainings in order to improve their understanding and ability in developing lesson plans and learning assessment techniques as well as able to conduct learning activities using the scientific approach.
- Learning implementation with scientific approach needs to be supervised by the school leader, in this case, the principal. In addition, the principal also needs to provide direction to the teachers at each stages of scientific learning implementation so that it continues to run in accordance with the applicable policies.

- Teacher as a subject who has an important role in the learning process is expected to be able to understand and carry out the entire process within scientific learning implementation, starting from lesson planning, learning implementation, and the learning assessment.

REFERENCES

Arikunto, 2014. *Evaluasi Program Pendidikan: Pedoman Praktis bagi Mahasiswa dan Praktisi Pendidikan.* Jakarta: Bumi Aksara.

Iryani, I., & Radiusni, R. 2017. Analisis Penerapan Pendekatan Saintifik pada Mata Pelajaran Ekonomi di SMA Nurul Falah Tahun Ajaran 2016/2017. *PEKA* 5(1), 75–84.

Kemendikbud. 2013a. *Peraturan Menteri Pendidikan dan Kebudayaan No. 81A Tahun 2013 tentang Implementasi Kurikulum.* Jakarta: Kemdikbud.

Kemdikbud. 2013b. *Peraturan Menteri Pendidikan dan Kebudayaan No. 65 Tahun 2013 tentang Standar Proses Pendidikan Dasar dan Menengah.* Jakarta: Kemdikbud.

Kemdikbud. 2014. *Peraturan Menteri Pendidikan dan Kebudayaan No. 103 Tahun 2014 tentang Pembelajaran pada Pendidikan Dasar dan Menengah.* Jakarta: Kemdikbud.

Kemdikbud. 2016. *Peraturan Menteri Pendidikan dan Kebudayaan No. 22 Tahun 2016 tentang Standar Proses Pendidikan Dasar dan Menengah.* Jakarta: Kemdikbud.

Kurniaman & Noviana. 2017. Penerapan Kurikulum 2013 dalam Meningkatkan Keterampilan, Sikap, dan Pengetahuan. *Jurnal Primary Program Studi Pendidikan Guru Sekolah Dasar Fakultas Keguruan dan Ilmu Pendidikan Universitas Riau* 6(2): 389–396.

Machali, I. 2014. Kebijakan Perubahan Kurikulum 2013 dalam Menyongsong Indonesia Emas Tahun 2045. *Jurnal Pendidikan Islam* 3(1): 71–94.

Musfiqon & Nurdyansyah. 2015. *Pendekatan Pembelajaran Saintifik.* Sidoarjo: Nizamia Learning Center.

Undang-Undang Nomor 20. 2003. *Concerning the National Education System.* Indonesia Government.

Borderless Education as a Challenge in the 5.0 Society – Abdullah, Adriany & Abdullah (eds)
© 2021 Taylor & Francis Group, London, ISBN 978-0-367-61960-2

The praxis of character-building education through children's literature in elementary school

D. Heryanto, T. Hartati, M. Darmayanti & W. Nurfalah
Universitas Pendidikan Indonesia, Bandung, Indonesia

A.H. Saputra
Universitas Terbuka, Jakarta, Indonesia

ABSTRACT: National character values are always inherent in the ideals of civic education and character-building education, having become an integral part of national culture. In the context of education, there are various interpretations of character-building practices: (1) character-building education manifested explicitly; (2) character-building education as part of dimensions of subjects contained in the curriculum; (3) character-building education implemented indirectly through the teaching and learning process in the classroom, thus also non-thematic; and (4) character-building education embedded in schools through an informal curriculum approach (hidden curriculum). One of the practices closely related to students' interaction in the pedagogical process is learning in a variety of the existing subjects, so both the planning and the process must refer to the children's character building. Children's literature in elementary school can incorporate the themes of character-building to introduce values in a fun way. Literature refines the character of the child, resulting in many literary works, especially children's poetry, being developed to direct the five aspects of character (religion, nationalism, independence, cooperation, and integrity). Therefore, this study aims to implement poetry learning in character-building education through cooperative learning (learning together).

1 INTRODUCTION

Education is the core of the nation's character-building strategy. In the education field, this can be carried out with teaching, learning, and facilities (The Ministry of National Education 2017). In the macro context, the implementation of character-building encompasses all activities of planning, organizing, implementing, and controlling quality involving all significant units within the national education system. The role of teaching is very strategic because it serves as the root of active national integration. Besides being influenced by political and economic factors, education is also influenced by sociocultural factors, especially relating to inclusion and social resilience.

There are four ways of interpreting character-building in the educational environment. First, character education is manifested explicitly by the creation of new subjects since teachers are more likely to realize character-building education in separate topics such as Character Education, Budi Pekerti Education or Ethics, and Pancasila Moral Education. Second, character-building is integrated into the curriculum, if the learning material used are arranged, so as to lead to specific character-building values. Even though there are no new subjects, this approach still uses the learning process from the existing theme. Separate questions can also be in the form of a grouping of students' character classes, such as religion, language, literature, civics, physical education, and health education. When integrated with this curriculum, character education can be done thematically according to the needs of the school. Third, character-building can be included as a dimension in the subjects contained in the curriculum indirectly through the teaching and learning process in the classroom, ain a way that is non-thematic so every teacher can be creative in providing enlightenment through the subject taught. Fourth, character-building can be instilled in schools through a, informal (hidden curriculum) approach. This is not explicitly done but occurs when friendly communication is established between teacher and students. What the children see at break time, the way they interact and communicate with teachers, the procedures and customs of courtesy in the classroom are essential tools for building student character. Elementary schools aim to have students gain literary experiences through literary appreciation learning activities, which are both receptive and expressive. The teaching of responsive literary appreciation directs students to understand and evaluate other people's literary works, such as through poetry reading activities. Meanwhile, the learning of strong literary appreciation leads students to expressing ideas, thoughts, and imaginative feelings, for example, through poetry writing activities.

Elementary school children are developing rapidly and children's literature can make a positive contribution to their development process. In their association with poetry, children get value for their development, namely (1) language development, (2) cognitive development, (3) personality development, and (4) social development (Huck 1987).

Student cognition also can be developed through literature because some aspects of the story, such as plots built based on events in a cause–effect relationship, cannot help but involve cognitive activities. Thus, children's literature has the potential to develop children's cognition. Based on this, this study aims to explain the process of developing children's poetry based on characters and to explain the implementation of children's poetry learning in character-building practices.

2 RESEARCH METHODS

This study uses the research and development method. This type of research method is used to produce specific products and test the effectiveness of these products, which makes this research different from other educational studies because it aims to develop products based on trials and then revise it to produce the outcome that is suitable for use. Borg and Gall state that development research is a process used to develop and validate products used in teaching and learning. Procedures are a series of steps that must be carried out in stages to achieve specific goals or complete a product (Sugiyono 2011).

The development phase of Dick & Carey (1996) adapted this research into four stages. The researcher tried to adjust the steps of the development of learning with the levels of developing the module. The participants of this study are students who take courses in Indonesian language and literature education in elementary schools. The four steps include analysis, design, validation and evaluation, and the final product.

2.1 *Requirement analysis*

This stage aims to review the objectives of the product, in this case a module. The first step is to conduct curriculum analysis to produce themes tailored to the Core Competencies (KI) and Basic Competencies (KD). The next step is to analyze the foundation of character education law in elementary schools, to determine the types of poetry products that are following curriculum demands.

2.2 *Product design*

The initial product design in this study is manifested in drawings and charts that can be used as a guide to assessing and making the product (Sugiyono 2011). The product design phase includes determining the module components, the concept of delivering and organizing the material, the types of tasks assigned, evaluation questions, drawings, articles, examples, and module layouts. This stage produces an initial product design in the form of a module that has previously carried out the preparation of product assessment instruments to use as guidelines in product design.

2.3 *Validation and evaluation*

This stage is the core stage in the form of a series of product development assessments. The pre-validation step was done by consulting the supervisor to get initial input. The pre-validation stage is useful for assessing product viability before being evaluated by the validator. Design validation is a process of activities to evaluate whether the product design will be more effective than the old one (Sugiyono 2011).

Validation of the initial design is done by asking experts to assess the product being designed (Sugiyono 2011). Experts evaluate and validate the product and make suggestions for product development. The results of the assessment and advice from experts are used to improve and revise the product being developed.

2.4 *Final product*

This stage produces the final product in the form of a draft character-based poem collection and learning tools such as modules that have been revised based on criticism and suggestions from the validation and evaluation stages. The final product is ready for mass production and is distributed as modules in the learning process.

3 RESULTS AND DISCUSSION

In elementary schools, poetry analyses are limited to the character analyses. Poetry development is carried out by analyzing the needs of characters in elementary schools. The researcher explains the role in the results of the study via renewed information and policies, namely the Strengthening of Character Education (SCE). Educational programs in schools aim to strengthen student character through the harmony of *olah hati, olah karsa, olah pikir,* and *olah raga* with the support of public engagement and cooperation between schools, families, and communities that are part of the National Movement Mental Revolution. The development of poetry in this research is to take steps to crystallize the character values developed: religious, nationalist, independent, cooperation, and integrity (Presidential Decree 2017).

Figure 1 becomes one of the references in developing poems for elementary school students, which are found in the curriculum of primary schools in Indonesia. Other than paying attention to character-building, researchers also pay attention to other essential aspects including the following:

- Students' cognitive levels divided into low and high classes in elementary school. The mindset of students from concrete to abstract.
- Contextual themes; the material is close to the students.
- The language style is simple but still attractive as literature work.
- The level of readability of the text using *judgment expert* testing and through the *fry graph.*

This poem development is carried out in collaboration with students who take courses in Indonesian language and literature education in elementary schools. The poetry developed with a variety of artistry approaches, including, through the description of an event, the omission of conjunctions, analogies, and symbolization of character themes and personal attitudes in doing work. This research produced 56 poems: 8 religious characters, 10 nationalism characters, 10 independent characters, 12 mutual poetry characters, and 16 poetry integrity characters. More detail is presented in the Table 1.

Here are two examples of poetry in a lower grade and upper grade in elementary school (Table 2).

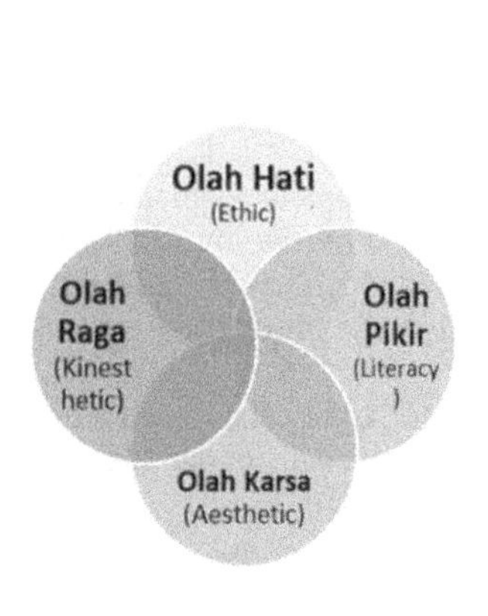

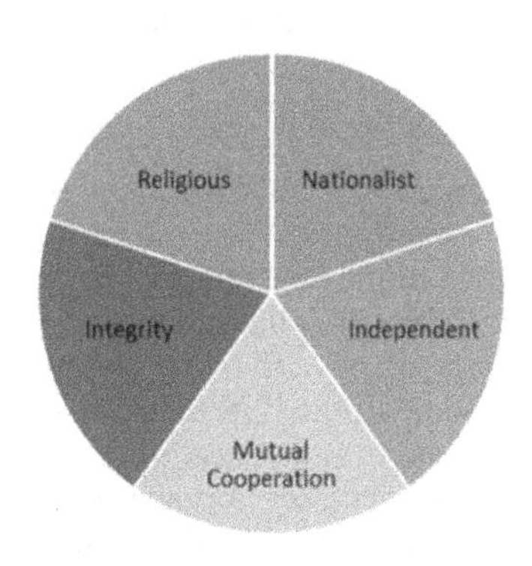

Figure 1. Character values development (The Ministry of National Education 2017).

Table 1. The results of the poem development of character-based.

Religious	Nationalism	Independence	Mutual Cooperation	Integrity
1. Heaven 2. God Is Accompanied 3. True Friends 4. Different 5. Life Color 6. Surrender 7. Mutual Respect 8. Peace	1. Heroes 2. Freedom 3. Playing Shots 4. Your blessings 5. My Indonesia 6. Take care of my Indonesia 7. Indonesia 8. Freedom! 9. Greeting the Youngest 10. That's Me	1. It can! 2. Life above the Sahara 3. Loneliness 4. Be motivated 5. I'm Not Scared 6. Sleeping hours 7. Posts 8. Trash 9. The Sunday 10. Bright Morning	1. Community Service 2. Working Together Community Service 3. Tired of Paid 4. Community Service 5. Mutual Cooperation 6. It's gone 7. Together 8. Community Service 9. Togetherness 10. Done House 11. My Housing Complex	1. Diligent Wisdom 2. Be honest 3. The High Value of Honesty 4. Corruptor 5. First Place 6. Travel Provisions 7. My country 8. Regents the Corruptor 9. The Key to Grasping the World 10. Don't Betray the Trust 11. I Am an Honest Child 12. My Indonesia 13. Go to School 14. My Teacher 15. The Gold Thing 16. Limitations

Table 2. Poetry in a lower grade and upper grade in elementary school.

Lower Grade	
Indonesia Version	English Version
Pagi yang Cerah	**Beautiful morning**
Di pagi yang cerah ini *Matahari tersenyum hangat* *Udara begitu segar* *Dan kicau burung terdengar bersahutan*	On this sunny morning The sun is smiling warmly The air is so fresh And birds chirping sounded friendly
Aku terbangun dari tidurku *Aku mandi dan sarapan pagi* *Badanku terasa segar,karena aku* *Tidur dengan lelap semalam*	I woke up from my sleep I take a shower and breakfast My body feels fresh, because of me Sleeping soundly last night
Aku bersiap – siap untuk pergi ke sekolah *Ya, sekolah. Tempat dimana aku menuntut ilmu* *Untuk mengejar cita – cita di masa depan* *Begitulah aku, di pagi yang cerah ini*	I'm getting ready to go to school Yes, school. A place where I study To pursue future goals So am I, on this sunny morning
Upper Grade **Mahalnya Nilai Kejujuran**	**The High Value of Honesty**
Segumpal hati di dalam dada *Berkecamuk antara dua pilihan* *Jujurkah atau dusta yang ada* *Jawaban menanti pada sebuah keputusan*	A lump of heart in the chest Raging between two choices Honest or lies The answer awaits a decision
Apakah kita jujur membawa mujur *Apakah kita jujur menjadi hancur* *Apakah kita dusta membuat bahagia* *Apakah kita dusta akan binasa*	Are we honest to bring good luck Are we honest being ruined Are we making happy lies? Are we lying will perish
Jujur... *Tak sedikit orang menjadi korban* *Akibat dari kepalsuan dan keserakahan* *Tak sedikit pula orang menjadi tenang* *Buah dari suatu kejujuran*	Honest... Not a few people become victims As a result of counterfeiting and greed Not a few people become calm The fruit of honesty

The researchers chose learning with the *Learning Together* Type Cooperative approach. Slavin (2008) revealed that David and Roger Johnson from the University of Minnesota developed the *Learning Together* model of cooperative learning. The model they examined involved students divided into groups of four or five students with different backgrounds working on assignments. These groups receive one assignment sheet as well as praise and awards based on the group's work.

The Learning Together model is almost the same as the STAD type, but the difference is that each group member is expected to be able to assess their own group's performance and help their group to be the best. In cooperative learning, each group must be able to show they are a compact group, both in terms of discussion and working on problems. Each member bears the responsibility to function optimally, but if the results are still not optimal, then they must improve the performance of the group.

Ghaith (2003) explains that the learning together model governs learning with the principle of positive interdependence, individual responsibility, interactive and collaborative steps, and group processing. Positive interdependence means that the achievement of goals is determined by the success of each group member in carrying out his personal responsibilities. Individual responsibility also means that besides working together, the performance of each member is assessed. Individual responsibilities are carried out along through interactive and collaborative steps according to group goals. Finally, team members carry out group processing to reflect how well the team is functioning and how its effectiveness can be improved. The steps of the *learning together* approach are:

- Orientation. The teacher delivers the lesson.
- Work in a group. Students form groups whose members consist of four or five students heterogeneously, both in terms of achievement, gender, ethnicity, etc. Each group receives an assignment sheet for discussion material and then completes it.
- Presentation. Several groups present the results of their work.
- They are giving praise and appreciation based on the work of the group. The rewards are assigned to groups based on the individual learning of all group members to increase student achievement and have a positive influence on the results (Slavin 2008).

Optionally, the teacher also can give a test or quiz to students as an evaluation tool to find out the extent to which the students' knowledge of poetry and writing poetry has increased.

Advantages of *Learning Together* Type of Cooperative Learning:

- Students become more active in learning because the teacher always gives them discussion material.
- Improved student cooperation in groups with the principle of joint learning.
- Students are trained to be brave and confident because they must step forward to present the results of their discussion in front of the class.
- The teacher is not too tired and busy because he only acts as a motivator and facilitator in the teaching and learning process.
- Students become more creative.

4 CONCLUSION

The practice of character-building through children's literature, especially poetry, in elementary schools is accomplished through the development of poetry that has character themes tailored to the cognitive, affective, and psychomotor development of students. The development of character values from Ki Hadjar Dewantara's Character Philosophy, *Olah Hati* (Ethics), *Olah Pikir* (Literacy), *Olah Karsa* (Aesthetics), and *Olah Raga* (Kinesthetic) into 18 character values and for character praxis is known as a crystallization of character values (religius, nationalists, independent, mutual cooperation, and integrity). The process of developing character values in poetry lies in contextual themes, selection of diction that is appropriate to the characteristics of the child, and safe/explicit messages for lower-grade students (classes I, II, and III), and implicit messages for upper-grade students (classes IV, V, and VI). After the analysis of essential competencies, a collection of poems was made that could facilitate material on children's literature (poetry) for classes I to IV.

Implementation of poetry learning for character-building through *Learning Together* Type of Cooperative Learning had the following core steps: (1) Orientation – the teacher delivers the lesson; (2) Group work; (3) Students form small groups heterogeneous in terms of achievement, gender, ethnicity, and others; (4) Each group receives an assignment sheet for discussion material and then completes it; (5) Presentations are done where several groups present the results of their work; and (6) Providing praise and appreciation based on the work of the group. The rewards are given to the group based on the individual learning of all group members, to increase student achievement, and to have a positive influence on the results. Optionally, the teacher can also give a test or quiz to students as an evaluation tool to find out the extent to which the students' knowledge of poetry and writing poetry increased.

In this research, the development of various poems with the theme of national character is still in the form of text and book drafts. Future research can be developed from fine arts to design and show more exciting display poetry. Integration and continuity of learning Bahasa

Indonesia and literature with other subjects, of course, can be accomplished by involving experts among the researchers.

The practice of character-building education through children's literature will not be optimal if it not supported by good literacy habituation. If there is standardized monitoring and evaluation, then it can be used as research material.

REFERENCES

Dick, W. 1996. The Dick and Carey model: Will it survive the decade? *Educational Technology Research and Development* 44(3): 55–63.

Ghaith, G. 2003. Effects of the learning together model of cooperative learning on English as a foreign language reading achievement, academic self-esteem, and feelings of school alienation. *Bilingual Research Journal* 27(3): 451–474.

Huck, C. 1987. *Children Literature in the Elementary School.* Chicago: Rand McNally College Publishing Company.

Presidential Decree. 2017. *Presidential Decree No. 87 of 2017 about Strengthening Character Education.* Jakarta: Presidential Decree.

Slavin, R. E. 2008. *Cooperative Learning: Theory, Research and Practice*, Terj. Nurulita, Bandung: Nusa Media.

Sugiyono. 2011. *Statistics for Research Revised Edition/Statistika Untuk Penelitian Edisi Revisi.* Bandung: CV. Alfabeta.

The Ministry of National Education. 2017. *National Policy on Nation Culture and Character Development in 2010–2025*. Jakarta: Kemendikbud.

Borderless Education as a Challenge in the 5.0 Society – Abdullah, Adriany & Abdullah (eds)
© 2021 Taylor & Francis Group, London, ISBN 978-0-367-61960-2

Adaptation of oral language skills program in hearing impaired children at SLBN Cicendo Bandung

T. Hernawati, I.D. Aprilia & D. Gunawan
Universitas Pendidikan Indonesia, Bandung, Indonesia

ABSTRACT: The impact of hearing loss may be directly associated with impaired language skills thus causing delay and difficulty in communication, especially verbally both receptively and expressively. Therefore, language development services for children with hearing impairments must be provided programmatically to minimize the impact. However, language development programs conducted by teachers in the classroom have not been integrated and structured according to the needs of children. Therefore, this research-based article draws attention to formulating oral language skills as a guide for teachers in training children with hearing impairments so they can communicate effectively. Conducted through descriptive qualitative method, the subjects on the current project were class teachers and students of grade 3 at Special School Negeri Cicendo Bandung. The results showed that oral language skills included the development in speech reading and speaking. Reading utterances include receptive semantics (picture vocabulary), receptive phonology (word discrimination), and receptive syntax (sentence comprehension). Speech development includes expressive semantics (oral vocabulary), expressive phonology (phoneme pronunciation), and expressive syntax (sentence imitation). The program can be implemented through three strategies: Integrated in thematic learning; integrated in the field of development of sound and rhythm perception communication; and special articulation programs. This project provides for an opportunity to gain insight into the recommendation of such program as a guide by teachers to improve oral language skills of children with hearing impairment, especially in lower primary classes.

1 INTRODUCTION

A direct impact of hearing loss is obstacles in language development, especially on spoken languages, which ultimately leads to obstacles in verbal communication. The ability to communicate verbally includes the ability to speak and listen or understand the conversation partner. Not only does hearing loss result in a lack of speech development, the greatest impact is on the occurrence of language poverty (Uden 1977). Limitations in language as a direct result of hearing loss will also have an inhibitory impact on the development of various other potentials such as academic, social, and emotional abilities, etc. With such terrible barriers, children with hearing impairment have special needs to minimize such effects.

In special schools for hearing impairment in children, there is a special program called the Development of Sound and Rhythm Perception Communication. In general, the program seeks to enable children with hearing impairment to be able to communicate both non-verbally and verbally and extend the sensitivity of their hearing function and the feeling of vibration to understand the meaning of various sounds, especially the sounds of language, with or without the use of hearing aids (Depdiknas 2000). The development of communication emphasizes the practice of verbal communication – both receptive and expressive – while the development of sound and rhythm perception emphasizes the practice of optimizing hearing ability to perceive various sounds including the sound of the language. The PKPBI program should be implemented through balanced and integrated material. However, in reality, teachers still focus on learning the practice of perceiving sounds instead of applying effective approaching methods in communication development, especially oral communication. Therefore, the preliminary study suggests that oral language skills in children with hearing impairment in SLB Negeri Cicendo have not been developed optimally. Development of oral communication includes both receptive and expressive aspects. In the receptive aspect, children with hearing impairment need to be trained to understand verbal language through effective methods, the Reflective Maternal Method (MMR), as the results of Prasetyo et al.'s (2017) research that suggest that MMR is effective in increasing students' interest in learning and understanding language. Meanwhile, children need both individual and group approaches for the development of expressive verbal language skills (speaking).

PKPBI is not enough to improve verbal language skills in children with hearing impairments, therefore integrated school programs, especially curricular programs, are necessary. In an effort to address the situation, current implementation is still not well programmed, thus the learning outcomes produce less-than-optimal results. Problems faced by schools in improving the oral language skills of children with hearing impairments should be able to be anticipated through designating appropriate interventions through systematic, collaborative, and sustainable learning programs that meets the children's needs.

2 LITERATURE REVIEW

2.1 *Impact of hearing obstacles on oral language ability*

"Hearing barrier" is a general term that refers to hearing loss – from mild to severe (Hallahan & Kauffman 1986, Hardman et al. 1990). The existence of these hearing impairments has an impact on the complexity of disabled people's lives. The direct impact of hearing impairment is impaired language skills, especially on spoken language both receptively and expressively. The ability to speak verbally is closely associated with the ability to capture what the speaker is saying, while the ability to speak verbally expressively is related to the ability to convey a message through speaking activities. Obstacles in oral language skills can ultimately affect socialization with the environment of the listening community.

2.2 *Development of oral language skills with hearing obstacles*

Oral language skills include listening and speaking. Listening skill is the ability to accurately receive and interpret messages in the communication process. Listening can be defined as an activity that includes listening to language sounds and identifying, looking at, and reacting to the meanings contained in the material (Tarigan 1991a). In the case of hearing impairment in children, listening activities are done through reading the utterances, such as capturing the movements of the articulation organs and the speaker's expression. Speaking is an ability to pronounce articulation sounds or words to express and convey thoughts, ideas, and feelings (Tarigan 1991b). Arsjad & Mukti (1993) suggested that speaking skill is the ability to say sentences to express and convey thoughts, ideas, and feelings. Listening and speaking skills are closely related and reciprocal.

The development of language skills in children with hearing impairments requires special strategies, one of which is through specific services, known as the Development of Communication and Sound and Rhythm Perception. In addition, the development of oral language skills can be integrated in thematic learning. In its implementation, it can use the reflective maternal method, which is learning with conversation as its axis. The stages of the reflective maternal method are heart-to-heart conversations (talking about what students experience); visualization of the conversation (writing what is said in direct sentences); deposit reading or video-visual (making a reading section as the result of conversation and visualizing a conversation using direct and indirect sentences); linguistic conversation (discussing important words or grammar); and reflection (asking questions about the content of the reading).

3 METHODS

The ultimate goal of this research is to formulate an oral language skills program adapted to the students' needs and school conditions. The program is a hypothetical program built on data that emphasizes the perspective and interpretation of participants. Based on the data generated, the method used is qualitative descriptive research method. This is in accordance with Bogdan and Taylor's opinion that qualitative research is one of the research procedures that produces qualitative descriptive data in the form of written or oral words from people and observed behaviors (Moleong 2012).

This research was conducted at SLB-Negeri Cicendo Bandung, with research subjects consisting of teachers, PKPBI teachers, and grade 3 students of SDLB. The school was chosen for several reasons: (1) The school is the largest special school for children with hearing impairment in Bandung; (2) There are a number of teachers in the school who have speech therapy expertise, but it doesn't function maximally; (3) Some teachers in the school, including teachers in grade 3 of SDLB, have implemented the Reflective Maternal Method (MMR), although not yet optimally.

4 RESULTS AND DISCUSSION

The program formulation was developed based on the initial ability (baseline) of spoken language of children with hearing impairments and special service activities developed by the teacher. Through initial data processing and analysis results from various theories, the following is an adaptation of the program.

4.1 *Development of language skills programs in hearing impaired children*

4.1.1 *Thought basis*

Oral language skills are very important for individuals, including children with hearing impairments, to communicate with their environment. Therefore, language skills are the main target of learning development in special schools for the deaf (SLB-B). For this learning service to be effective, a systematic and integrated pattern or program that can guide teachers in implementing oral language skills programs in children with hearing impairments is needed.

4.1.2 *Program objective*

It is conducted to improve oral language skills in children with hearing impairment.

4.1.3 *Scope*

Oral language skills include listening or reading utterances and speaking. The listening or reading aspects of speech include receptive semantics (picture words), receptive phonology (word discrimination), and receptive syntax (sentence comprehension). The aspects of speech developed include expressive semantics (oral vocabulary), expressive phonology (phoneme pronunciation), and expressive syntax (imitation of the sentence).

4.1.4 *Subject and users*

The chosen subjects with regard to the development of oral language skills program were students in grade 3 of SLBN Cicendo Bandung. Such oral language skills programs can be utilized by class teachers or PKPBI teachers in primary classes, especially in grade 3.

4.1.5 *Oral language skills program for children with hearing impairment*

Oral language skills program of hearing impaired students of grade 3 can be seen in Table 1.

4.1.6 *Program implementation guide*

Language skills programs can be implemented through three strategies: first, integrated in thematic learning; second, integrated in the field of specialization, under the Development of Sound and Rhythm Perspective Ability (PKPBI); third, special articulation programs.

Table 1. Oral language skills program of hearing impaired students of grade 3.

Basic Competence	Indicator	Material	Approach/Method	Evaluation	Source
Able to understand words	Able to show pictures of words correctly	Reading the words; glass, pineapple, chair, spoon, pencil, banana, hand. Sleep, eat, read, and mop	Classical/individual Reflective Maternal Method	Performance test: picture vocabulary	Bunawan & Yuwati (2000), Rusyani (2017), Sumardjo, Heribertus (ed). (2013)
Able to distinguish words	Able to discriminate the utterances of words containing phonemes b and p; j and r; h and m, and d and g.	Discrimination in the expression of words containing phonemes b and p; j and r; h and m; k and g, and d and s	Classical/individual Reflective Maternal Method	Oral test: word discrimination	
Able to understand sentences	Able to read utterances of simple sentences with the response designation of the image.	Read simple sentence sentences	Classical/individual Reflective Maternal Method	Performance test: sentence comprehension	
Able to verbally explain the meaning of words.	Able to explain the meaning of words verbally with more than one word	Oral Vocabulary	Classical/Reflective Maternal Method	Oral test: oral vocabulary	
Able to pronounce phonemes	Able to pronounce bilabial, dental, labiodental, laryngal, palatal and velar phonemes/ consonants.	Phoneme/consonant practice of bilabial, dental, labiodental, laryngeal, palatal and velar	Individual approach Method: - Global words - Synthetic analysis - Speech sound - Multi-sensory - Phonetic placement	Performance test: phoneme pronunciation	
Able to mimic oral utterances	Able to mimic phrases. Able to imitate simple sentences	Practice speaking/imitating simple phrases and sentences	Approach: Classical/individual Method: - MMR - Multi-sensory - imitation - recitation	Oral test: imitation of the sentence	

– *Integrated in Thematic Learning.*

The formulated strategy learning is done classically by classroom teachers using the Reflective Maternal Method (MMR), a language learning method through conversation as its axis. The core stages of the method are the heart-to-heart conversation stage, the visualization of the conversation, the preparation of deposit/video-visual reading, linguistic conversation, and reflection. Before implementing the program, the teacher makes a learning plan consisting of three stages: pre-learning, learning, and post-learning/reporting stages. The learning design is flexible in the sense that the stages can be applied to various materials. In this regard, it can be described as follows.

Learning Design for Children's Oral Language Skills with Hearing Obstacles

School Name: SLB Negeri Cicendo
Educational Unit: SDLB
Type of Obstacle: Hearing Obstacle
Class: III/A
Time allocation: 6 hours of study (6 x 30 minutes)

- Stage Pre-Learning Activities
  At this stage the teacher makes learning plans that relate to:
  a) Determine the themes & sub themes to be discussed in learning.
  b) Prepare the media that is expected in accordance with the conversation that will occur.
  c) Plan what form of stimulation will be given so that the child finds the material or object to discuss.
- Learning Activity Stage (Table 2).
- Post-Learning Activity Stage

After learning has been conducted, the teacher makes a journal or current-learning report, which describes the activities that have been carried out. If the teacher has not already made a lesson plan, the journal/learning can be used as a basis in the preparation of the lesson plan (for official reports).

– *Integrated in PKPBI Specialization Field*

The Development of Sound and Rhythm Perception Communication (PKPBI) is an area of specialization for children with hearing impairments, which is carried out via four hours of learning and implemented by PKPBI special teachers. Under its implementation, PKPBI is divided into two fields, the field of communication development and that of sound and rhythm perception development, with two hours of learning each.

The oral language skills program is integrated in the field of communication development, with two hours of study per week. The spoken language skills program in this strategy is more directed at developing phonological abilities both receptively and expressively. Such a breakthrough program is implemented through the Individualized Educational Program (IEP) based on the results of the assessment. The implementation can be in-class or individually. The components of an IEP are: student identity; current level of student ability (based on assessment results); basic competence; indicators; description of services that include material, methods, media, learning activities (initial activities, core, and closing); evaluation.

- Special Articulation Program

Oral language skills programs, especially in terms of phoneme pronunciation can be carried out individually or in groups. PKPBI teachers can take turns practicing articulation with students in special spaces individually, while other students study with class teachers. Evaluation is carried out to determine the success of program implementation. Evaluation of results is intended to determine the effectiveness of program, which can be conducted through a second assessment.

4.2 *Discussion*

As stated in the introduction, children with hearing impairment experience obstacles in oral language, yet they actually have the potential to develop their oral language skills. Based on

Table 2. Learning activity stage.

Activity Stage	Activity	Time (minutes)
a. Preliminary Activities	Student Conditioning: students sit in a semicircle, the teacher sits in front of students in the middle position, continuing to pray. Conversation stimulation.	10
b. Core Activity		
Heart to Heart Conversation Stage (Perdati)	Students converse with the teacher and other students about events that are happening (situational), or previous experiences. Conversations can be aligned with themes in the curriculum. The teacher applies the catch technique and multiple roles, namely capturing messages or thoughts expressed by students through improvised vocalization or sign language, then expressing them, and responding to them.	30
Conversation Visualization Stage	The teacher and students review the conversation that just happened and write the conversation on the blackboard and use direct sentences (conversation visualization). The teacher gives curvilinear phrases or groups of accents on the visualization results of the conversation. Students are guided by the teacher to read and interpret sentences with demonstrations or pictures.	40
Stages of Making Deposits & Video-visual Reading	The teacher makes interesting reading based on the results of the visualization of the conversation, equipped with the appropriate picture. Students together with the students read deposit readings, with the guidance of the teacher, then give the title of the reading. Students are individually guided to read sentence-by-sentence and correct the speech.	40
Linguistic Conversation Stage (Percali)	The teacher points or writes words or sentences containing grammatical topics, then discusses them.	20
Reflection Stage	Students are asked to answer questions about the content of the reading. Students are asked to read words, then show pictures/writing accordingly. Students are asked to read the utterances of two words, then convey whether the two words are the same or different.	30
c. Closing Activities	The teacher and students summarize the learning outcomes and end the learning with prayer and greetings.	10

results obtained in this line of research, it is demonstrated that such programs are tailored to needs and abilities according to the results of the assessment. Assessment activities are necesarily important especially for children with special needs as a basis for adaptive education or learning programs.

Broadly speaking, the scope of the program consists of developing the ability to read utterances and speaking, both of which cover semantic, phonological, and syntactic aspects. This is consistent with the opinion of McLoughlin & Lewis (2008), which suggests that a comprehensive test to measure the ability of spoken language is the Test of Language Development-primary, measuring receptive and expressive aspects in the semantic phonology and syntax dimensions.

The program produced from this research can be implemented through three strategies as explained in the program implementation guidelines. One of them is integrated with thematic learning that is carried out using reflective maternal methods. Learning with the reflective maternal method is learning using conversation as its axis. Each stage of learning is based on conversation. The conversation in question is a conversation between students and between students and the teacher talking about what happened to students. Language understanding is expected to be easier because what is said is related to what is thought, felt, and experienced. Because this learning is very situational, the learning design is different from the learning design commonly used in regular schools. Learning Design by applying reflective maternal methods is more general or flexible, so that it can be applied in a variety of learning. This choice is supported by the results of research by Prasetyo et al. (2017) that the application of reflective maternal methods is effective in increasing student interest in learning and improving language understanding. In addition, the results of Zulmiyetri's research (2017) show that by using the Reflective Maternal Method (MMR) and the use of attractive media, the verbal skills of children with hearing impairments have increased.

5 CONCLUSION AND RECOMMENDATION

5.1 *Conclusion*

Based on results obtained in this line of research, we conclude that developing oral language skills means developing the skills and knowledge that go into listening and speaking, all of which have a strong relationship to reading comprehension and to writing, and shall be adapted to the needs of students based on the results of the assessment.

5.2 *Recommendation*

- This program can be used as a guide for teachers in improving language skills in children with hearing impairment.
- This program is only a hypothetical program, and therefore, it is necessarily fundamental to test the effectiveness of the program.
- The program is focused on oral language skills. It is also necessary to conduct more comprehensive research that includes verbal communication (listening, speaking, reading, and writing).

REFFERENCES

Arsjad, Maidar, G., & Mukti, U.S. 1993. *Pembinanan Keterampilan Berbicara Berbahasa Indonesia*, Jakarta: Erlangga.

Bunawan, L., & Yuwati, C. S. 2000. *Penguasaan bahasa anak tunarungu*. Jakarta: Yayasan Santi Rama.

Depdiknas. 2000. *Pedoman Guru Pengajaran Bina Persepsi Bunyi dan Irama untu Anak Tunarungu SLB Bagian B*. Jakarta.

Hallahan, D. P., & Kauffman, J. M. 1986. *Exceptional Children: Introduction to Special Education*. New Jersey: Prentice-Hall, Inc.

Hardman, M. L., Drew, C. J., Egan, M. W., & Wolf, B. 1990. *Human Exceptionality* (third ed.). Massachusetts: A Division of Simon & Schuster Inc.

McLoughlin, J. A., & Lewis, R. B. 2008. *Assessing Students with Special Needs* (Seventh ed.). New Jersey: Pearson Prentice Hall.

Moleong. L. J. 2012. *Metode Penelitian Kualitatif*. Bandung: PT Remaja Rosdakarya.

Sumardjo, H. 2013. *Didaktik Metodik Pemerolehan Kemampuan Berbahasa Anak Tunarungu Sekolah Dasar*. Jakarta: tanpa penerbit.

Tarigan, D. 1991a. *Pendidikan Bahasa Indonesia 1*. Jakarta: Departemen Pendidikan dan Kebudayaan.

Tarigan, H. G. 1991b. *Pengajaran Kosakata*. Bandung: Angkasa.

Uden, V. 1977. *A Word of Language for Deaf Children: Basic Principles: a Maternal Reftective Method*. Amsterdam: Rotterdam University Press.

Prasetyo, Asrowi, & Sunardi. 2017. The Using Reflective Maternal Method to Improve Language Learning and Understanding of Hearing Impairment Students in Grade 2 Pemalang State Extraordinary Schools-SLB Negeri Pemalang, Indonesia. *European Journal of Special Education Research* 2(2): 25–34.

Rusyani. 2017. *Artikulasi dan Optimalisasi Fungsi Pendengaran*. Bandung: Departemen Pendidikan Khusus FIP-UPI.

Zulmiyetri. 2017. Metoda Maternal Reflektif (MMR) untuk Meningkatkan Kemampuan Bahasa Lisan Anak Tunarungu. *Jurnal Konseling dan Pendidikan* 5(2): 62–67.

Borderless Education as a Challenge in the 5.0 Society – Abdullah, Adriany & Abdullah (eds)
© 2021 Taylor & Francis Group, London, ISBN 978-0-367-61960-2

The views of teachers concerning the curriculum of the special education program at Universitas Pendidikan Indonesia

E. Ratnengsih, B. Susetyo, N. Warnandi, E. Heryati & K.A. Fajrin
Universitas Pendidikan Indonesia, Bandung, Indonesia

ABSTRACT: In this article, the authors explore the views of teachers concerning the curriculum of the special education program at Universitas Pendidikan Indonesia. Participants in this study are special education teachers who graduated from the special education department of Universitas Pendidikan Indonesia (UPI). The teachers participated in filling out open-ended questionnaires about the curriculum that focus on the composition of achievement, content, special skills, and competency needs that have not been met yet. The study identified issues with the current practices, collaboration, competency, and training needed for special teachers. Recommendations are for further studies and suggestions for the development of the curriculum in the future of special education programs.

1 INTRODUCTION

The main milestone in the implementation of educational services, especially in formal education, is focused on teachers. Teachers must not only fit a "trusted and figured" slogan, more than that teachers must have some competencies as determined in existing regulations in Indonesia, including pedagogical, personal, social, and professional competence. These further consist of 10 pedagogical competency indicators, 5 indicators of personality competence, 4 indicators of social competence, and 5 professional competency indicators.

Law Number 14 of 2005 concerning teacher and lecturer article 5 paragraph (1) states that the profession of teacher and lecturer is a specialized field of work that requires professional principles (Undang-undang Republik Indonesia nomor 14 2005), namely: (1) has talents, interests, vocations, and idealism; (2) has educational qualifications and background in accordance with their field of work; (3) has the competencies needed in accordance with their area of work; (4) adheres to the professional code of ethics; (5) has the rights and obligations in carrying out the tasks; (6) earns a specified amount of income based on his or her work performance; (7) has the opportunity to develop his or her profession in a sustainable manner; and (8) obtains legal protection in carrying out their professional duties. The manifestation of teacher professionalism must meet several criteria, including: (1) Bachelor or Diploma IV academic qualifications; (2) pedagogical, personal, social, and professional competencies; (3) educator certificates; (4) physical and mental health; and (5) ability to realize the goals of national education. The implication of this law is that a teacher must have a Bachelor's academic qualification, where the process to achieve these requirements must be obtained through the level of formal education in higher education or we are often familiar with the Institute of Educators and Education Personnel.

The process of achieving academic qualifications for prospective teachers in university should refer to the competencies that support the main tasks of teachers governed by the law: in the Law of teacher and lecturer mentioned it is stated that the teacher is a professional educator with the main task of educating, teaching, guiding, directing, training, assessing, and evaluating students in early childhood education through formal education, basic education, and secondary education. Furthermore, each individual is to reach the level of maturity as the ultimate goal of the educational process undertaken (Uno 2009). Therefore, it is important for universities to design the curriculum that can produce undergraduate teachers who can carry out their main tasks as a teacher.

Teachers of children with special needs or known as special education teachers, are required to have the same prerequisites and competencies set out in the teacher and lecturer law, even though the competencies listed contain different descriptions, adjusted for the context of students with special or unique needs. At minimum, a special education teacher must have competencies to (1) Make learning plans tailored to the needs of students through curriculum modifications; (2) Conduct learning processes carried out in accordance with the conditions and abilities of students with reference to the principle of flexibility; (3) Conduct assessments with referring to the principle of flexibility in accordance with the abilities and obstacles of students; and (4) Supervise learning in collaboration with parents and the community (Garnida 2015, Hamalik 2011, Setiawan & Sitorus 2017).

The curriculum designed by the university is expected to contribute greatly to the realization of professional teachers. In addition to curriculum development, it must also contribute to improving the quality of education in general, including university covering 4 areas, which are: the quality of inputs, processes, outputs, and outcomes. Those are (1) Educational input declared of good quality; (2) Quality education process able to create an active, creative, and fun atmosphere, (3) Output is declared good quality if the learning outcomes in the academic and non-academic fields of students are high; and (4) Outcome is declared of good quality if the graduates are quickly absorbed in the world of work, a fair wage, and all parties acknowledge the competence of, and satisfaction with the graduates.

For the Department of Special Education UPI that generates special education teachers, curriculum development to achieve quality inputs, processes, and outputs can be done at the management level of university. However, the outcome aspect needs data that requires an empirical search process related to the level of absorption of graduates in the world of work, a fair wage, and satisfaction with the graduates. All aspects revealed in the outcome will be good when graduates embed the teacher competencies needed by employment.

In general, the implementation of education in several countries has shown poor results. This condition is caused by lack of resources, overcrowded classrooms, diverse student and teacher organizations, reduced funding, pedagogical decision-making based on market needs, and an emphasis on high-risk accountability focused on teacher performance (Cordingley 2013).

The views of special education teachers, especially those who have worked in educational services for children with special needs, can compare and even evaluate the competencies needed at the place where they are currently working compared to the knowledge and skills they gained during formal education at the university level. The views of teachers on the design of the special education curriculum will be beneficial for future curriculum development.

The purpose of this study is to discover teacher perceptions or views about the curriculum of the special education department at Universitas Pendidikan Indonesia.

2 METHODS

The study was conducted using a survey method involving 23 participants. Data processing was done in a descriptive way. Data in this study were collected from 20 teacher participants. All of these teachers are graduates of specialized undergraduate programs in Indonesian Education universities who work in both public and private special schools. They have worked in a special school for children with visual impairments, hearing impairments, intellectual disabilitys and physical motor impairment. Participants consisted of 5 men and 15 women. They have a working experience span of 5–20 years. All were participants who took part in the socialization of the new curriculum at the UPI Special Education Department.

Data collection in this study required all teacher participants to fill out questionnaires after explanation of the new curriculum of the UPI Special Education department. The explanation covered achievement of competencies, curriculum structure, and composition of courses.

Instrument development was carried out to explore the suitability of the curriculum in aspects of competence, content, and composition with the competencies and needs that are really relevant for teachers in a special education school. The questionnaires were openly presented.

3 RESULTS AND DISCUSSION

Based on the data collection, the results explored include achievement of competencies, curriculum structure, and learning models provided in lectures. The results of filling out the questionnaire obtained the following data (Tables 1 and 2):

Table 1. Scoring of achievement of competence and curriculum structure.

	Scoring	
Aspects	Accordance	Not Accordance
Achievement of Competence	6%	94%
Curriculum Structure	12%	88%

Table 2. Scoring of learning model.

	Scoring	
Aspect	Good	Poor
Learning Model	100%	0%

The results of data collection on the aspects of achievement of competence showed that 94% of competency outcomes set by the special education department stated that they were in accordance with the competencies needed by teachers for the purpose of assessing children with special needs at schools. However, 6% of participants were not accordance with it.

In the aspect of curriculum structure – that is, the distribution of courses that will be taken by special education students both general and special courses, and courses that must be taken by all students and courses that are required in all specialties (specialization of vision, hearing, intelligence, physical motives and behaviors) –88% of participants answered that they were already in accordance with the competencies required by special education teachers in the field. The remaining 12% answered that they were not accordance with the requirements yet.

In the aspect of the learning model applied during classes in the Special Education Department aiming to achieve the competency of students in the form of knowledge and skills, 100% of participants stated that the classroom process in the Special Education Department was good.

The results of the survey in the form of an open-ended questionnaire obtained several findings as follows:

- Special education teachers need to have competency in certain subjects (mathematics, science, etc.). Some participants said that they had difficulty teaching related to deeper content, because they did not have skills and knowledge to master the content in certain subject areas.
- Special education teachers should have specific vocational field–specific competencies. Some of the participants argued that often students with special needs did not have a clear direction in terms of vocational skills after graduating from school because of the limited ability of teachers in the vocational field. The often led to children returning to school after graduating.

- Education development competencies for multiple disabilities need to get a large enough portion of practice. Some participants working in special schools stated that currently the students are mostly in the moderate to severe category. In fact, there are so many cases included in the category of compounded obstacles, that all cases taken in the classroom should be in the category moderate, heavy, and compound. Some special education teachers assert that children with special needs with moderate categories should attend more regular or inclusive schools.
- PPL should be done in some inclusive organizing schools or at resource centers. Some participants, especially respondents from regular schools, suggest some field practices in special education should be carried out in regular school settings. Special education teacher resources in inclusive school settings are seen as very limited.
- Special education teachers must develop ICT use/competence for children/individuals with special obstacles. Some students with special types of needs are greatly helped by the presence of ICT in learning. The inability of teachers to use/develop ICT causes learning to be less than optimal.
- There should be a special competency training program for teacher capacity building. Some participants saw the importance of the UPI special education department as expanding the capacity-building program for teachers through special skills training, which was legalized in the form of special certifications on specific skills.
- There should be increased collaboration between practitioners and academics in other fields in research and teaching. Researches conducted by lecturers should be carried out together with the teacher, so that the results of the study could really have a greater value for practitioners, especially teachers.

Based on the data obtained, there are several aspects of competency achievement, including curriculum structure, which are indicated as not meeting the competencies required by special education teachers, especially for the development of learning processes in regular or inclusive schools. Some of the problems that arise in schools are the large number of children with special needs who must be dealt with, and knowledge and the need to learn more about their characteristics (McSheehan et al. 2006). In addition, special education teachers lack understanding of how to adjust class-level content for children with special needs whose learning capacity is limited in order to work on their abilities in inclusive schools (Ruppar et al. 2011). In this regard, the collaboration of special education teachers and general teachers is very important for the success of inclusive education, and the readiness of teachers to collaborate with each other is needed for learning and student success. Collaboration is thought to contribute to more positive beliefs in teachers and predict fewer teachers' concerns about children with special needs in inclusive schools.

The design of UPI's special education curriculum is expected to increase the competencies needed for handling students in inclusive school settings. A teacher is required to be able to provide quality learning for all students in the class, including children with special needs. One way to support the quality of learning for children with special needs is to increase the capacity of personnel preparation programs and professional development activities to ensure teachers have the expertise to implement effective practices for children with special needs in inclusive school settings (Ryndak et al. 2013). The result should be that teacher competencies that are formulated in a good curriculum will be able to produce special education teachers who are ready to be placed at any time in either segregated or inclusive settings.

4 CONCLUSION

The competency goals, structure, and learning model designed in the curriculum of the special education department, in general, have fulfilled the competencies needed by the teacher. Some notes on competencies and learning models and also general input become valuable data for further study in order to develop the curriculum of the special education department in the future.

REFERENCES

Cordingley, P. 2015. The contribution of research to teachers' professional learning and development. *Oxford Review of Education* 41(2): 234–252.

Garnida, D. 2015. *Pengantar Pendidikan Inklusif*. Bandung: Refika Aditama.

Hamalik, O. 2011. *Proses Belajar Mengajar*. Jakarta: Bumi Aksara.

McSheehan, M., Sonnenmeier, R. M., Jorgensen, C. M., & Turner, K. 2006. Beyond communication access: Promoting learning of the general education curriculum by students with significant disabilities. *Topics in Language Disorders* 26: 266–290.

Ruppar, A.L., Dymond, S.K., & Gaffney, J.S. 2011. Teachers' perspectives on literacy instruction for students with severe disabilities who use augmentative and alternative communication. *Research and Practice for Persons with Severe Disabilities* 36: 100–111.

Ryndak, D., Jackson, L.B., & White, J.M. 2013. Involvement and progress in the general curriculum for students with extensive support needs: K-12 inclusive-education research and implications for the future. *Inclusion* 1: 28–49.

Setiawan, D., & Sitorus, J. 2017. Urgensi Tuntutan Profesionalisme dan Harapan Menjadi Guru Berkarakter (Studi Kasus: Sekolah Dasar dan Sekolah Menengah Pertama di Kabupaten Batubara). *Cakrawala Pendidikan*, (1): 122–129.

Undang-undang Republik Indonesia nomor 14. 2005. *tentang Guru dan Dosen*. Jakarta: Indonesia Government.

Uno, H. B. 2009. *Profesi Kependidikan: Problema, Solusi, dan Reformasi Pendidikan di Indonesia*. Jakarta: Bumi Aksara.

Borderless Education as a Challenge in the 5.0 Society – Abdullah, Adriany & Abdullah (eds)
© 2021 Taylor & Francis Group, London, ISBN 978-0-367-61960-2

Teachers' self-efficacy in the implementation of Indonesia 2013 curriculum: An analysis of senior high school teachers' lesson plan

L. Dewi & Y. Rahmawati
Department of Curriculum and Instructional Technology, Universitas Pendidikan Indonesia, Bandung, Indonesia

ABSTRACT: Indonesia implemented a new curriculum in 2013. The curriculum was developed based on the global need for 21st century skills. The value of this revised curriculum is to improve students' character through the learning process. The new curriculum mandates that the learning process should focus on Student Active Learning (SAL), which is based on 21st century skills. In order to achieve SAL in Senior High School (SHS), the learning process should use a scientific approach with a variety of instructional models. Teachers have to instill the values, knowledge, and working skills needed in the future through the learning process. This research aims at analyzing teachers' lesson plans to find out whether the lesson plans contain critical thinking, creativity, collaboration, and communication skills (4C), information literacy, character building, and Higher Order Thinking Skills (HOTS) in order to investigate whether the learning process is in line with the curriculum mandate. These values and skills should be written and implemented by teachers. The lesson plan analysis was conducted on sixteen subjects in SHS. The findings revealed that the lesson plans had used a scientific approach and a variety of instructional models, but the values and skills still needed to be improved. The research suggests that it is important to help teachers improve the learning process with 4C, information literacy, character building, HOTS through the lesson plans designed.

1 INTRODUCTION

Shifting to the 2013 Curriculum in senior high school (SMA, *Sekolah Menengah Atas*) was based on several considerations, including the importance of improving the values of character, knowledge, and 21st-century skills known as 4C, namely critical thinking, creativity, collaboration, and communication skills (Partnership 21st Project 2008). These changes were also intended to solve the problem of Indonesian students' low level of critical thinking skills. According to the results of PISA and TIMMS research in early 2000, Indonesia ranked 39 out of 41 countries involved in PISA. However, the 2015 PISA report revealed that Indonesia had experienced periodic improvements in student skills in the aspects of science, literacy, and mathematics, even though it was still ranked 62 out of 69 countries (Pratiwi 2019). This ongoing curriculum change is geared toward helping graduates, as candidates for Indonesian human resources in the future, to be able to compete with other human resources.

Changes in the curriculum also require changes in implementation of the learning process. Thus, all elements involved should get ready for the change, and the teachers should be prepared to face and implement these changes. In addition, they should have the knowledge and the ability to conduct the learning process well, fulfilling the demands of the change (Schwarzer & Hallum 2008).

To carry out a learning process that is in line with the applicable mandate and the demands of the community, the teachers should equip themselves so that learning can be achieved according to the objectives. Self-efficacy becomes an important aspect for the teachers to prepare themselves to carry out the learning process. Self-efficacy is reflected in the individuals' ability to organize and take action in a program or implementation of change (Bandura 2013, Lee & Lee 2014, Perepa et al. 2019). The 2013 Curriculum mandates that the learning process in high school be carried out using a scientific approach through a variety of student-oriented learning models. Its implementation requires readiness, beliefs, knowledge, and motivation in the individual as the implementer (Al Rasyidin & Sinaga 2017, Bandura 2013, Lee & Lee 2014, Perepa et al. 2019). The aspects measured to investigate teachers readiness in carrying out learning based on the 2013 Curriculum include the ability of teachers to prepare the learning process, carry out the learning, and evaluate the learning itself. Lesson planning important to implementing learning. Through careful lesson planning preparation, the learning process will meet expectations (Derri et al. 2014, Janjai 2012, Nesusin et al. 2014).

Therefore, this research aims at investigating lesson plans designed by teachers in high school by referring to the Regulation of Ministry of Education and Culture of the Republic of Indonesia No. 22 of 2016 concerning education process standards. The element analyzed was the compatibility of existing components, the application of values to be taught in the learning process as a provision for students' lives in the future. The values included in the learning process are scientific approach, character values, and 21st century skills including critical thinking skills, creativity, collaboration, communication, information literacy values, and higher-order thinking skills (HOTS), which are reflected in the learning process and assessment (Dhewa et al. 2017). Based on the results of the research, teachers modified implementation of the scientific approach in 2013 Curriculum-based learning.

2 METHODOLOGY

This research employed a document-analysis method for 16 subjects in high schools in Bandung and Cimahi. The instrument in this research was a lesson plan assessment guide based on the guidelines for the preparation of the lesson plan. The subjects studied are shown in Table 1:

Table 1. Distribution of subjects.

No	Subjects	Numbers of Lesson Plan
Natural Sciences Group		
1	Biology	1
2	Physics	1
3	Chemistry	2
Social Sciences Group		
4	Economics	1
5	History	3
6	Sociology	2
Compulsory Subjects		
7	Mathematics	2
8	Bahasa	3
9	Islamic education	1

3 RESULTS AND DISCUSSION

As explained earlier, this research analyzed the lesson plans developed by teachers following guidance in the 2013 Curriculum. The lesson plans included compulsory subject groups such as Bahasa and Islamic Education and Mathematics. The subjects in the Natural Sciences group were represented by Biology, Physics, and Chemistry. Meanwhile, social studies subjects were represented by Economics, History, and Sociology.

Table 2. Compatibility of lesson plan components.

No	Component	Compatibility		
		S	KS	TS
1–5	Subject identity, including: School name; Name of subject; Class/semester; Subject matter; Time Allocation	100%		
6	Learning objectives are formulated based on Basic Competency (KD, *Kompetensi Dasar*), using operational verbs.	100%		
7	Basic competence and indicators of achievement of competence.	87.5%	12.5%	
8	Subject material contains relevant facts, concepts, principles and procedures.	100%		
9	The learning method is used by educators to create a learning atmosphere and learning process so that students reach the basic competencies to be achieved.	62.5%	37.5%	
10	Learning media are in the form of learning process aids to convey subject matter. Learning resources are in the form of printed and electronic media books, the environment, or other relevant learning resources.	81.3%	18.7%	
11	The learning steps are carried out through the stages of preliminary activities, core activities and closing activities.	93.7%	6.3%	
12	Assessment of learning outcomes uses an authentic approach.	93.7%	6.3%	

S= *Sesuai* (Appropriate); KS= *Kurang Sesuai* (Less appropriate); TS= *Tidak Sesuai* (Not appropriate).

According to the Minister of Education and Culture Regulation of the Republic of Indonesia No. 22 of 2016 concerning education process standards, components of the lesson plan were generally following the compulsory component of the lesson plan, detailed in Table 2.

3.1 *Compatibility of lesson plan components*

In designing lesson plans, teachers should follow the existing regulations. The research found that the lesson plan component was in accordance with applicable regulations. The Regulation of Minister of Education and Culture of the Republic of Indonesia No. 22 of 2016 suggests that the components of a lesson plan are (1) school identity; (2) subject identity; (3) class; (4) subject matter; (5) time allocation; (6) learning objectives; (7) basic competencies and indicators; (8) learning material; (9) learning methods; (10) media & learning resources; (11) stages of learning; and (12) assessment. Table 2 explains the results of the analysis of the compatibility of the lesson plan components.

An analysis of the lesson plan compatibility components is needed to assess the consistency between what is written and what should be written. The focus of this compatibility analysis was consistency between the formulation of basic competencies, indicators, learning objectives, subject matter of learning, learning steps, and evaluation techniques (Janjai 2012, Nesusin et al. 2014). The formulation of consistent and systematic lesson plans aid teachers to teach in class. Even though learning is situational, if the lesson plan has been designed appropriately beforehand, the learning process will go according to plan (Derri et al. 2014, Janjai 2012, Nesusin et al. 2014). Lesson plans can help teachers design learning that is interactive, inspiring, fun, challenging, efficient, and that fosters students' motivation to participate more actively.

3.2 *Compatibility with the principle of lesson planning*

Lesson plan preparation should not only pay attention to the applicability of the components but also to compatibility between the components and the principles mandated in the regulations following the demands of current and future needs. To become reliable human resources candidates, students need to be prepared to face the times. Nowadays, students should have particular skills to deal with various 21st-century challenges. In addition, during the 21st century, the Industrial Revolution 4.0 was born, meaning more digital tools are used in operating various equipment (Derri et al. 2014, Janjai 2012, Nesusin et al. 2014).

In line with these challenges, the education process should implement various approaches, strategies, and learning methods that lead to the acquisition of 21st-century skills. This was conducted to prepare honest, tough, responsible, disciplined human resources with excellent skills who are ready to compete with the outside world. Thus, the lesson plan should include character values, 4C skills, information literacy, and HOTS. The following are the results of the analysis of the lesson plans designed by teachers based on the values and skills students should have.

3.3 *Character values*

Table 3 is the result of a review of the lesson plans in terms of the availability of character values.

Based on these data, the character values written in the lesson plan, especially in the initial learning stages and the core activities, were religious values including the habit of praying before learning, studying the Qur'an, discipline, responsibility, self-confidence, mutual cooperation, and communication skills. In a number of the lesson plans assessed, the character values were explicitly written out. Character values that need to be instilled in students based on the 2013 Curriculum consist of 18 character values that become a Strengthening Character Education Program, namely religious, honest, tolerance, discipline, hard work, creative, independent, democracy, curiosity, national spirit, love of the motherland, respect for achievement, friendly/communicative, love peace, love to read, care about the environment, care about society, and responsibility (Agboola & Tsai 2015, Agung 2011, Dhewa et al. 2017). The character values are grouped into five major groups namely religious values, nationalism, integrity, independence, and mutual cooperation. The purpose of the application of character education is to re-improve heart-processing, thought-processing, taste-processing, and kinesthetic-processing (Agboola & Tsai 2015, Agung 2011, Bialik et al. 2015). These character values are not taught but are grown and internalized, ideally through a process of guidance and habituation, so that they can impact behavior and action in the future (Al Rasyidin & Sinaga 2017, Kaimudin 2014, Pal 2011, Sari 2013). For this reason, habituation at the initial and core stages of learning needs to be continually developed.

3.4 *Learning leading to improving 4C*

Education is one way to prepare generations to survive and compete in the 21st century. According to the mandate of the Indonesian Government concerning education process standards, the learning process should pay attention to individual differences, increase active

Table 3. Analysis of character values.

No	Subjects	Findings
Natural Sciences Group		
1	Biology	Based on the analysis of the three subjects in the natural sciences group, the value of strengthening character education appeared in lesson plan and was written explicitly in the initial and core stages of learning. The values listed are religious values such as gratitude. Other values that are written explicitly are discipline, communication, curiosity, and responsibility. There is one subject that did not explicitly input these character values.
2	Physics	
3	Chemistry	
Social Sciences Group		
4	Economics	In the social sciences subject group, the character values written explicitly in the initial stages and the core activities of learning were gratitude, discipline, mutual cooperation, confidence, responsibility, and curiosity.
5	History	
6	Sociology	
Compulsory Subjects		
7	Mathematics	In the compulsory subject group, the character values were written explicitly and included religious values such as praying, reading the Qur'an, and disciplinary values.
8	Bahasa	
9	Islamic education	

Table 4. Studies on the main activities learning.

No	Subjects	Findings
Natural Sciences Group		
1	Biology	In the science group, the scientific approach had been used with discovery learning as the learning method used in all three subjects. The learning steps compiled were in accordance with the selection of learning methods.
2	Physics	
3	Chemistry	
Social Sciences Groups		
4	Economics	In this subject group, the scientific approach had been applied using discovery learning as one of the learning methods used. In almost all three subjects, discovery learning was a favorite learning method used.
5	History	
6	Sociology	
Compulsory Subjects		
7	Mathematics	In these three subjects, the scientific approach had been applied using learning methods that support student learning activities. The learning methods used in the lesson plan were discovery learning, project-based learning, and cooperative learning. The steps of learning at the core stage also seemed consistent with the stages of learning in the chosen learning method.
8	Bahasa	
9	Islamic education	

participation, and provide learner-centered education (Educational for All Global Monitoring Report 2006, Organisation for Economic Co-operation and Development 2003, Partnership 21st Project 2008, Seftiawan 2019, Yen & Halili 2015). The scientific approach, a learning model adapted to the learning needs of students, is expected to be used by the teacher in the implementation of learning. Table 4 presents the core stages of learning from the analysis of the use of scientific approach through a variety of learning models that fit the learning needs of students. The 4C should emerge in the learning activities that use a scientific approach (Partnership 21st Project 2008, Yen & Halili 2015).

The analysis of the lesson plan document of the three different subject groups revealed that there was a tendency for the use of the same learning method, namely discovery learning. Only one subject used cooperative learning. Discovery learning methods can indeed be applied in various subjects but it was not the only learning method used (Janjai 2012, Mauliate et al. 2019, Nesusin et al. 2014). For example, there was one subject that included two learning methods, namely discovery learning and project-based learning. Both learning methods have different characteristics and orientations; they cannot be combined in one learning process.

It can be concluded that, in general, teachers had arranged the lesson plan well by applying a scientific approach and learning methods supporting the achievement of 4C. However, it is necessary to have a deeper understanding related to the use of varied learning methods, so that the determination of the learning method can be adjusted to the competencies to be achieved and the material to be delivered (Bialik et al. 2015, Partnership 21st Project 2008, Yen & Halili 2015).

3.5 *Application of literacy skills in the learning process*

The lesson plans prepared by the teachers implemented literacy skills. However, the teachers did not specifically mention literacy skills that were appropriate to the characteristics of the subjects. Thus, it seemed that the preparation of the lesson plan in participating the literacy program was merely a discourse. The lesson plan should have shown clear activities. For example, in watching a video, the teacher should mention what videos the student will see, what the student should do after watching the video, and another specific activities.

Literacy skills developed by teachers were still centered on basic literacy. Some subjects had shown efforts to increase scientific literacy through the activities of collecting, processing, and reporting the information (Educational for All Global Monitoring Report 2006, Organisation for Economic Co-operation and Development 2003, Seftiawan 2019, Yen & Halili 2015). The application of digital, financial, cultural, and citizenship literacy was still not explicitly

Table 5. Analysis of literacy skills implementation.

No	Subjects	Findings
1	Biology	In the lesson plan, the teacher had explicitly explained literacy activities in the
2	Physics	learning process, such as:
3	Chemistry	Basic literacy: students were directed through the activity of viewing video and
4	Economics	images; observing the Worksheet; reading the book; and listening to the infor-
5	History	mation conveyed by the teacher;
6	Sociology	Science literacy: discuss, ask questions, process information and data
7	Mathematics	Digital literacy: not yet visible
8	Bahasa	Financial literacy: none
9	Islamic Education	Cultural and citizenship literacy: none

mentioned. Table 5 shows the result of lesson plan analysis related to the implementation of literacy skills in schools.

Literacy skills consist of basic literacy, numeracy, scientific literacy, digital literacy, financial literacy, and cultural and citizenship literacy (Educational for All Global Monitoring Report 2006, Organisation for Economic Co-operation and Development 2003). In 2015, the results of the PISA research showed that the level of student literacy, especially in reading, was still in a low category. Since then, the government continued to promote various activities to support the improvement of student literacy. A local newspaper, pikiranrakyat.com, on April 25, 2019 reported that the government, represented by the Ministry of Education and Culture, stated that student literacy in Indonesia had increased. This was based on the results of research on 6,500 10th-graders in 34 provinces in Indonesia (Seftiawan 2019). This increase was certainly based on the efforts designed by the government to grow literacy skills through reading interest programs.

3.6 *Higher Order Thinking Skills (HOTS)*

HOTS is a high-level thinking skill that trains students' cognitive abilities through specific learning processes to provide understanding, reasoning, analysis, problem-solving, and critical and creativity thinking (Bialik et al. 2015, Yen & Halili 2015). The teaching and learning process that applies HOTS is reflected in the formulation of indicators at C3 level (comprehension) and above, the use of learning methods that lead to student active learning and the implementation of appropriate learning, and the determination of the learning assessment process that uses HOTS questions. In addition, HOTS learning can also be seen from the use of the 4C approach and literacy. The results of the research indicated that the design stages of learning had shown HOTS-oriented learning processes (Bialik et al. 2015, Yen & Halili 2015).

This research focused more on the formulation of indicators, the selection of evaluation types, and the form of questions used. A more detailed explanation is presented in Table 6:

Table 6. Analysis of the implementation of HOTS.

No	Subjects	Findings
Natural Sciences group		
1	Biology	**Formulation of Indicators:** Of the three types of subjects analyzed, two subjects had
2	Physics	used operational verbs at the C3 learning level and above. One subject still uses
3	Chemistry	operational verbs under C3.
		Selection of the type of assessment: The lesson plan had used various types of assessment for aspects of attitude, knowledge, and skills. The test used was a written test

(*Continued*)

Table 6. (*Continued*)

No	Subjects	Findings
		in the form of description and multiple-choice, oral test, attitude assessment, process evaluation, and performance evaluation.
Social Sciences group		
4 5 6	Economics History Sociology	**Formulation of Indicators:** Of the three types of subjects analyzed, two subjects have used C3 level operational verbs and above. One of the subjects still used verbs in the learning phase under C3. **Selection of types of assessment:** Teachers had used various types of assessment for aspects of attitudes, knowledge, and skills. The test used was a written test in the form of description and multiple-choice, oral test, attitude assessment, process evaluation, and performance evaluation.
Compulsory Subjects		
7 8 9	Mathematics Bahasa Islamic Education	**Formulation Indicator:** Of the three types of subjects analyzed, two subjects have used verbs at the learning stage level C3 and above. One of the subjects still used verbs in the learning phase under C3. **Selection of types of assessment:** Teachers had used various types of assessment for aspects of attitudes, knowledge, and skills. The test used was a written test in the form of description and multiple choice, oral test, attitude assessment, process evaluation, and performance evaluation.

Based on the lesson plans analyzed, the teachers had compiled indicators that lead to the formulation of HOTS level indicators strengthened through the preparation of the learning stages that also lead to HOTS learning. At the end of the stages the teacher had compiled a complete assessment tool on aspects of attitudes, knowledge, and skills (Derri et al. 2014, Mauliate et al. 2019). If the lesson plan has included four components (character values, 4C, literacy, and HOTS), the implementation of the learning process carried out by the teacher has also led to the expected direction. Therefore, the learning process will produce students who are accustomed to behave well, have knowledge, and have good skills in accordance with the stated education level.

4 CONCLUSION

The understanding and self-efficacy of teachers in understanding the implementation of the 2013 Curriculum is reflected through the evidence of lesson plans that are in line with expectations set in various rules and guidelines for curriculum implementation. Based on the mandate delivered, the current national curriculum is one that aims at producing students who can survive and compete in the global era with various challenges faced. The rapid development of information, knowledge, and technology should be welcomed with various innovations designed in learning. This is to invite students to also have innovative thoughts and actions in the future.

The education process in high school has been prepared in such a way as to produce graduates who have the expected skills. This expectation is realized through the inculcation of character values, the application of critical thinking skills, creativity, collaboration, and also communication, cultivation of literacy culture, and HOTS learning. Therefore, in the future, Indonesia will have human resources ready to adapt, compete, and take part in the field they are engaged in and become successful in living their lives.

Teacher understanding has been reflected in the preparation of lesson plans that include these four components. However, there is still a need to increase understanding and also the application of learning methods that vary from the four components, so that it will give a different and innovative nuance. This research only analyzed the dimensions of written documents (lesson plans). Further research is expected to analyze the implementation of the learning process in accordance with the lesson plan that has instilled character values, critical thinking skills, creativity, collaboration, communication, literacy culture, and HOTS learning.

ACKNOWLEDGMENTS

This research was funded by an internal university fund from Universitas Pendidikan Indonesia through a Competitive Research grant with contract number 293/UN40.D/PP/2019. The author would like to thank Universitas Pendidikan Indonesia for providing the funds to carry out this research.

REFFERENCES

Agboola, A., & Tsai, K. C. 2015. Bring character education into classroom. *Euroean Journal of Education Research* 1(2): 167–70.

Agung, L. 2011. Character education integration in social studies learning. *International Journal of History Education* 12(2).

Al Rasyidin, Z., & Sinaga, A. I. 2017. Implementasi Nilai-Nilai Pendidikan Karakter Pada K13 Dalam PAI dan Budi Pekerti di SMP Islam Al Amjad Medan Sunggal. *EDU-RILIGIA: Jurnal Ilmu Pendidikan Islam dan Keagamaan* 1(4).

Bandura, A. 2013. Exercise of personal and colletive efficacy in changing. In *Self Efficacy in Changing Society* (edited by Albert Bandura). New York: Cambridge University Press. (Publish 1995, Transferred to digital printing 2009).

Bialik, M., Bogan, M., Fadel, C., & Horvathova, M. 2015. *Character Education for 21st Century: What Student learn?* Boston: Center for Curriculum Redesign.

Derri, V., Papamitrou, E., Vernadakis, N., Koufou, N., & Zetou, E. 2014. Early professional development of physical education teachers: Effects on Lesson Learning. ERPA Congress. *Procedia: Social and Behavioral Sciences* 152(2014): 778–783.

Dhewa, K. M., Rosidin, U., Abdurrahman, A., & Suyatna, A. 2017. The development of Higher Order Thinking Skill (HOTS) instrument assessment in physics study. *IOSR Journal of Research & Method in Education (IOSR-JRME)* 7(1): 26–32.

Educational for All Global Monitoring Report. 2006. *Understandins of Literacy*. Chapter 6. [Online] http://www.unesco.org/education/GMR2006/full/chapt6_eng.pdf. Downloaed 25-04–2019.

Janjai, S. 2012. Improvement of the Ability of the Student in an Education Program to Design the Lesson Plans by Using an Instruction Model based on the Theories of Constructivism and Metacognition. I-SEE2011. *Prcedia Engineering* 32(2012): 1163–1168.

Kaimudin. 2014. Implementasi Pendidikan Karakter dalam Kurikulum 2013. *Jurnal Dinamika Ilmu* 14(1).

Lee, Y., & Lee, J. 2014. Enhancing pre-service teachers' self-efficacy beliefs for tehnology integration through lesson planning practice. *Journal of Computers & Education* 73(2014): 121–128.

Mauliate, H. D., Rahmat, A., & Wachidah, S. 2019. Evaluation the lesson plan of English language learning in junior high school, Seraphine Bakti Utama West Jakarta. *Internatonal Journal of Scientific Research and Management (IJSRM)* 7(7): 1078–1086.

Nesusin, N., Intrarakhamhaeng, P., Supadol, P., Piengkes, N., & Poonpipathana, S. 2014. Development of lesson plans by the lesson research approach for the 6th grade students in social research subject based on open approach innovation. *5th World Conference on Educational Sciences – WCES 2013. Procedia – Social Behavioral Sciences* 116(2014): 1411–1415.

Organisation for Economic Co-operation and Development. 2003. *Literacy Skills for the World of Tomorrow. Executive Summary*. [Online] http://www.unesco.org/education/GMR2006/full/chapt6_eng.pdf. Downloaded 26-10–2019.

Pala, A. 2011. The need for character education. *International Journal of Social Sciences and Humanity Studies* 3(2).

Partnership 21st Project. 2008. *21st Century Skills*. 21st Century Skills.

Perera, H. N., Calkins, C., & Part, R. 2019. Teacher self-efficacy profiles: Determinants, outcomes, and generalizability across teaching level. *Contemporary Educational Psychology* 58: 186–203.

Pratiwi, I. 2019. Efek program PISA Terhadap Kurikulum di Indonesia. *Jurnal Pendidikan dan Kebudayaan* 4(1).

Sari, N. 2013. The Importance of Teaching Moral Values to The Students. *Journal of English and Education* 1(1): 154–162.

Schwarzer, R., & Hallum, S. 2008. Perceived teacher self-efficacy as a predictor of job stress and burnout: Mediation analyses. *Applied Psychology: An International Review* 57: 152–171.

Seftiawan, D. 2019. *Literasi di Indonesia*. Bandung: Pikiran Rakyat. tersedia pada pikiranrakyat.com tanggal 25 April 2019.

Yen, T., & Halili S. 2015. Effective teaching of higher-order thinking (Hot) in education. *TOJDEL* 3(2): 41–47.

Borderless Education as a Challenge in the 5.0 Society – Abdullah, Adriany & Abdullah (eds)
© 2021 Taylor & Francis Group, London, ISBN 978-0-367-61960-2

Reading comprehension skills through use of child-language approaches for children with intellectual disability

M.A. Saepulrahman
Department of Special Education, Universitas Pendidikan Indonesia, Bandung, Indonesia

ABSTRACT: By definition, intellectual disability affects a child's level of intelligence and this will have an impact on their ability to understand the contents of reading. This is because the words used in the reading text are different from the words used by children every day, as their vocabulary is small. This study aims to determine the influence of children's language when used in reading texts. This study uses a combination research method with a balanced mixed-model combination type (concurrent triangulation strategy). Quantitative research methods used are true-experimental design with the characteristics of the control group experimental sample taken randomly from certain populations. The design is pretest–posttest equivalent group design. The results showed an increase in reading comprehension in children with intellectual disability through effective child-language approaches. Thus it can be recommended that learning material in the source book should use children's language. First pay attention to the language used by children every day, then use that in the reading text.

1 INTRODUCTION

Over 1% of all students aged 6 to 21 years have some developmental delay (US Department of Education 2005 in Slavin 2011), resulting in limitations in several aspects of development, including physical, language, emotions, social adjustment, and personality. Somantri (2005) argues, "Language development is closely related to the development of cognition, both of which have a reciprocal relationship. The development of cognitive impairment in children with intellectual disabilities, therefore language development will also be hampered." The language development of children with intellectual disabilities is slower than children in general or the same as typical children with MA (mental age) younger than CA (chronology age). "Many mentally retarded children speak fluently but lack vocabulary" (Amin 1995). Bowey (1986) in Ormrod (2009) argues, "The linguistic ability of children in general aged six years as many as 8000 to 14,000 words and for children in grades six to eighth grade as much as 50,000 words." Vocabulary mastery of children with mild intelligence barriers means the child is able to read, but has difficulty in understanding reading from books, because the language used is not understood. It is feared that this obstacle causes the child to be unable to understand the lesson that is being taught.

Hallahan and Kauffman (1982) argued, "Speech problems (difficulties in the formation of sound, such as articulation errors) occur more frecuently in retarded than in normal children. The exact incidence of speeh and language problems is difficult to determine, and estimates vary widely." Research of Hallahan and Kauffman (1982) in Abdurahman (2017) stated, "The prevalence among retarded children in special classes was lower (8 to 26 percent), but still higher than in the general population." Rochyadi (2010) argued, "However, the Beta Coefficient it was is found that linguistic awareness is 0.72, while the coefficient path value for linguistic awareness is 0.25. This means that linguistic awareness is influencing position as the prerequisite of early reading skill compared to the visual perception aspect."

Soendari (2008) argued, "Reading is an auditive and visual activity to obtain meaning from symbols in the form of letters or words. This activity includes two processes, namely the decoding process, also known as technical reading or beginning, and the process of understanding." The ability to read with comprehension is the ability to read at a higher level than the initial reading. Children with mild intelligence obstacles need help in reading comprehension, needing to be guided in developing reading comprehension skills. If left untreated, it can be difficult for children to learn the subject matter. Reading comprehension skills require subject matter that can be read easily by children. The source books used in learning use a lot of language that has not been mastered by children, so it is difficult for children to understand.

Children will understand the subject matter they read when using language common in their everyday life. This research provides learning material using the approach of children's language experience (learning experience approach). The teacher still gives the lesson material using the source book as a whole. Before being given subject matter through the child-language approach, children are first given subject matter in the form of reading texts that children learn in school. Many children with intellectual disability have difficulty understanding the contents of the text in the textbooks. This study is to determine the effectiveness of reading comprehension skills through child-language approaches for children with intellectual disability.

2 RESEARCH METHODS

This study uses a combination of balanced mixed model (concurrent triangulation strategy) methods. Gall et al. (2003) argued, "some researchers believe that quantitative and qualitative research are incompatible because they are based on different epistemological assumptions." Sugiyono (2011) argues, "with this method the results of research will be more complete, valid, reliable, and objective; because by using triangulation data collection techniques, the weaknesses of one data collection technique will be overcome by other data collection techniques." Creswell (2009 in Sugiyono 2011) states, that "This method is a popular method among another combination method. Because both methods are used at the same time, it is more efficient in terms of time."

Quantitative research methods used are true-experimental design with the characteristics of the control group experimental sample taken randomly from certain populations. The design is a pretest–posttest equivalent group design. The research procedure consists of three stages: introductory research on children's language skills, the planning stage of learning based on children's language, and the stage of implementing reading learning through the child-language approach.

3 RESULTS AND DISCUSSION

The results comparing children's ability to read pretest and posttest can be seen as follows (Figure 1).

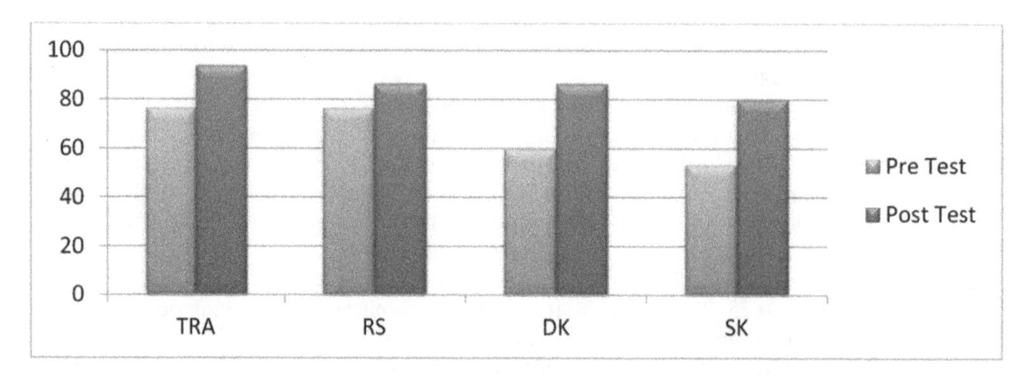

Figure 1. Improvement of experimental group reading comprehension ability.

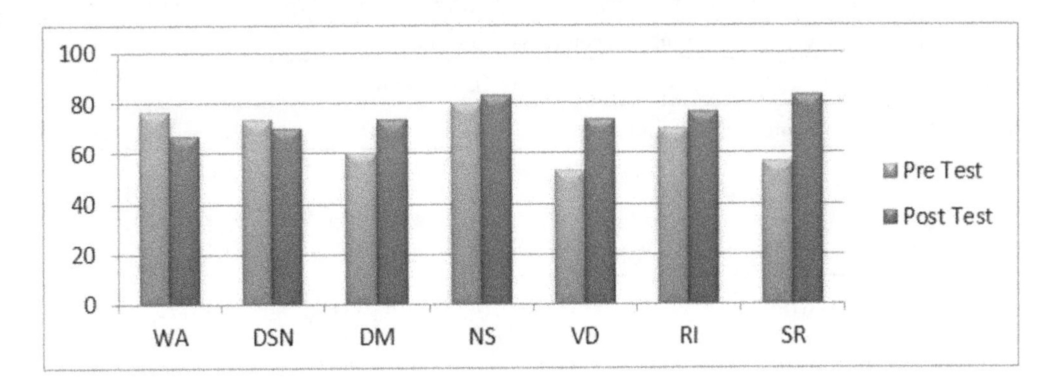

Figure 2. Improved reading group comprehension ability.

There was a significant difference in the experimental group. The average ability at pretest was 66.6% and the posttest score using the child-language approach reached 85.8%, an increase of 19.2%. Increased ability to listen was seen in the control group in Figure 2.

This shows that not all subjects in the control group experienced the same increase as in the experimental group. The pretest results reached 67.07% and the posttest results reached 76.6%, an increase of 9.53%. Thus, students who were not exposed to subject matter through the child-language approach, showed a smaller increase in ability to read for comprehension than students who were exposed. Results of pretest and posttest experimental group and control group are shown in Figure 3.

The above figure shows that the experimental group has an average reading comprehension ability compared to the reading comprehension ability of the control group (see Table 1).

Mann Whitney test for the first sample calculation.

$U_1 = n_1 n_2 + \frac{n_1(n_1+1)}{2} - R_1$

$U_1 = 4 \text{ x } 7 + \frac{4(4+1)}{2} - 38$

$U_1 = 28 + \frac{20}{2} - 38$

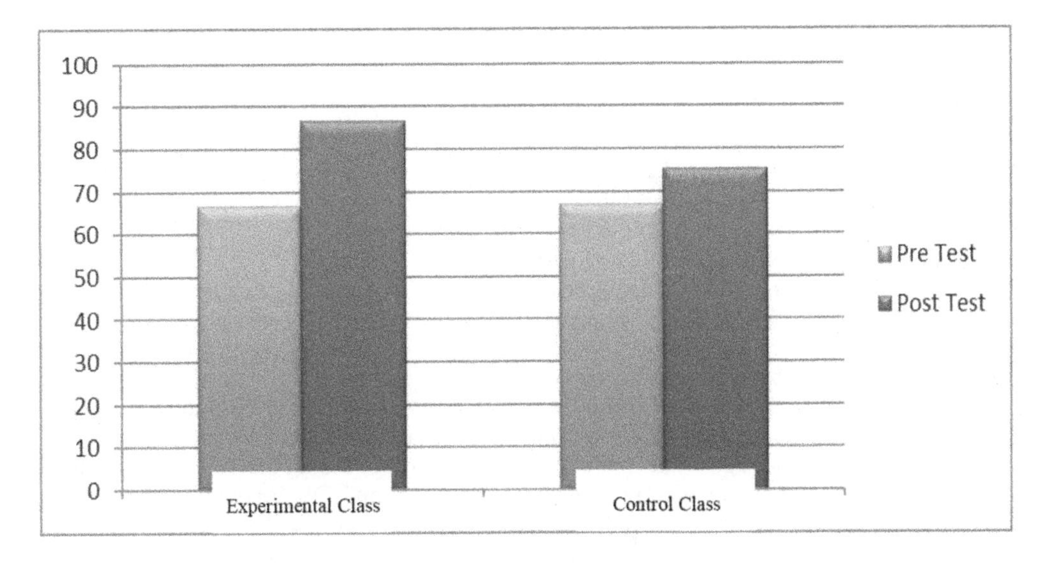

Figure 3. Pretest and posttest scores.

Table 1. Ranking of experiment group and control group scores.

Experiment Group				Control Groups			
No.	Subject	Score	Ranking	No.	Subject	Score	Ranking
1	TRA	93.3	11	1	WA	66.6	1
2	DK	86.6	9.5	2	DSN	70.0	2
3	SK	80.0	8	3	DM	73,3	3.5
4	RS	86.6	9.5	4	NS	83,3	6.5
$n_1 = 4$		$R_1 = 38$		5	VD	73.3	3.5
				6	RI	76.6	5
				7	SR	83.3	6.5
				$n_2 = 7$		$R_2 = 28$	

$U_1 = 28 + 10 - 38$

$U_1 = 0$

Calculation for the second sample.

$U_2 = n_1 n_2 + \frac{n_2(n_2+1)}{2} - R_2$

$U_2 = 4 \text{ x } 7 + \frac{7(7+1)}{2} - 28$

$U_2 = 28 + \frac{56}{2} - 28$

$U_2 = 28 + 28 - 28$

$U_2 = 28$

Smallest statistical computation:

$U_{\text{smallest}} = n_1 n_2 - U_{\text{biggest}} =$

$0 = (4)\,(7) - 28 =$

$0 = 28 - 28 = 0$ (correct calculation)

The calculation results have the smallest U value, namely U = 0

Criteria for rejection of H0 for one side if the score of Ucount Utable or opportunity (p) ≤ probability is real level (α), and vice versa. The price U = 0, for n1 = 4 is the chance (p) = 0.003, because p > probability is the real level or 0.003 > 0.05, then H0 is rejected, H1 is accepted. Thus, it can be concluded that there is a significant difference between the reading ability of the experimental group and the control group showing the positive effects of teaching subject matter through the child-language approach.

Based on the decision-making criteria, H0 is accepted if Ucount ≤ Utable, it is obtained Utable, 0.082 > (α = 0.05). Thus, H0 is rejected and H1 is accepted, meaning that learning using subject matter through the child-language approach is significant in increasing the reading comprehension ability of children with mild mental retardation, compared to that not given subject matter through the child language approach.

The ability of children with mild intelligence obstacles to read before being taught a lesson through the child-language approach is as follows. TRA, RS, and DK adequately understand the reading; SK has inadequate understanding. After teaching reading material through child-language approaches, reading comprehension for TRA, RS, and DK becomes very adequate; SK becomes adequate. In all cases, reading comprehension becomes better.

It is known that the ability to read for children in the control group is relatively low. There are two children with the lowest ability to read in the group as well. Therefore, the teachers made an application of reading comprehension which refers to the lesson plan, teaching material, and teaching and learning processes. The Ability to Read the Control Group's Understanding in the Pre-Test Stage, it is known that the ability to read the understanding of the control group before being given adequate treatment. There are also inadequate two children. Application of reading comprehension made by teachers regarding RPP, material, and process.

The thing that distinguishes the application of reading comprehension between the experimental group and the control group is in terms of the process alone because the lesson plans and materials used are the same, based on study of the lesson plan and seeing the learning process, which is delivered to students in accordance with the existing lesson plan.

Based on the data, it is known that the ability of children with mild intelligence barriers to read for comprehension before being given subject matter through the child-language approach is inadequate. The other three students are adequate. Reading comprehension carried in relation to the lesson plan illustrates active student learning. Study material is sourced from books. The learning process is in accordance with the lesson plans. The child-language approach to the material is based on the results of the experimental group students' interviews with the same number of paragraphs, sentences, and words as the subject matter from the source book.

Components and the number of questions in each component is the same, and the process of learning to read is the same as before: it only differs in the material conveyed. The reading comprehension results obtained by one student are adequate and for the other three students are very adequate. The results obtained by students in the experimental group show the application of reading comprehension skills through child-language approaches is effective in improving understanding reading skills for children with mild developmental delay.

Subject matter presented through child-language approaches can improve reading comprehension for children with intelligence barriers. Tarigan (2008) in Novitasari (2011) says, "the general ability of spoken language contributes to a background of beneficial experience and skills for teaching reading." The children have shown an increase in reading comprehension. This is because the language used is easier to read and be understood by children. Among the children who did not get treatment in the form of reading material containing children's language, some have increased and decreased. The increase was not evenly distributed. The ability of children with mild intelligence barriers to read for comprehension will increase if the subject matter is presented through a child-language approach. The potential of this child can still be optimized, holding on to constructivist theories that emphasize experience, as children's language is part of experience. The child's ability to read starts from saying what is on his or her mind. The ability to read for comprehension is needed in the learning process, because by understanding the contents of the reading the child will obtain information from what is read. What is learned by children with intelligence barriers must be meaningful. Broto (1975) in Abdurrahman (2003) argues, that reading does not only speak the written language or symbols of language sounds, but also responds to and understands the contents of written language.

The ability of children with mild intellectual delay to read with understanding can be optimized by providing subject matter through the child-language approach. This approach can be applied to other subjects, in connection with the application of reading comprehension in the teacher's previously made lesson plans.

In providing subject matter, the teacher chooses material that is in accordance with the existing SK and KD. The choice of material does not always originate in the same class as the child and might be taken from a lower class. The reading material can be sourced from source books, but for children with mild intellectual difficulties the material provided is in a language understood by the child. Reading comprehension in children with mild intellectual difficulties improves through the child-language approach, that is, is the provision of reading material that uses language derived from children. Based on the data obtained, children with mild intellectual difficulties who receive treatment in the form of child-language reading material develop the ability to better read for understanding. It can be seen from the results of the posttest that the experimental group's reading comprehension skills are very adequate. In the control group there was an increase and those who experienced a decrease, although both were given treatment. This is related to the opinion of Mursell (2008). Whether or not a lesson is successful depends on the level of meaning it contains for the child. Successful teaching must be based on lessons that contain as much meaning as possible for children, not on mechanical routines. Teaching will fail if the child is forced with threats or penalties to memorize things that have almost no meaning for the child.

4 CONCLUSION

In general, it can be concluded, reading comprehension of mildly delayed children with mental age of seven to nine years can be improved by providing subject matter through a child-language approach. The application of reading comprehension in mildly delayed children through the child-language approach is related to the material and the process of learning to read by the steps of exploring children's language first. The results obtained from the child are identified from the vocabulary, language structure, and emerging words. The preparation of reading material for the number of paragraphs, sentences, and words is equated with the subject matter through the source book, as are the components and the number of questions in order to have the same level of difficulty. After the reading material has been validated, it is then used with the experimental group students. The use of child-language material with mildly delayed children has shown to improve reading comprehension skills.

REFERENCES

Abdurahman. 2017. Effectivity of teaching model mediated learning based to improve speaking ability of student with intellectual disability. *Proceeding The 2nd International Conference on Special Educational Needs 2017.* Bandung: UPI Press.

Abdurrahman. 2003. *Children's Education Learning Difficulties.* Jakarta: PT Rineka Cipta.

Amin. 1995. *Children's Orthopedic of Children with Intellectual Disability.* Bandung: Ministry of Education and Culture.

Gall, M. D., Gall, J. P., & Borg, W. R. 2003. *Educational Research.* Boston: Pearson Education, Inc.

Hallahan, D. P., & Kauffman, J. J. 1982. *Exceptional Children: Introduction to Special Education*, New Jersey: Prentice Hall, Inc.

Mursell. 2008, *Teaching with Success (Successful Teaching).* Jakarta: Earth Literacy.

Novitasari. 2011. *Use of the Language Experience Approach to Improve Beginning Reading Ability for Children with Difficulty in Learning to Read*, Thesis on UPH PKh. Bandung: Unpublished.

Ormrod, J. 2009. *Educational Psychology, Helping Students Grow and Develop.* Sixth Edition. Jakarta: Erlangga.

Rochyadi, E. 2010. Effects of linguistic awareness and visual perception awareness on the ability to read: Beginning of children with developmental ability, *Proceedings First Annual Inclusive Education Practices Conference.* Bandung: Rizqi Press.

Slavin, R. 2011. *The Psychology of Educational Theory and Practice*, Ninth, volume 2. Jakarta: PT Index.

Soendari. 2008. *Teaching Module for Assessment of Children with Special Needs.* Bandung: Unpublished

Somantri. 2005. *Extraordinary Child Psychology*, Bandung: PT Rineka Aditama.

Sugiyono. 2011. *Quantitative, Qualitative and R & D Research Methods.* Bandung: CP Alfabeta.

Borderless Education as a Challenge in the 5.0 Society – Abdullah, Adriany & Abdullah (eds)
 ISBN 978-0-367-61960-2

Implementation of gamification strategy to improve the learning motivation of children with emotional and behavioral issues

N. Warnandi & R.N. Istighfar
Universitas Pendidikan Indonesia, Bandung, Indonesia

ABSTRACT: This study aims to determine the effect of applying gamification strategy to increasing the learning motivation of children with emotional and behavioral barriers. This research is an experimental research with one-group pretest/posttest design. Data collection was carried out using observation instruments and questionnaires regarding learning motivation. Sampling in this study used probability sampling with lottery sampling techniques. Obtained samples are six children with emotional and behavioral barriers of class. Data analysis used non-parametric statistical techniques using the Wilcoxon Test. The results showed that gamification had an effect on increasing the learning motivation of children with emotional and behavioral barriers. Gamification can be applied as an alternative in increasing the learning motivation of children with emotional and behavioral barriers.

1 INTRODUCTION

The learning process includes how the teacher modifies the material, modifies the process, and designs various activities to help students understand the material. It is no less important that teachers understand the characteristics of students and class routines in the learning process, especially for children with emotional and behavioral barriers who have different learning characteristics from children in general (Soemantri 2007). Emotional and behavioral barriers can have an impact on learning situations, such as limited concentration power, lack of ability to learn from experience, lack of motivation, and so on (Setiawan 2009).

Observations have found low motivation to learn in children with emotional and behavioral barriers. This can be seen from the behavior exhibited by students in class during the learning process: showing less enthusiastic attitude to learning, often in and out of class; disturbing other classes; disturbing friends when studying in class; and reluctant to do the assignments given by the teacher. Hatomi (2016) explains that learning motivation in children with disabilities tends to be low compared to children in general because the will, desires, and learning goals do not appear to follow learning. Children with emotional and behavioral barriers often refuse to learn and when the teacher asks them to do exercises, the child tends to find excuses (Santrock 2007). However, if the child is invited by the teacher to play games during recess, the child is very enthusiastic. The low motivation of these students is thought to be the learning strategies used by class teachers in teaching less interested, less motivated, and less engaged students. Teachers in schools feel uneducated about the learning strategies used and tend to use traditional teaching strategies such as drill questions and lectures. One alternative that can increase the learning motivation of children with emotional and behavioral barriers is to apply gamification, which is a game-based learning strategy where the elements contained in the game are applied to non-game activities, one of which is in the learning process. The elements consist of points, challenges, badges, levels, leaderboards, and several other game elements. Gamification is defined as the use of game mechanics, aesthetics, and game thinking to engage people, motivate action, promote learning, and solve problems (Kapp 2012).

Gamification is considered appropriate to increase the learning motivation of children with emotional and behavioral barriers, because the concept of game-based learning is in accordance with the characteristics of children who prefer playing games to learning in the classroom.

Gamification is expected to be able to accommodate the needs and learning styles of the 21st-century generation that are very close to the game world. The combination of the elements contained in the game is a learning strategy effort to make the delivery of learning more interesting and enjoyable so as to increase the learning motivation of children with emotional and behavioral barriers. The results of the study of the application of this gamification can be one solution in increasing learning motivation.

2 METHODS AND DESIGN

The method used in this study is a mixed method, using two techniques at the same time, quantitative and qualitative. This study involved all students with emotional and behavioral barriers in SLB E Handayani. The class is given a pretest (initial test) to find out the extent of the learning motivation of children with emotional and behavioral barriers before teaching with a gamification strategy, including teaching mathematics and science through gamification. Finally, we conducted posttest (final test) on the class to see the development of the learning motivation of students with emotional and behavioral barriers. At the end of the study, we also conducted a distribution of learning motivation questionnaires to see student responses related to learning with gamification strategies. The research design used in this study was concurrent triangulation designs as submitted by Creswell (2010): "research collects quantitative and qualitative data concurrently (at one time), then compares these two databases to find out whether there are convergences, differences, or some combination."

In this form of research design, researchers simultaneously collect quantitative and qualitative data. The data that has been collected is analyzed by quantitative and qualitative data analysis methods then results are interpreted to provide a better understanding of interesting phenomena at the stage of interpretation and discussion.

The learning motivation of children with emotional and behavioral barriers in this study will be explained in two phases, namely before and after being treated via mixed methods. In a quantitative research approach, researchers will formulate the problem based on the researchers' point of view through the analysis of observational data before and after being treated. Whereas in the qualitative approach, the problem is formulated by the researcher based on the perspective of the subject or participants studied through filling out the learning motivation questionnaire after being given treatment to strengthen quantitative data so that the mixed methods can be carried out properly.

3 RESEARCH RESULTS

Data obtained from the results of this study are quantitative data and qualitative data. Quantitative data is observational data of the learning motivation of children with emotional and behavioral barriers. The data was obtained from the pretest and posttest results. Qualitative data were obtained from the results of the learning motivation questionnaire given to each student. Research on increasing motivation to learn in children with emotional and behavioral barriers through this gamification strategy was designed in a Learning Implementation Plan, and was conducted for six meetings.

3.1 *Quantitative data description*

3.1.1 *Learning motivation score of children with emotional and behavioral barriers before being given learning with gamification (Pretest)*

Pretest score data of learning motivation of children with emotional and behavioral barriers before being given learning with gamification. Data shows the pretest scores of six research samples. The first sample with the inititals RW received a score of 42 out of a maximum score of 75, while the second sample with the initials DV obtained a score of 26 from a maximum score of 75. The total score obtained by AR from all four aspects was 34 out of a maximum score of 75, MN reached a score of 21, RR a score of 16, and, AS a score of 19 from a maximum score of 75.

3.1.2 *Learning motivation score of children with emotional and behavioral barriers after being given learning with gamification (Posttest)*

After six sessions of learning with gamification, a posttest was conducted. The first sample with the provincial RW received a score of 69 out of a maximum score of 75, while the second sample with the initials DV obtained a score of 36 out of a maximum score of 75. The total score obtained by AR from all four aspects was 61 out of a maximum score of 75, MN achieved a score of 57, RR a score of 56, and AS a score of 56 from a maximum score of 75.

The results of calculations using the Wilcoxon test formula are as follows:

The hypothesis proposed in this study is:

H1: Application of gamification increases the learning motivation of children with emotional and behavioral barriers in SL Handayani

H0: Application of gamification has no effect in increasing the motivation to learn of children with emotional and behavioral barriers in SL Handayani.

To test the hypothesis there are criteria for decision making namely:
H0 is rejected if J (count) ≤ J table
H0 is accepted if J (count) ≥ J table

Based on the results of the Wilcoxon test results that can be seen in Table 1, there is no negative differences, so all subjects were given a positive sign. From the calculation results obtained J (count) = 0 and based on the critical value of the Wilcoxon test at a significance level of 0.05 with the number n = 6, then the J table = 0 is obtained, then H0 is rejected because 0 ≤ 0. Thus the hypothesis proposed in the study this is accepted: it shows that gamification influences the learning motivation of children with emotional and behavioral barriers.

4 DISCUSSION

The results of this study indicate that before gamification, most children with emotional and behavioral barriers have low motivation. This is indicated by the acquisition of quantitative data at the pretest stage, which is then interpreted as one of three learning motivation criteria, namely high, medium, and low. The findings obtained from the pretest data, showed that most children with emotional and behavioral barriers have low motivation in learning. This is in line with the theory expressed by Hatomi (2016) that "learning motivation in children with disabilities tends to be low compared to children in general this is because the will, desires and goals for learning do not appear indicated by the inability of children with Tunalaras to participate in learning."

The first sample, RW, was shown to have better learning motivation than the others, as shown by having the highest pretest score among the group of six. RW is a child who is quite easygoing about accepting lessons, but he often shows verbal aggression to the teacher and his friends by

Table 1. Table with the Wilcoxon test on learning motivation.

No	Research Samples	Score					
		Pretest	Posttest	Different	Rank	(+)	(-)
1	RW	42	69	27	2.5	2.5	
2	DV	26	36	10	1	1	
3	AR	34	61	27	2.5	2.5	
4	MN	21	57	36	4	4	
5	RR	16	56	40	6	6	
6	AS	19	56	37	5	5	
Total						20	0

saying harsh words. RW tends to be easily bored and arrogant. Therefore, in the aspect of learning motivation related to affective domain achievement, he gets low ratings. Before implementing learning with a gamification strategy, RW tended to underestimate the teacher when explaining the lesson by saying, "What the heck are you not really clear", "Ah, that I already understand, ma'am." RW was also prone to annoying his friends when working on a task he had completed. He also often mocks his friend who gets a lower value than him. However, after the implementation of learning with the gamification strategy, RW began to show changes. He no longer shows verbal aggression to his teacher and friends. When studying, he is no longer busy bothering his friends and even sometimes gets the highest level in learning mathematics with a gamification strategy and becomes a mentor to friends who are at a level lower than him. Although when explaining to his friend, RW is sometimes upset with his friend who has difficulty understanding the material by saying, "Ah, you can't do this!" In carrying out the missions in gamification when learning science, RW is also better able to concentrate than before. RW shows a significant change in learning motivation, especially in the aspect of choice of tasks or interests. RW no longer plays and talks when working on a given task. This is consistent with the statement on the student motivation questionnaire which states that the learning that has been followed is interesting, not boring, encouraging students to complete missions in the game and encouraging more active involvement. But RW does not agree with the statement that many friends help him in understanding the material. According to RW, he feels that he is the one who helps his friends more in understanding the material.

In the second sample, DV, the pretest results show that learning motivation is still lacking in all four aspects. DV is the only female in class VII. DV who tends to be quiet and shy, which makes her less active in learning and easily discouraged if she encounters difficulties in learning and is reluctant to ask the teacher or her friends when she does not understand the material. DV rarely expresses opinions or responds to teachers who ask questions. As a result, DV has the lowest learning achievement among the group and gets unsatisfactory pretest results. Observation results show an increase in learning motivation as seen from behavioral changes during the process of learning mathematics and science. The most visible improvement in learning motivation is in the aspects of effort and persistence. During the learning process, DV would readily despair when unable to do tasks that are considered difficult and tended to remain silent. After the implementation of learning with gamification strategies, there was behavioral change: when learning Mathematics with a gamification strategy, DV tried to complete challenges at each level and was willing to try again when she failed to advance to the next level. DV even wants to ask friends and teachers for help on how to solve challenges in the game, although motivation to learn in the achievement aspect still needs to be improved. In the questionnaire, DV agreed that the learning that had been followed was fun and felt guided by the teacher. However, DV feels that the learning that has been followed does not encourage her to be more active, even though from observations that have been carried out DV looks more active during learning. This shows that there is a mismatch between the results of observations and the results of student learning motivation questionnaires.

The third sample, AR, is the child who is boldest against the teacher and often fights with other friends. He also often argues with teachers and classmates, especially with RW. AR is often seen as being unable to control his emotions and impulsive behavior. AR's learning motivation before learning with gamification was low. He often shows displeasure in participating in learning even skipping class hours. In addition, low motivation to learn is also indicated by low learning achievement. However, after the application of gamification, there was a change in AR. AR previously opposed the teacher several times saying, "Ah, ma'am, if I know that the lesson is difficult, it's better if I skipped class": after the application of gamification, he showed a significant change in behavior. In fact, after learning with gamification five times, he said "Ma'am learned rich yesterday again, ma'am, exciting!" Although the learning of mathematics with gamification includes children who are slow in achieving the level, AR does not easily give up. Children who get a high enough power acquisition in the game become a "Math Legend." The results of the questionnaire also showed that AR liked to follow the learning and he felt learning with gamification was not boring but he stated that he still had difficulty understanding the material.

The fourth sample, MN, tends to have high anxiety and often lies, which makes him less liked by his friends. MN is also easily bored in learning and often sleeps in class when learning takes place. According to the results of observations that have been made, MN tends not to ask at all if he does not understand about material. When working on assignments, MN is also often careless because he is easily discouraged if he encounters difficulties in learning. MN is classified as a child with high creativity. MN has a hobby of making graffitI, which is not unusual when children are learning. This is done because children feel bored with the usual learning. After the application of gamification, children begin to show an increase in learning motivation. This is evident from the results of observations that show MN begin to want to ask the teacher and his friend about material that is not yet understood. In addition, when doing assignments, MN also looks excited and when he encounters difficulties MN, is willing to try again with various strategies and even ask teachers and friends for help. However, there is a mismatch between the results of the learning motivation questionnaire and MN observation results. MN's questionnaire showed the same answer for all the question, both positive and negative. Even the "class" column, he answered "TK", even though he was currently sitting in class VII. This shows that MN was just careless in filling out the questionnaire.

The fifth sample, AS, showed that learning motivation increased from before the application of gamification. Gamification is able to accommodate the characteristics of AS, who has limited powers of concentration and is easily distracted. However, after the implementation of gamification, AS can more easily focus because AS really likes games and learning gamification is learning like playing a game. AS seemed to enjoy learning and at the end of the lesson AS even said "Try all the lessons learned like this. It's cool." AS's increased learning motivation can also be seen from the break time, which is usually used for playing and snacks but with gamification AS spends time working on missions in unfinished gamification. In accordance with the results of the questionnaire obtained, he stated that he agreed that the learning was fun and made him addicted to learning. But AS's behavior of damaging goods has not changed, as is evidenced when the writer prepares learning media for gamification, AS damages the media.

The sixth sample, RR, tends to be easily distracted which often results in low achievement. His attitude of opposing the teacher is also often shown during class learning. Before the implementation of gamification, RR's learning motivation was in the low category, and observations show that he did not ask the teacher or friends about the material he did not understand at all. However, after learning with gamification strategy, RR even asked the teacher and his friends five times. In addition, RR also showed changes of achievement, namely in the affective, cognitive, and psychomotor domains.

Overall, children with emotional and behavioral barriers who became the study sample experienced an increase in learning motivation after learning with a gamification strategy. This increase can be seen from the results of the pretest and posttest. Based on the results of the study, it is evident that gamification has an influence on the learning motivation of children with emotional and behavioral barriers of class VII. The results of the study are supported by studies such as Alsawaier (2018), Filippou et al. (2018), and Kusuma et al. (2018). "Game design techniques and game elements such as stories, points, and challenges [gamification] can motivate learners and change behavior."

REFERENCES

Alsawaier, R. 2018. The effect of gamification on motivation and engagement. *International Journal of Information and Learning Technology*.

Creswell, J. W. 2010. *Research Design Qualitative & Quantitative Approaches*. United States.

Hatomi, A. 2016. Efektivitas Model Value Clarification Technique (VCT) Dalam Meningkatkan Motivasi Belajar Anak Tunalaras Di Kelas 3 SDLB Bhina Putera Surakarta. *JASSI Anakku* 17(1): 34–39.

Filippou, J., Cheong, C., & Cheong, F. 2018. A model to investigate preference for use of gamification in a learning activity. *Australasian Journal of Information Systems* 22.

Kapp, K. 2012. *The Gamification of Learning and Instruction*. United States: Pfeiffer.

Kusuma, G. P., Wigati, E. K., Utomo, Y., & Suryapranata, L. K. P. 2018. Analysis of gamification models in education using MDA framework. *Procedia Computer Science* 135: 385–392.

Santrock, J. W. 2007. *Psikologi Pendidikan*. Jakarta: Prenada.

Setiawan, A. 2009. Mengembangkan Motivasi Belajar pada Anak Tunalaras. *JASSI ANAKKU* 8(1): 54–60.

Soemantri, T. S. 2007. *Psikologi Anak Luar Biasa*. Bandung: Refika Aditama.

Borderless Education as a Challenge in the 5.0 Society – Abdullah, Adriany & Abdullah (eds)
© 2021 Taylor & Francis Group, London, ISBN 978-0-367-61960-2

Burnout in a cheerful environment: Student stress survey in early childhood teacher education program, Universitas Pendidikan Indonesia

A.D. Gustiana, H. Yulindrasari, M. Agustin & O. Setiasih
Early Childhood Teacher Education Program, Universitas Pendidikan Indonesia, Bandung, Indonesia

ABSTRACT: To a certain degree, stress can play a motivating role in learning. However, too much pressure can create burnout in students, which, in turn, will inhibit student learning. Early detection of stress is vital to navigating situations that can perpetuate student burnout. ECTE curriculum involves a lot of music and art classes. The learning environment is also built on fun learning principles that are the core of early childhood education. Exposure to fun learning principles and activities does not prevent students from experiencing stress. This study surveys potential tension in the learning environment of early childhood teacher education (ECTE) program, Universitas Pendidikan Indonesia. We surveyed 49 students of the class of 2014–16 using questionnaires. The result shows that 60% of students experienced burnout. The symptoms that they experienced were loss of motivation and cognitive exhaustion. This research also identifies factors by which the students felt the most pressure. Lack of English language proficiency and the high financial cost of study are perceived to be the most significant causes of students' stress leading to burnout.

1 INTRODUCTION

Stress during an academic journey is very common among students. From the perspective of school psychology, students' stress could have two contradictory effects. On the one hand, stress could play a motivating role for success. On the other hand it could also sabotage the students' learning. Thus early detection of student stress is necessary to prevent the unwanted effect of stress. Unmanaged stress could develop into burn-out and depression that could lead to suicidal thoughts.

There are extensive studies about student burnout in the context of medical students (Brazeau et al. 2010, Dyrbye et al. 2006, Dyrbye et al. 2009) and technical-college student (Yang 2004, Yang & Farn 2005). Burnout in early childhood teacher education students is under-researched. This research will contribute to understanding student burnout in a perceived less-pressure discipline, such as early childhood education.

Stress is subjective, thus individual perceptions on the object of stress or the cause of stress are influenced by his/her experience in dealing with it (Lazarus & Folkman 1984). Students' interpretation of the demand of the coursework determines how the student cope with the demand. Some student may see the demand as too much, some may think that it is moderate, and some other may feel no pressure at all. Despite the subjectivity, early detection of students' potential distress is very important to help them cope with the stressors effectively. This research is part of our attempt to help our students overcome stress they endure during their study to become early childhood teachers. We aim at mapping out which stressors contribute to students' stress and burnout to inform our intervention in dealing with students' stress.

To provide understanding in the context of our research, the following section clarifies what we mean by student stress and burnout. It will be followed by a brief explanation of research methods we used for the research. Then, we will briefly explain our findings.

1.1 *Student stress and burnout*

Student burnout is defined as an emotional condition when students feel mentally exhausted from academic demands (Lightsey & Hulsey 2002, Skovholt & Trotter-Mathison 2014). The feeling of exhaustion and fatigue arrives from hard work, irrational expectations, lack of time to finish the tasks, and lack of leisure time to release stress. Psychologically, the symptoms of burnout range from feeling guilty, helpless, hopeless, trapped, sad, or ashamed, which produce a cycle of discomfort and finally result in resentment toward learning and the tasks that should be accomplished.

We can group learning/student burnout into three categories (Syah 1999): (a) sensory associated fatigue, such as hearing and vision problem; (b) physical fatigue that relates to physiological aspects; and (c) mental fatigue. Mental/psychological fatigue is the disturbance that contributes most to learning failure. The socioemotional learning environment and unorganised habits also contribute to student burnout (Dierkes et al. 2003).

1.2 *Contributing factors to student burnout*

There are three major factors contributing to student burnout: (a) student personality; (b) learning environment; (c) socioemotional relation with others within the learning community, such as with other students, teachers, and administrators (Jacobs & Dodd 2003).

Student personality includes gender and personality types. Men are more susceptible to stress and burnout then women. Women are more flexible then men, because women tend to be able to handle great stress due to the socialisation process that teaches women to embrace emotions, empathy, and compassion (Lightsey & Husley 2002). Most of ETCE students are women; hypothetically, they should be more resilient to burnout. Personality -wise, idealistic, enthusiastic, and obsessive students are prone to burnout, because they set a high learning goals and tend to push themselves to reach the goals; they invest a lot of time and energy in their learning up to the point that they get tired physically and mentally (Jacobs & Dodd 2003). When the result is not as they expected, they will get disappointed and take the failure seriously. Therefore, they are vulnerable to burnout.

Learning environment also contributes to student burnout. Excessive study load, extensive study hours, overwhelming responsibilities, routine and non-routine work, and other administrative work that exceeds students' capacity and ability are conditions that lead to student burnout. A negative relationship with peers, teachers, and other parties could also perpetuate burnout. Social conflict, harassment, and ineffective communication between students and teachers, and/or among students are risk factors to burnout.

2 METHODS

This research is a survey that is quantitative in nature. The survey took place in June 2017 with a total sample of 49 students from class of 2014 (17 students), 2015 (15 students), and 2016 (17 students). This research used simple random sampling. We used two questionnaires: the Burnout Symptoms Questionnaire (BSQ) that consists of 98 items and the Risk Factors of Burnout Questionnaire (RFBQ) that consists of 40 items. We used simple descriptive statistics to identify the burnout profile of ETCE students.

3 RESULTS AND DISCUSSION

3.1 *Level of burnout and the symptoms*

Figure 1 shows that 60% of ECTE students experienced a high level of burnout, and 40% of students had low levels of burnout. This shows that there is a serious problem that the ECTE program, as an institution, has to solve to overcome the burnout phenomenon. The following graphic explains in which burnout symptoms most students experienced.

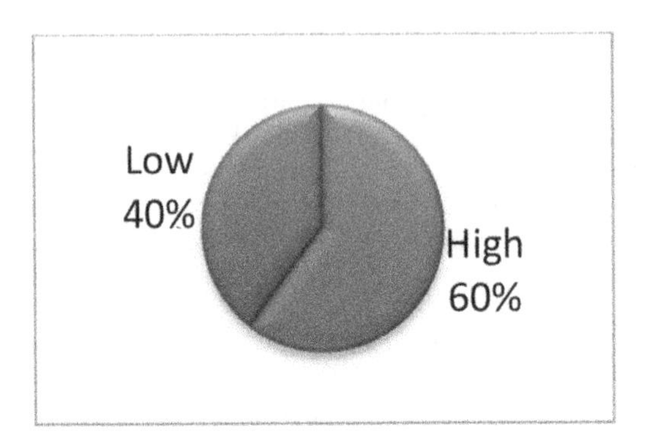

Figure 1. ECTE students' burnout level.

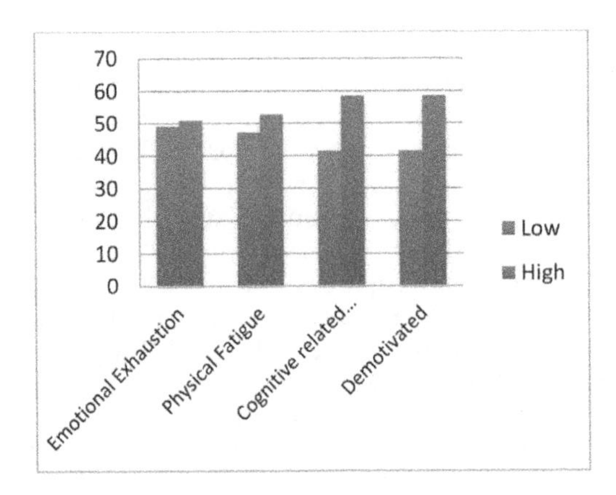

Figure 2. ECTE students' burnout symptomS.

Figure 2 shows that the students experienced demotivation and other cognitive-related symptoms such as inability to think clearly, lack of understanding of the learning content, and difficulty solving the academic problem that they need to complete. They also experienced emotional and physical exhaustion,though less commonly than the other two symptoms.

3.2 *The potential stressor leading to burnout*

We also studied potential stressors that might put our students at risk of burnout. We found two major categories of potential stressors: first, are factors that come internally from the students, or what we call personality issues; second, are factors that come externally, or external factors. The following sections will describe our findings in more detail.

3.3 *Personality factors*

There are four personality factors that, according to the students, may cause them stress. First, communication skill apparently plays an important role in a student's ability to cope with stress situations. Some 67.92% of students claimed that difficulty in making effective communication either with the teachers or with other students contributes to their stress. The difficulties include lacking direct communication with the teachers when they need direction in relation to their learning. Second, 58.59% students admitted that procrastination eventually made them stress because they were finally rushing to finish the task at the last minute. Third, being a perfectionist also brings disadvantages in learning. Perfectionists tend to have high expectations and set the target high. When they fail to reach the target or do not perform as they wanted to perform they are easily disappointed and stressed out. There were 37.74% students who claimed that wanting to be perfect in every task stressed them out. Forth, inability to set the possible outcome or setting the standard too high and not being able to achieve it led to learned helplessness, which is one of the most important symptom of burnout. Some 50.94% of the students felt that the gap between what they thought they could do and what they could actually do generates feelings of inadequacy and helpless.

3.4 *External factors*

We will list the external factors we found contribute to students' burnout in ECTE program. First, students of the sample mentioned that lack of consultative meeting with lecturers caused them stress, especially for students in the last year of their studies. Some 83.02% students said that lack of consultative meeting prolonged their study and made them unmotivated to continue what they were working on. Second, since English is foreign language for most Indonesian students, references written in English become challenging for them, and 98.11% of the students complained that they could not understand the books or journal articles written in English. Therefore, when the lecturer asked them to read the English source, they were likely to feel stress. Third, monotonous methods of teaching are also a problem, and 62.26% students got bored with the conventional teaching method that only requires them to listen to the lecturer. Fourth, financial factors. There are a lot of assignments in ECTE program that require many costly materials. In addition, not all students come from a middle-class family, therefore, the financial demand of the assignments became one of their sources of stress. Fifth, 54.72% of the students stated that their non-academic activities or personal problems also contributed to academic burnout they experienced.

The survey found that all the subdimensions of academic burnout have a high correlation with academic procrastination. High positive significance has been found between procrastination and study burnout, burnout caused by family, and boredom with teacher's behavior. Solomon and Rothblum (1984), in a survey that they took with university students, state that procrastination behaviour appears in 46% of the students in homework/project writing, 30% doing their reading homework weekly, 28% preparing for the exams, 23% attending the courses regularly, and 11% in administrative acts (paying the tuition, returning to the library). The fact that high school students in Indonesia have to deal both with academic studies and preparation for the university exam creates significant pressure on them. Although it seems an unhealthy application to utilize academic procrastination as a defensive mechanism, for teenagers who have to overcome the burnout caused by this problem, it may seem a creative adaptation.

4 CONCLUSION

There are four important conclusions we drew from the survey: first, students of early childhood teacher education programs are not immune to study-related burnout, although their study involves creating fun learning environments for young children. Second, knowing that lack of student–lecturer consultative session could perpetuate student stress and burnout, PGPAUD studies programs could encourage the lecturers (academic advisors) to increase their availability for student consultation. Third, literatures written in English are considered the culprit of student stress. PGPAUD studies programs could encourage students to take English courses or create an English club to enhance their English skills. Fourth, the high cost of assignments and other financial problems also contribute to stress. It is difficult to provide extra funding for students to cover their expenses for assignments, therefore we recommend ECTE lecturers be considerate about costs in giving assignments.

REFERENCES

Brazeau, C. M., Schroeder, R., Rovi, S., & Boyd, L. 2010. Relationships between medical student burnout, empathy, and professionalism climate. *Academic Medicine* 85(10): S33–S36.

Çakıra, S., Akçab, F., Kodazc, A F., & Tulgarerd S. 2013. The survey of academic procrastionation on high school students with in terms of school burn-out and learning styles. *Procedia: Social and Behavioral Sciences* 114(2014): 654–662.

Dierkes, M., Antal, A. B., Child, J., & Nonaka, I. (Eds.). 2003. *Handbook of Organizational Learning and Knowledge*. Oxford University Press, USA.

Dyrbye, L. N., Thomas, M. R., Harper, W., Massie Jr, F. S., Power, D.V., Eacker, A., Szydlo, D. W., Novotny, P. J., Sloan, J. A., & Shanafelt, T. D. 2009. The learning environment and medical student burnout: A multicentre study. *Medical Education* 43(3): 274–282.

Dyrbye, L. N., Thomas, M. R., Huntington, J. L., Lawson, K. L., Novotny, P. J., Sloan, J. A., & Shanafelt, T. D. 2006. Personal life events and medical student burnout: A multicenter study. *Academic Medicine* 81(4): 374–384.

Jacobs, S. R., & Dodd, D. 2003. Student burnout as a function of personality, social support, and workload. *Journal of College Student Development* 44(3): 291–303.

Lazarus, R. S., & Folkman, S. 1984. *Stress, Appraisal, Anti Coping*. New York: Springer.

Lightsey, R.O. Jr., & Hulsey, C. D. 2002. Impulsivity, coping, stress, burnout and problem gambling among university students. *Journal of Counseling Psychology* 49(2): 202–211.

Skovholt, T. M., & Trotter-Mathison, M. 2014. *The Resilient Practitioner: Burnout Prevention and Self-Care Strategies for Counselors, Therapists, Teachers, and Health Professionals*. London: Routledge.

Solomon, L. J., & Rothblum, E. D. 1984). Academic procrastination: Frequency and cognitive-behavioral correlates. *Journal of Counseling Psychology* 31: 503–509.

Syah, M. 1999. *Psikologi Pendidikan dengan Pendekatan Baru*. Remaja Rosdakarya: Bandung.

Yang, H. J. 2004. Factors affecting student burnout and academic achievement in multiple enrollment programs in Taiwan's technical–vocational colleges. *International Journal of Educational Development* 24(3): 283–301.

Yang, H. J., & Farn, C. K. 2005. An investigation the factors affecting MIS student burnout in technical-vocational college. *Computers in Human Behavior* 21(6): 917–932.

Developing educational competencies (Teachers) for education in the industry 4.0 Era

Borderless Education as a Challenge in the 5.0 Society – Abdullah, Adriany & Abdullah (eds)
© 2021 Taylor & Francis Group, London, ISBN 978-0-367-61960-2

The description of the ability and difficulties faced by preservice (PLP) teacher in conducting Classroom Action Research

M. Darmayanti, N.D.C. Anasta, A.R. Riyadi & T. Hartati
Universitas Pendidikan Indonesia, Bandung, Indonesia

ABSTRACT: Reflective action is an inseparable part of Classroom Action Research. Conducting Classroom Action Research becomes an essential part for teachers to improve their teaching quality and develop their profession. This research aims to describe teachers' abilities in executing Classroom Action Research and elaborate difficulties faced by the teachers. There are 4 indicators of teachers' abilities in conducting the Classroom Action Research (CAR), namely listing the background of problems, formulation of the problems, collecting, processing and displaying data, and writing the results of CAR. This research uses qualitative approach with descriptive method. The subjects of the research are 50 teachers serving as advisers of the preservice teaching program in Elementary Teacher Education Program (PGSD). The result of the research shows that teachers are qualified enough in compiling background of problems and designing the CAR thoroughly. However, two indicators indicate that the teachers still face some difficulties. The major difficulty is formulating problems, and the rests are collecting, processing, and displaying data.

1 INTRODUCTION

A teacher must have the competence described as a professional educator. Four competencies established are pedagogic, personality, social, and professional competencies. In professional competence, one thing teachers need to have is to develop professionalism in a sustainable manner by taking reflective actions. The competencies mentioned are: (1) continuously reflecting on one's own performance; (2) utilizing the results of reflection to increase professionalism; and (3) conducting Classroom Action Research to improve professionalism. Based on the explanation, it appears that teachers must have the ability to reflect, use, and conduct CAR (Menteri Pendidikan Nasional 2007).

Based on preliminary data obtained, there were difficulties faced by teachers in implementing CAR. Quoted from Kompas.com media, the fact that many teachers do not understand CAR makes them think that CAR is difficult. Furthermore, based on research, teacher literacy on research implementation is still insufficient and teachers have difficulty in finding related literature (Trisdiono 2015). The research focuses on the analysis of a teacher's difficulties in terms of linguistic and theoretical study aspects but does not reveal the ability and difficulties of teachers as a whole in conducting CAR. Therefore through this research, the researcher wants to reveal the Interpretation of teacher's ability and difficulty in conducting Classroom Action Research.

The role of a professional teacher in the learning process is very important as a key to a successful student-learning process. Professional teachers are teachers who are competent in building good learning processes so that they can help students develop all their potential. This makes the teacher a component of the focus of attention of the central government and regional governments in improving the quality of education, especially concerning the teacher competence. The development of teacher professionalism through the Teacher Educators program is an effort to increase teacher competency, one of which is through reflective action.

Reflective action is an important part of teachers who want the quality of learning that they manage to improve while increasing their professional development. Reflection should be carried out in an effort to evaluate and make self-introspection of the whole learning process that has been carried out. Reflection on learning and Classroom Action Research (CAR) is some forms of teacher's reflective actions that can improve the quality of learning as well as a bridge to develop himself. Reflection occupies a central place is participatory action research cycles of 'look, think and act' (Koch et al. 2005).

There are several forms of reflection in learning. This is due to the fact that reflections on learning are complex and involve various aspects of the implementation. There are three forms of reflective practices that include reflection in action, reflection on action, and reflection about action. Reflection in action is related to the decision-making process by the teacher when actively involved in learning. Reflection on action is a reflection carried out before and after the action is carried out. Reflection on action is a relatively comprehensive reflection activity, by taking a broader and deeper perspective and being critical of its learning practices by examining it from various other aspects, such as ethic, moral, politic, economic, sociology, etc. (Payong 2011).

CAR is one of the most appropriate researches used to follow up the results of learning reflection. The findings obtained from the reflection of learning determine the corrective action needed, and then the action is applied in learning process as well as in the form of CAR. Teacher reflection is not simply a teacher's self-description of what took place in his/her classroom. Critical self-reflection is a documented response of the application of ideas in the classroom (Pultorak 1996). CAR is a study of a social situation with a sight intended to improve the quality of action in it (Elliot 1991). This is consistent with what was stated that CAR is how a group of teachers can organize the conditions of their learning practices, and learn from their own experiences. They can experiment an idea of improvement in their learning practices (Wiraatmadja 2008).

A double result will be obtained by a teacher in the CAR, both improving the quality of learning and developing his professionalism. It is this fact that builds awareness that quality improvement or the effort to make the most appropriate learning improvements is improvements made in the real context by the actors themselves in their daily practice through systematic and documented efforts that are none other than through Action Research.

2 METHODS

The research method used is a case study. The case study method of research is used because it is relevant to research objectives, which is to obtain a specific and comprehensive picture of the ability and difficulties faced by preservice teacher in conducting Classroom Action Research. The participant of this study were 50 teachers who were teacher educators in the Primary School Teacher Education Program. The data were collected through an interview and a questionnaire. The qualitative data obtained were processed using the qualitative data processing techniques of the Miles and Huberman models, which included data reduction, data presentation, drawing conclusions and data verification.

3 RESULTS AND DISCUSSION

3.1 *Interpretation of abilities and difficulties in setting up background of problems*

Based on a result of the data collection, the ability of arranging the background of problems of almost half of the teachers is in the proficient category while the other half needs further guidance (Figure 1).

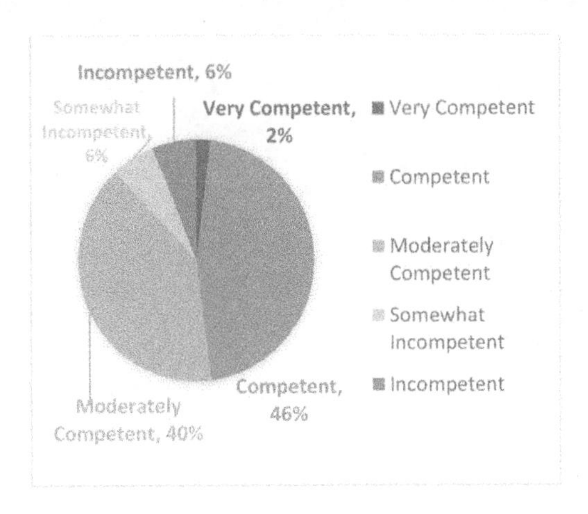

Figure 1. Interpretation of abilities and difficulties in arranging background of problems.

Based on the conducted interviews, teachers have difficulties, especially in several matters such as, analyzing core problems and systematic substance background, using written language, linking the main problem with the solutions taken, and determining the priority of problems experienced by students in learning.

Teachers' difficulties in analyzing problems is due to lack of understanding of learning reflection. Reflection activities are inseparable from the teacher's experience in teaching, namely evaluating activities by finding weaknesses and strengths of the learning that has been carried out. The process of reflection is an activity to reveal the meaning of learning from experience. This is an important activity; that experience is the basis for knowledge, and it leads to wise actions (Raelin & Coghlan 2006).

Some teachers experience obstacles in relating the subject matter to the solutions taken. That is due to the lack of conducting literature studies. This is in accordance with research, which states that teachers are still lacking in studying theories due to difficulties in finding literature (Trisdiono 2015). This gives the impact of the difficulty of connecting theory with practice.

3.2 *Interpretation of the abilities and difficulties in compiling the formulation of the problem*

Referring to Figure 2, almost half of the teachers found it quite difficult to compiling the formulation of the problem. Things that are considered difficult related to the formulation of the problem which comes in unclear format, confusion in collecting data and determining the priority of students' problems which are considered complex, and the adjusting of the background.

The process of making the formulation of the problem cannot be separated from the background because it relates to the priority of the problem to be solved by efforts to improve it. Therefore, it cannot be separated from connecting practice with theory. Making connections between theory and practice has proved elusive, as researchers have come to realize the complex set of skills teachers require to reflect critically on their practice in order to address the needs of students in schools (Darling-Hammond & McLaughlin 1995). Action research has more relevance to teacher practice because the problem under investigation is real and personal than what the student teacher will investigate in situations of temporary and artificial teaching practice (Volk 2010).

The appointment of the formulation of the problem leads to efforts to improve learning. Action research looks for answers to the question, 'how do I improve my work?' (Kitchen & Stevens 2008). It is in accordance to Kemmis and McTaggart (1988) who suggested that action research is generally defined as a form of educational research wherein a professional,

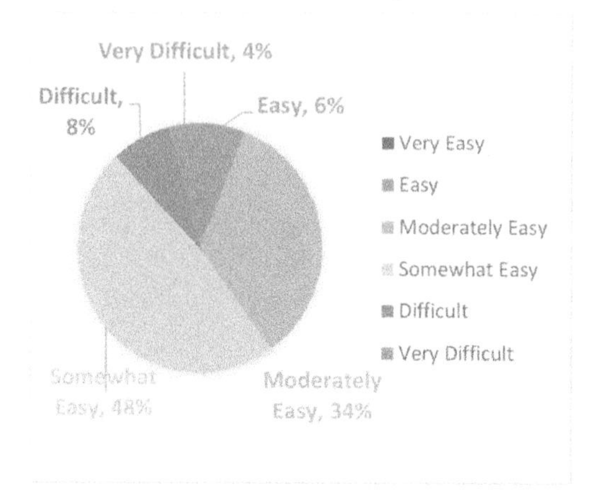

Figure 2. The Interpretation of abilities and difficulties in formulating problems.

actively involved in practice, engages is systematic, intentional inquiry into some aspect of that practice for the purpose of understanding an improvement. Analogized ACR benefits as an activity "to shed light of the democratic schools (Schmuck 1997).

3.3 *Interpretation of abilities and difficulties in collecting, processing, and displaying data*

An important capability in conducting CAR is to collect, process, and display data. Based on Figure 3, most teachers are insufficient in these abilities. These difficulties are in the process of elaborating data to the report (Chapter IV results and discussion), distinguishing ways of obtaining qualitative and quantitative data, processing data, having different formats, and lacking of reference books to compare field results with findings.

When teachers experience difficulties in collecting, processing and displaying data, it certainly becomes a big obstacle in implementing CAR. Action research is a scientific activity, meaning that the data obtained must be based on an investigation that has been carried out. The inquiry process involves data gathering, reflection on the action as it is presented in the data, generating evidence through the data, and making claims to knowledge based on conclusions drawn from validated evidence (McNiff 2002).

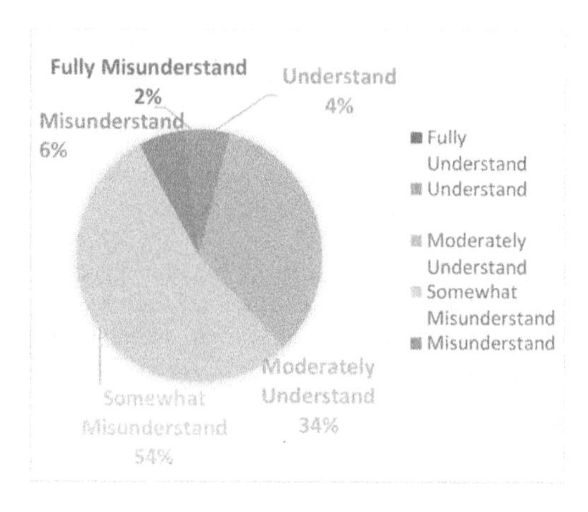

Figure 3. Interpretation of abilities and difficulties in collecting, processing, and displaying.

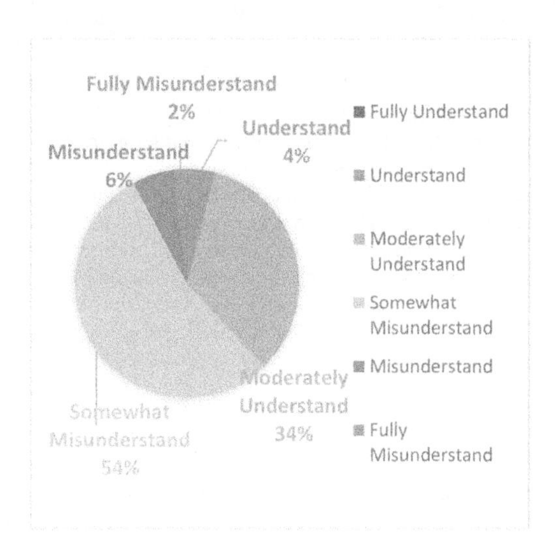

Figure 4. Interpretation of abilities in CAR.

3.4 *The interpretation of abilities and difficulties in writing ACR research*

The last stage in conducting CAR is making a report. Teachers' ability to report CAR can be seen in Figure 4. Judging from the clarity of the CAR report format design, more than half of teachers felt the format was clear or confident of the format, while the rest were less sure and some even felt the reporting format was unclear so that they were not sure whether CAR that they made would be accepted or not by the assessors. It was revealed that the conditions that caused this were confusing systematics, different guidelines, no examples of role models, and unknown CAR criteria.

The final stage after the teacher takes action is to report it as evidence in improving the competence of educators. In reporting action research, some teachers have difficulty because of inadequate amount of information about the format of the report appropriate for the assessors. This decreases teachers' motivation because they are unsure about the assessors' acceptance over the report. There are obstacles such as insufficient amount of time, motivation and confidence to do action research as more of an integral part of the teacher's professional portfolio (Best & Kahn 2006). Based on interviews, the teacher had difficulty in relating theory to the results of the study. The great difficulty actions researchers face is to bridge the two worlds of theory and praxis, but if they try to avoid these difficulties, they will be reduced to either consultants or academic scientists (Miller et al. 2003). By making research reports, teachers will strengthen their existence as professional educators. CAR can be a useful framework for teachers who work with marginalized youth for its potential to both highlight ways in which teachers are marginalized by association and help to improve their situations by giving them professional respect and voice in research (Stapleton 2018).

4 CONCLUSION

Action research is one effort to improve the quality of learning and teacher professional development. In general, teacher understanding and skills in conducting action research are in the sufficient category and tend to be insufficient. In compiling the background of the problem and making a full CAR design proficiently, the restrain is the different report formats. As for capabilities that must be further improved in formulating problem, difficulties are found in collecting, processing and displaying data.

REFERENCES

Best, J. & Kahn, J. 2006. *Research in education*. Boston MA: Pearson.

Darling-Hammond, L. & McLaughlin, M.W. 1995. Policies that support professional development in an era of reform. *Phi delta kappan* 76(8): 597–604.

Elliot, J. 1991. *Action Research for Educational Change*. Philadelphia: University Press.

Kemmis, S. & McTaggart, R. 1988. *The action research planner, 3rd*. Victoria: Deakin University.

Kitchen, J. & Stevens, D. 2008. Action research in teacher education: Two teacher-educators practice action research as they introduce action research to preservice teachers. *Action Research* 6(1): 7–28.

Koch, T., Mann, S., Kralik, D. & van Loon, A.M. 2005. Reflection: Look, think and act cycles in participatory action research. *Journal of Research in Nursing* 10(3): 261–278.

McNiff, J. 2002. *Action research for professional development: Concise advice for new action researchers (3rd ed.)*. Dorset, England: Author.

Menteri Pendidikan Nasional. Peraturan Menteri Pendidikan Nasional Nomor 16 Tahun 2007 Tentang Standar Kualifikasi Akademik Dan Kompetensi Guru. Jakarta: Kementrian Pendidikan dan Kebudayaan. 2007.

Miller, M.B., Greenwood, D., & Maguire, P. 2003. Why Action Research. 1(1): 9–28.

Payong, M.R. 2011. *Sertifikasi Profesi Guru: Konsep Dasar, Problematika, dan Implementasinya*. Jakarta: PT Indeks.

Pultorak, E.G. 1996. Followling the Developmental Process of Reflection in Novice Teachers: Three Years of Investigation. *Journal of Teacher Education* 47(4): 283–291.

Raelin, J.A. & Coghlan, D. 2006. Developing managers as learners and researchers: Using action learning and action research. *Journal of Management Education* 30(5): 670–689.

Schmuck, R. 1997. *Practical action research for change*. Arlington Heights, IL: Skylight.

Stapleton, S.R. 2018. Teacher participatory action research (TPAR): A methodological framework for political teacher research. *Action Research* pp. 1–18.

Trisdiono H. 2015. *Analisis Kesulitan Guru Dalam Melaksanakan Penelitian Tindakan Kelas*. Yogyakarta: LPMP Yogyakarta.

Volk, K.S. 2010. Action research as a sustainable endeavor for teachers: Does initial training lead to further action?. *Action research* 8(3): 315–332.

Wiraatmadja, R. 2008. *Metode Penelitian Tindakan Kelas*. Bandung: PT Remaja Rosdakarya.

Diversity in education

Borderless Education as a Challenge in the 5.0 Society – Abdullah, Adriany & Abdullah (eds)
© 2021 Taylor & Francis Group, London, ISBN 978-0-367-61960-2

The urgency of social support and being productive people: Public service reports

D.Z. Wyandini, S. Maslihah & H.M.E. Kosasih
Department of Psychology, Universitas Pendidikan Indonesia, Bandung, Indonesia

ABSTRACT: Mental disorder is a disease that interfers with the process of thinking and/or behavior that impact decision-making in daily life. If not handled properly, the disease will have a negative impact on people with mental disorders, their family, and society. The article aims to describe the social service practice concerning mental disorders rehabilitation in the Psychology Department of Universitas Pendidikan Indonesia in Bandung, Indonesia. The activity was aimed to provide psycho-education to attempt to solve problems related to mental disorders. Participants in this activity were 15 people with mental disorders, 5 social workers, and 15 local civilians. The activities included a presentation about the urgency of social rehabilitation for people with mental disorders, discussion with the local civilians, and observation of the social-life situation of people with mental disorders. The results show that the presentation increased knowledge about the urgency of social rehabilitation. The observation describes how the rehabilitation center provides treatment in the form of workshop making musical instruments such us acoustic and electric guitars, and others instruments from wood and bamboo. The interview shows that the workshop's wood/bamboo musical instruments can make people with mental disorders calmer and more productive in the rehabilitation center. The implication of this social service practice provides insight to reduce negative stigma about people with mental disorders.

1 INTRODUCTION

People with mental disorders (in Bahasa known as ODGJ: *Orang Dengan Gangguan Jiwa*) are one of the four main health problems in developed, modern, and industrial countries. Although the mental disorder is not considered a disorder that causes death directly, the severity of the disruption means that the inability will hamper development, because the individuals are not productive or efficient (Hawari 2001, Lestari & Wardhani 2014). Mental disorders interfere with the process of thinking and/or behavior that impacts decision-making in daily life (National Mental Health Association 2000). According to the American Psychiatric Association (2013), mental disorders are symptoms or patterns of clinically apparent psychological behavior that occur in a person and are associated with a state of distress (painful symptoms) or disability (disruption in one or more areas of important functions) which increases the risk of death, pain, disability, or loss of freedom.

ODGJ is not a condition where cause is easily determined. Many interrelated factors can cause mental disorders. Psychological factors (personality), mindset and ability to cope with problems, the existence of brain disorders, speech disorders, the condition of foster care, not being accepted in the community, and the presence of two problems and failure in life may be factors that can cause mental disorders. There are many factors that can trigger mental disorders (Lestari et al. 2014).

Experts estimate that 15% of the global population will have an ODGJ problem in 2020 (Harpham et al. 2003). Other data shows that the number of ODGJ in West Java increased sharply in one year. Based on data from the Division of Prevention and Transmission of Non-Communicable Diseases (P2TM), ODGJ in West Java in 2017 amounted to 11,360.

In 2018, the number was 16,714 according data from the West Java Health Office, 2018 (Sarasa 2019).

People with mental disorders (ODGJ) not only experience the effects due to symptoms and illness, but also stigmatization (Kapungwe et al. 2010). Some 75% of ODGJ experience stigma from the community, government, health workers, and the media (Hawari 2001). The main cause of stigmatization in the community and among health workers is the result of ODGJ violent behavior. Stigma can make treatment not continue, even though it is very important to know the severity of mental disorders.

The success of psychiatric therapy lies not only in psycho-pharmaceutical drug therapy and other types of therapy but also family knowledge and patient participation in treatment (Hawari 2001, Lestari & Wardhani 2014). There is no cost in treatment but the burden borne by families who live with people with mental disorders further worsen the situation. Therefore, there is a need for comprehensive treatment to help cure ODGJ, particularly social rehabilitation.

Social rehabilitation for ODGJ is a process of re-functionalization and development to enable a person to be able to maintain his recovery and carry out social functions properly. Social rehabilitation aims to help restore patient social functioning. Related to this, the role of mental rehabilitation employees, psychologists, or social workers is very important in carrying out the process of social rehabilitation. Patients are said to recover when they are able to communicate and interact well, as determined by tests conducted by the medical team.

2 RESEARCH METHODS

2.1 *Participants*

Participants were social workers and prospective volunteers at a social rehabilitation center, and people with mental disorders (ODGJ): a total of 15 ODGJ, 15 local residents, and 5 social workers. The rehabilitation center is located in Bandung, Indonesia.

2.2 *Programs*

The program includes presentations on community contributions to ODGJ rehabilitation presented by clinical psychologists; observation; and interviews about the situation at the rehabilitation center.

2.3 *Procedure*

The program began with a presentation on community contributions in the rehabilitation of ODGJ. This presentation explains ODGJ data in Indonesia, problems faced by ODGJ, alternative solutions for ODGJ rehabilitation, therapeutic communication, active listening, and family therapy. After that, the team members observed ODGJ activities in the rehabilitation center, therapies provided for ODGJ, and facilities in the rehabilitation center such as the condition of the room, bedroom, and sanitation. Other team members conducted interviews with ODGJ, volunteers, social workers, and the local residents who were present at the time of the activity. The questions were regarding local resident responses to the presence of ODGJ in rehabilitation center, therapies provided for ODGJ, the effects of therapy, knowledge of rehabilitation center workers about ODGJ, and participants' opinions about the presentation.

3 RESULTS AND DISCUSSION

Table 1 shows that one of the daily activities is chatting with employees (such as social workers) of the rehabilitation center. Conversations with ODGJ are a form of therapeutic communication. According to Fite et al. (2019), therapeutic communication is a purposeful

Table 1. Observation.

Aspect	Description
ODGJ activities in the rehabilitation center	ODGJ's daily activities: sleep, exercise, eating, bathing, routine checkups by health workers, taking therapy to make musical instruments, chatting with other patients and with employees at the rehabilitation center.
Therapies provided for ODGJ	ODGJ's therapy is to make musical instruments from wood and bamboo. They make musical instruments that are marketed to several regions in Indonesia, and also receive orders to make custom musical instruments.
Facilities in the rehabilitation center	One bedroom houses 4–6 people; there are 4 bathrooms; there are three meals a day cooked by a rehabilitation center employee and 1 patient, good room lighting, but lack of air circulation.

Table 2. Interview.

Question	Description
Local resident responses to the presence of ODGJ in rehabilitation center	Locals still don't have social awareness about the existence of the mental rehabilitation centre; locals still don't know how to deal with people with mental disorders in the rehabilitation centre; locals still have negative stigma about people with mental disorders.
Therapies provided for ODGJ	One of the main therapies provided at the rehabilitation center is making musical instruments from wood and bamboo.
The effects of therapy	The therapy can make ODGJ calmer and more productive.
Knowledge of rehabilitation center workers about ODGJ	Most rehabilitation center employees already have sufficient knowledge about ODGJ while prospective volunteers still do not have knowledge about ODGJ.
Participants' opinions about the presentation	The presentation increased knowledge about the urgency of social rehabilitation.

interaction between health professionals and patients that helps to achieve positive health outcomes. Conversations with fellow patients can create a helping relationship between two or more individuals or groups that provide and receive help or support to meet basic needs throughout life.

Table 2 shows that locals still have negative stigma about people with mental disorders. This fact is consistent with Kapungwe et al. (2010): 75% of people with mental disorders experience stigma from the public, government, health workers, and the media.

4 CONCLUSION

Through a public service program with the theme "Community Contribution to People with Mental Disorders (ODGJ)" the rehabilitation center is expecting to increase public knowledge about people with mental disorders (ODGJ). It is hoped this will reduce the negative stigma that arises in relation to ODGJ; it is further expected that people will be able to actively participate in the rehabilitation process of ODGJ healing.

REFERENCES

American Psychiatric Association. 2013. *Diagnostic and Statistical Manual on Mental Disorders, fifth edition (DSM-V)*. Washington, DC: American Psychiatric Press.

Fite, R. O., Assefa, M., Demissie, A., & Belachew, T. 2019. Predictors of therapeutic communication between nurses and hospitalized patients. *Heliyon* 5(10): e02665.
Harpham, T., Reichenheim, M., Oser, R., Thomas, E., Hamid, N., Jaswal, S., ... & Aidoo, M. 2003. Measuring mental health in a cost-effective manner. *Health Policy and Planning* 18(3): 344–349.
Hawari, D. 2001. *Pendekatan Holistik Pada Gangguan Jiwa Skizofrenia*. Jakarta: Gaya Baru.
Kapungwe, A., Cooper, S., Mwanza, J., Mwape, Sikwese, L. A., Kakuma, Lund, R.C., & Flisher, A. J. 2010. Mental illness: Stigma and discrimination in Zambia. *African Journal of Psychiatry* 13: 192–203.
Lestari, P., Choiriyyah, Z., & Mathafi. 2014. Kecenderungan atau Sikap Keluarga Penderita Gangguan Jiwa terhadap Tindakan Pasung. *Jurnal Keperawatan Jiwa* 2(1).
Lestari, W., & Wardhani, Y. F. 2014. *Stigma dan Penanganan Penderita Gangguan Jiwa Berat Yang Dipasung*. Surabaya: Pusat Humaniora, Kebijakan Kesehatan dan Pemberdayaan Masyarakat, Badan Litbang Kemenkes RI.
National Mental Health Association (2000). *People with mental disorder*. [Online], https://www.mhanational.org/people%20with%20mental%20disorder, Accessed by: 29 April 2019.
Sarasa, A. B. 2019. *16.714 Warga Jabar Menderita Gangguan Kejiwaan Berat*. [Online], https://daerah.sindonews.com/read/1379303/174/16714-warga-jabar-menderita-gangguan-kejiwaan-berat-1550314454, Accessed by: 29 April 2019.

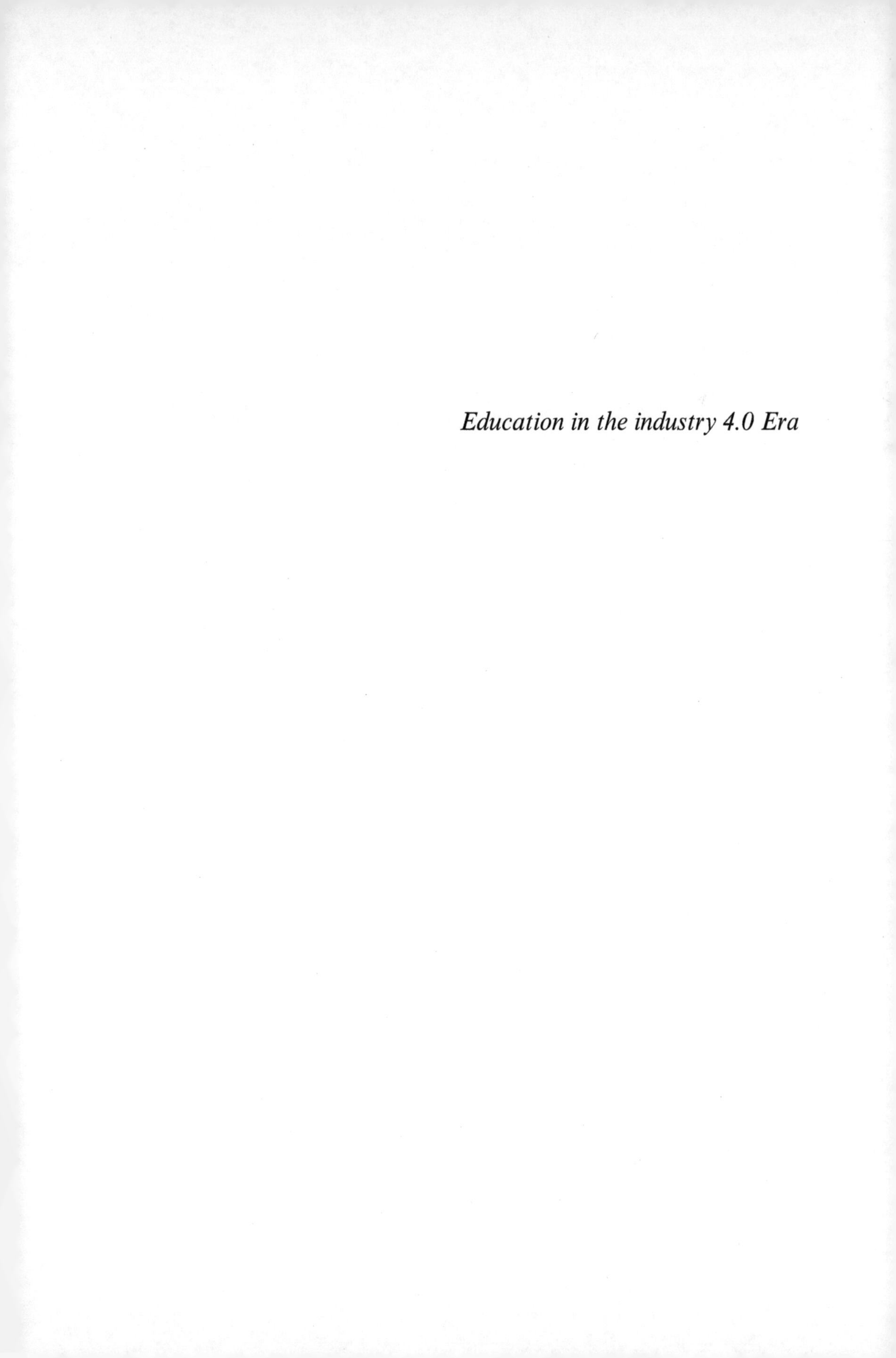

Education in the industry 4.0 Era

Borderless Education as a Challenge in the 5.0 Society – Abdullah, Adriany & Abdullah (eds)
© 2021 Taylor & Francis Group, London, ISBN 978-0-367-61960-2

Augmented Reality: The potential in mathematics education

M.I.S. Guntur & W. Setyaningrum
Universitas Negeri Yogyakarta, Yogyakarta, Indonesia

ABSTRACT: Technological developments encourage teaching and learning activities in the classroom to adapt. One of the technologies developed to collaborate with class learning is Augmented Reality (AR). AR is a technology that combines cyberspace and the real-world using smartphones, tablets, or laptops. This study aims to explain mathematical material that is suitable to be combined with AR and how AR can help teachers explain the material in the classroom. The research method used is a literature review and the data analysis used is a scoping review study based on Arksey and O'Malley (2005). The findings of the scoping review are discussed concerning multiple dimensions that are explored under research questions. The results showed that AR is very suitable to be developed in geometrical material but it does not close the possibility to be developed in different materials such as vectors, algebra, and others.

1 INTRODUCTION

Technological developments change the way people do things, ranging from how they search for the latest information to how they seek knowledge. For example, changes that occur in the world of education by technology are the use of computers to replace papers on national examinations (Retnawati et al. 2017). The use of instructional media in education is becoming increasingly diverse and interactive. One current change is the utilize Augmented Reality (AR). AR is a technology that can change the environment around the user into a digital interface that cannot be seen and felt by others by placing virtual objects in the real world in real-time – be it images, animations, sounds, or smells (Nugroho & Ramadhani 2015). With the ability to add virtual objects into the real world, this AR technology opens up unlimited opportunities (Farisi & Pratamasunu 2018).

The development of this technology is very rapid with increasingly large market growth encouraging researchers to continue to develop this technology (Serio et al. 2013). This is also supported by the fact that the hardware needed to implement this AR technology is only a smartphone camera and a marker. There are many researchers into the potential of AR in the world of education (Furió et al. 2013, Ifenthaler & Eseryel 2013, Majid & Majid 2018, Martín-Gutiérrez et al. 2015, Spector et al. 2014, Sylla et al. 2019). The research aims to determine suitable mathematical material to be developed using Augmented Reality. This research will be very useful for teachers or practitioners who want to develop AR in learning Mathematics.

1.1 *Augmented Reality*

Augmented Reality (AR) is a breakthrough in the interaction technology between humans and machines, which can cause the effect of computer-animated images in the real world (Hanan et al. 2018). According to Azuma et al. (2001), AR is a technology that combines two-dimensional or three-dimensional cyberspace into a three-dimensional real environment and then projects these virtual objects in a real environment. According to Pamoedji and Maryuni 2017, AR is a technique that combines two-dimensional and three-dimensional virtual objects

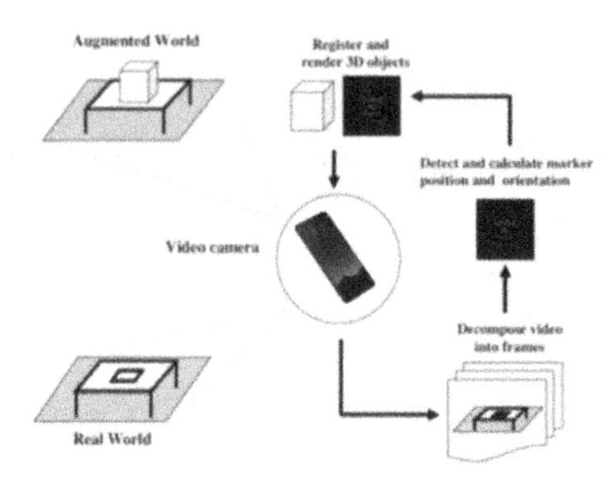

Figure 1. AR process in smartphone.

into a real three-dimensional scope and then projects the virtual objects into real-time. From these explanations it can be concluded that AR is a technology capable of adding both two-dimensional and three-dimensional virtual objects to real-world objects in real-time.

The working principle of Augmented Reality is tracking and reconstruction. At first, the marker is detected using a camera. Detection methods can involve a variety of missal edge-detection methods, or other image processing algorithms. Data obtained from the tracking process is used in the reconstruction of coordinate systems in the real world (see Figure 1) (Mustaqim & Kurniawan 2017).

The key to the success of AR technology is how realistic the virtual objects that are added blend with the real environment appear. This must be supported by a good digital-image-processing algorithm. The software must be able to calculate the position of real-world coordinates so that it can determine the exact coordinates to add virtual objects that have been prepared (Pamoedji & Maryuni 2017).

1.2 *Marker*

Marker is a special patterned image that has been recognized by the ARToolkit Memory Templates, where these markers function to be read and recognized by the camera then matched with the ARToolkit template. After that, the camera will render 3D objects above the marker. In general, the markers that can be recognized by ARToolkit are only square-shaped markers with black frames inside. However, as time goes by, many AR developers can create markers without black frames (see Figure 2) (Chari et al. 2008).

The virtual coordinates of the marker function to determine the position of the virtual object that will be added to the real environment. The position of the virtual object will lie perpendicular to the marker. The virtual object will stand in line with the z-axis and perpendicular to the x-axis (right or left) and y-axis (front or back) of the virtual marker coordinates (see Figure 3).

2 RESEARCH METHODS

This study followed the five-phase process developed by Arksey & O'Malley (2005).

2.1 *Identifying the research questions*

The research aims to determine suitable mathematical material to be developed using Augmented Reality. Research questions were developed after a literature review was carried out for this purpose.

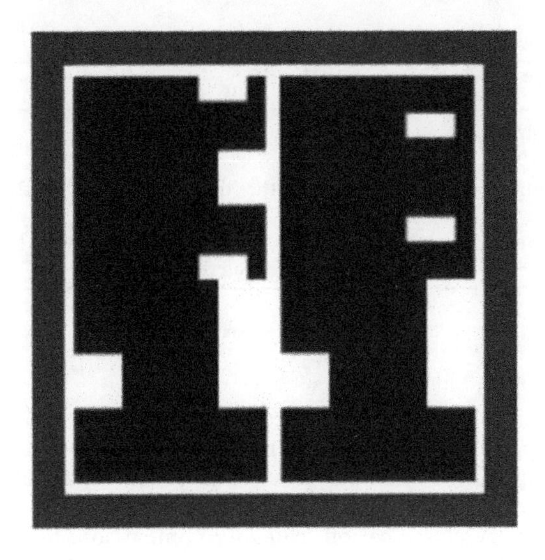

Figure 2. An example of a marker can be seen.

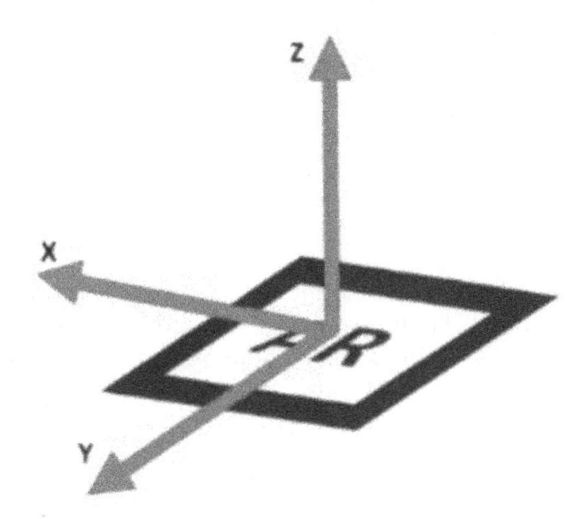

Figure 3. Illustration of the virtual marker coordinates.

2.2 *Identifying relevant studies*

The use of AR in education, especially mathematics, is a research topic that is currently developing. But the study of literature on AR in education, especially mathematics education, is limited (Akçayır & Akçayır 2017). The keyword used is Augmented Reality mathematics and found 1,106 articles.

2.3 *Study selection*

Not all data obtained can be used in this study. The next step was to reduce the articles that have been obtained. Google Scholar found 1,106 articles, which were then reduced to 5 articles. Criteria for selecting data as follows in Table 1.

Table 1. Inclusion and exclusion criteria.

Criteria	Inclusion	Exclusion
Period	The last 20 years (1999–2019)	Studies outside these dates or period
Study focus	Mathematics Education	Research other than the realm of Mathematics Education is not used
Research method	RnD, Development,	Qualitative, RnD, Quasi-experiment, Mix method Development, Survey Survei

2.4 *Charting of data*

In this phase, the article that was examined was first coded into the Microsoft Excel program with the help of ARF. It was observed that several articles included more than one sample. In the study, each sample group was coded separately. Data was checked with the help of content analysis methods. Content analysis is a method that includes text arrangement, category classification, category comparison and extraction of theoretical results (Cohen 2013).

2.5 *Collating, summarizing, and reporting findings*

The last phase included a comparison of results, summarising and reporting them. These can be found in the Results and Discussion section.

3 RESULTS AND DISCUSSION

Based on the Google Scholar article search, several articles fit the research objective criteria. All of the research results are described in Table 2.

Research conducted by Olief aims to use mobile augmented mentions as interactive learning media in cubes and cuboids. With this media, all combinations of cube nets and cuboids can be simulated in the user's Android smartphone without the need for long preparation and large space. The use of AR in learning also has a positive impact on students and teachers themselves (Farisi & Pratamasunu 2018). One of the features of OLief's AR is shown in Figure 4.

Table 2. Selected article.

Name	Years	Domain	Result
Olief	2018	Three-dimensional geometry	Mobile Augmented Reality as an interactive learning medium of cube and cuboid webs
Saiful Rizal	2018	Three-dimensional geometry	Development of mathematics learning media of cube and cuboid network based on Augmented Reality in class 5th basic school
Hannes Kaufmann	2003	Three-Dimensional geometry	Collaborative Augmented Reality in education
Achmad Buchori	2017	Three-dimensional geometry	Mobile Augmented Reality media design with waterfall model for learning geometry in college
Patricia Salinas	2013	Calculus	The development of a didactic prototype for the learning of mathematics through Augmented Reality

Figure 4. AR images developed by olief.

Media development carried out by Saiful Rizal is the Augmented Reality technology that can bring up 3D virtual objects in the form of cube nets and blocks directly on the user's device (Rizal & Yermiandhoko 2018). The difference in the development carried out by Saiful Rizal is that the cube nets have been opened and then students will determine what number is on the front side. This kind of knowledge is needed by students when taking the academic potential test, which is very common in Indonesia. One of the features of Saiful Rizal's AR is shown in Figure 5.

Construct3D is a three-dimensional geometry construction too. Construct3D is based on the Studierstube, which uses AR to allow multiple users to share a virtual space. They develop a system for the improvement of spatial abilities and maximization of transfer learning. Experimentation with geometric constructions improves spatial skills (Kaufmann 2002). This study aims to develop learning media to improve students' spatial skills. One of the features of Kaufman's AR is shown in Figure 6.

Achmad Buchori said that "students have difficulty if they are asked to determine the location of the diagonal in a three-dimensional object, especially if the diagonal is located beside or behind the objects. They find it difficult to imagine how the form of the diagonal is. However, by using AR-based geometry media, they can rotate the three-dimensional objects freely, by

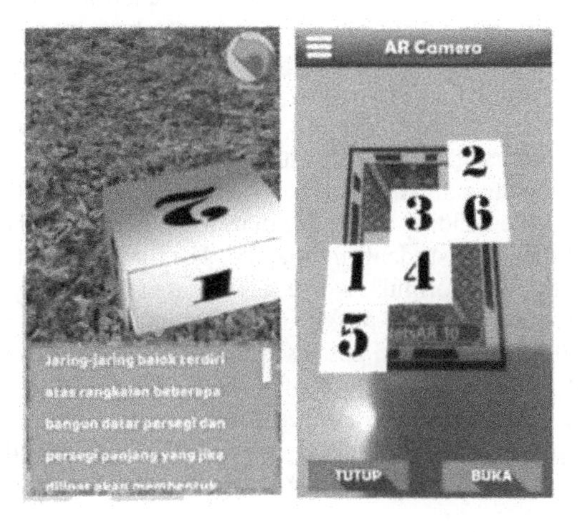

Figure 5. AR images developed by Saiful Rizal.

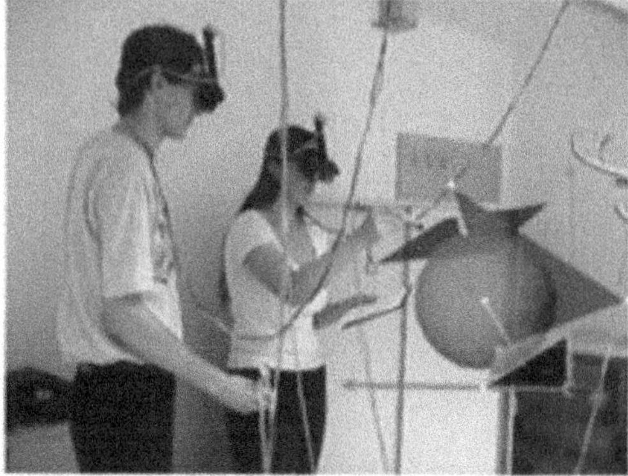

Figure 6. AR images developed by Kaufman.

Figure 7. AR images developed by Achmad Buchori.

merely moving the highlighted marker, rotating it, until they can find the part that they want to see" (Buchori et al. 2017). One of the features of Achmad Buchori's AR is shown in Figure 7.

In Patricia Salinas's research, this research focuses on displaying 3D objects from a function starting from parabola, ball, cuboid, or other 3D objects (Quintero et al. 2015). This study aims to display the 3D object of a function that will then be analyzed in calculus: for example, to determine the area or volume of the intersection of two functions. For example, the application of the concept of a limit in a function can also be displayed. This technology can replace similar applications such as GeoGebra or paint 3d on Microsoft with better capabilities because students can see the object from various sides. One of the features of Patricia Salinas's AR is shown in Figure 8.

In some of the research results previously discussed, AR can show an animation of how open cube nets open, which of course this will help students imagine the process of opening cube nets without having to make objects manually. A second research has shown that AR can visualize the distance between points in the building of a space that, when put in the classroom without any 3D objects, would be very difficult for students to imagine. The third is about AR's ability to visualize 3D object forms from a function formula, where this feature is still rarely found in other applications. Additional articles opt for the use of AR in learning mathematics but users are still limited to space and calculus. While the potential possessed by AR itself is widely publicized, as supported by Guntur et al. (2019), AR has a lot of potentials to be developed, as supported by Cheng & Tsai (2013). This is because AR has advantages besides being able to make three-dimensional objects visible through student smartphones. Augmented Reality can produce sounds and various other animations so that it can stimulate learning in the classroom to be more active.

Figure 8. AR images developed by Patricia Salinas.

4 CONCLUSION

Development of Augmented Reality to teach the geometry of three-dimensional objects is very possible. On the other hand, it also shows limited types of research that combine AR with mathematics education. Other material that might be developed using AR is a vector in which there are interrelated algebraic and geometrical analyses.

ACKNOWLEDGMENT

We would like to express ours to Yogyakarta State University for all the support given to the researchers. Through the service, we can write this article as one of the service out-put.

REFERENCES

Akçayır, M., & Akçayır, G. 2017. Advantages and challenges associated with Augmented Reality for education: A systematic review of the literature. *Educational Research Review* 20: 1–11.

Arksey, H., & O'Malley, L. 2005. Scoping studies: Towards a methodological framework. *International Journal of Social Research Methodology 8*(1): 19–32.

Azuma, R., Baillot, Y., Behringer, R., Feiner, S., Julier, S., & MacIntyre, B. 2001. Recent advances in Augmented Reality. *IEEE Computer Graphics and Applications* 21(6): 34–47.

Buchori, A., Setyosari, P., Dasna, I. W., Ulfa, S., Degeng, I. N. S., Sa'dijah, C., ... & Karangtempel, S. T. 2017. Effectiveness of direct instruction learning strategy assisted by mobile augmented reality and achievement motivation on students cognitive learning results. *Asian Social Science* 13(9): 137–144.

Chari, V., Singh, J. M., & Narayanan, P. J. 2008. *Augmented Reality Using Over-Segmentation*. Center for Visual Information Technology, International Institute of Information Technology.

Cheng, K. H., & Tsai, C. C. 2013. Affordances of Augmented Reality in science learning: Suggestions for future research. *Journal of Science Education and Technology* 22(4): 449–462.

Cohen, L., Manion, L., & Morrison, K. 2013. *Research Methods in Education*. Routledge.

Farisi, O. I. R., & Pratamasunu, G. Q. O. 2018. Mobile Augmented Reality sebagai media pembelajaran interaktif jaring-jaring kubus dan balok, *NJCA (Nusantara J. Comput. Its Appl.)* 3(2): 96–104.

Furió, D., GonzáLez-Gancedo, S., Juan, M. C., Seguí, I., & Costa, M. 2013. The effects of the size and weight of a mobile device on an educational game. *Computers & Education* 64: 24–41.

Guntur, M. I. S., Setyaningrum, W., Retnawati, H., Marsigit, M., Saragih, N. A., & bin Noordin, M. K. 2019. Developing Augmented Reality in mathematics learning: The challenges and strategies. *Jurnal Riset Pendidikan Matematika* 6(2).

Hanan, R. A., Fajar, I., Pramuditya, S. A., & Noto, M. S. 2018. Desain Bahan Ajar Berbasis Augmented Reality pada Materi Bangun Ruang Bidang Datar. In *Prosiding Seminar Nasional Matematika dan Pendidikan Matematika (SNMPM)* 2(1): 287–299.

Ifenthaler, D., & Eseryel, D. 2013. Facilitating complex learning by mobile Augmented Reality learning environments. In *Reshaping learning*: 415–438. Berlin, Heidelberg: Springer.

Kaufmann, H. 2002. Construct3D: An Augmented Reality application for mathematics and geometry education. In *Proceedings of the Tenth ACM International Conference on Multimedia*: 656–657.

Majid, N. A. A., & Majid, N. A. (2018). Augmented Reality to promote guided discovery learning for STEM learning. *International Journal on Advanced Science, Engineering and Information Technology* 8(4–2): 1494–1500.

Martín-Gutiérrez, J., Contero, M., & Alcañiz, M. 2015. Augmented Reality to training spatial skills. *Procedia Computer Science* 77: 33–39.

Mustaqim, I. & Kurniawan, N. 2017. Pengembangan Media Pembelajaran Berbasis Augmented Augmented Reality, *J. Edukasi Elektro* 1(1): 36–48.

Nugroho, N. A., & Ramadhani, A. 2015. Aplikasi pengenalan bangun ruang berbasis Augmented Reality menggunakan android. *Jurnal Sains dan Informatika* 1(1): 20–24.

Pamoedji, A. K., & Maryuni, R. S. 2017. *Mudah Membuat Game Augmented Reality (AR) dan Virtual Reality (VR) dengan Unity 3D*. Elex Media Komputindo.

Quintero, E., Salinas, P., González-Mendívil, E., & Ramírez, H. 2015. Augmented Reality app for calculus: A proposal for the development of spatial visualization. *Procedia Computer Science* 75: 301–305.

Retnawati, H., Kartowagiran, B., Arlinwibowo, J., & Sulistyaningsih, E. 2017. Why are the mathematics national examination items difficult and what is teachers' strategy to overcome it? *International Journal of Instruction* 10(3): 257–276.

Rizal, S., & Yermiandhoko, Y. 2018. Pengembangan media pembelajaran matematika materi jaring-jaring kubus dan balok berbasis Augmented Reality pada kelas v sekolah dasar, *J. Penelit. Pendidik. Guru Sekol. Dasar* 6(6).

Serio, Á. D., Ibáñez, M. B., & Kloos, C. D. 2013. Impact of an Augmented Reality system on students' motivation for a visual art course. *Computers & Education* 68: 586–596.

Spector, J. M., Merrill, M. D., Elen, J., & Bishop, M. J. (Eds.). 2014. *Handbook of Research on Educational Communications and Technology* (pp. 413–424). New York, NY: Springer.

Sylla, J., Cristina, & Cerquira. 2019. Learning basic mathematical functions with Augmented Reality, *ICST Inst. Comput. Sci. Soc. Informatics Telecommun. Eng.* 265: 508–513.

Global citizenship education

Borderless Education as a Challenge in the 5.0 Society – Abdullah, Adriany & Abdullah (eds)
© 2021 Taylor & Francis Group, London, ISBN 978-0-367-61960-2

Content analyses on peace education on curriculum 2013 in junior secondary schools in Indonesia

D. Wahyudin, T. Ruhimat, L. Anggraeni & Y. Rahmawati
Universitas Pendidikan Indonesia, Bandung, Indonesia

ABSTRACT: Peace education is a hot issue in schools in every country. However, its implementation is varied among school curricula. It is a process of promoting the knowledge, skills, attitudes, and values needed to bring about behavior changes that will enable children, youth, and adults to prevent conflict and violence, both overt and structural; to resolve conflict peacefully; and to create the conditions conducive to peace, whether at an intrapersonal, intergroup, national, or international level. Conflict and violence in the school social climate can have a negative impact on student learning. The current state, college school climate is not always peaceful and safe, because conflicts often occur in schools. Minor conflicts on schools among students can take many forms. The method used is research and development (R & D), simplified in three steps: preliminary study, model development, and model validation. The results showed that: (i) peace education in schools has cognitive, affective, and psychomotor aspects; (ii) implementation of peace education in schools has fostered a culture of peace, which can be seen in terms of peaceful negative and positive peace. This includes integrated approaches in subjects as well as extra curricularly (Pramuka/Scouts).

1 BACKGROUND

In every country, peace education is a hot issue. There has been discussion and development of topics to be introduced and mastered by citizens, especially the younger generation. It is also implemented in school curricula as subject matter or as topics integrated in many subjects. However, the implementation of peace education is varied in school curriculum among countries (UNESCO 2015). It is a multifaceted educational program that encompasses different approaches capable of transforming the behavioral patterns of people through the inculcation of desired knowledge, attitudes, and skills for effective contribution to the cultural, social, economic, and political development of their countries (Kartadinata et al. 2015, UNESCO 2015). In addition, Hicks as quoted by Alimba (2013) described peace education as activities that develop the knowledge, skills, and attitudes needed to explore concepts of peace, to enquire into the obstacles to peace (both in individuals and societies), to resolve conflicts in a just and nonviolent way, and to study ways of constructing a just and sustainable alternative future. Similarly, peace education is the process of promoting the knowledge, skills, attitudes, and values needed to bring about behavioral changes that will enable children, youth, and adults to prevent conflict and violence, both overt and structural; to resolve conflict peacefully; and to create the conditions conducive to peace, whether at an intrapersonal, intergroup, national, or international level (Fountain 1999, UNESCO 2012). In addition, peace education hopes to create a commitment to peace in the human consciousness (Harris 2009).

On the other side, school as a social system is a place that should have a climate conducive to supporting the learning process. A peaceful and pleasant environment is very conducive to facilitate a better learning process. Conversely, conflict and violence in the school or college social climate can have a negative impact on the learning process of students.

If we look at the current condition, college school climate is not always peaceful and safe. This is because conflicts often occur in schools and college schools, both simple and more serious conflict, either horizontally or vertically. They are horizontal conflicts such as interpersonal conflict among student groups in school, or between students in other schools, or between teachers and teachers. Vertical conflicts can occur between students and teachers, between teachers and college leaders, and between students and college leaders. Whatever the form, if present, conflict will at least interfere with the learning process and then will weaken the process and impact student achievement.

2 LITERATURE REVIEW

The curriculum can be seen as an instrumental strategic input in educational programs. Oliva (1988) and Wahyudin (2017) confirmed that the curriculum should be an instrument of reconstruction of knowledge, systematically developed to control managerial educational institutions; curriculum as that reconstruction of school and university to enable the learners to increase his or her control of knowledge and experience. However, there should be coherence between curriculum and learning knowledge and experience systematically developed under the auspices of the institution (Wahyudin 2016). First, the curriculum rests on purposes or goals – curriculum objectives to be achieved. Likewise, when the curriculum is conceived as the transmission of cultural heritage, the curriculum should serve to transfer cultural heritage to the next generation. Second, the curriculum is based on the context of the curriculum used. A curriculum based on, for example, essentialism, is seen as the transmission of cultural heritage by teaching the younger generation in order to prepare tham for a better life in the future. Third, the curriculum is based at strategic vantage points on the chosen curriculum development (Wahyudin 2017, Wahyudin 2018). Development also can't be separated from the processes, which have better teaching strategies and teaching techniques (Darling-Hammond & Bransford 2007, Hunkins & Ornstein 2016, Oliva 1988).

Peace education is described by UNESCO (2015) as the process of promoting the knowledge, skills, attitudes, and values needed to bring about behavioral changes that will enable children, youth, and adults to prevent conflict and violence, both overt and structural; to resolve conflict peacefully; and to create the condition conducive to peace, whether at an intra-personal, interpersonal, intergroup, national, or international level. In addition, Harris (2009) stated that peace education trys to inoculate students against the evil effects of violence by teaching skills to manage conflict.

Furthermore, a peaceful world was the dream of those who drafted the Charter of the United Nations in 1944: "We the peoples of the United Nations have resolved to save the succeeding generations from the scourge of war'; and UNESCO's Constitution: "to build peace in the minds of men" (UNESCO 1945a,b).

To recognize equal human dignity, it is essential to live peacefully together, in brotherhood and otherness. However, before long, aid was substituted by loans, cooperation for development by exploitation, multilateralism by plutocratic groups, and, even worse, the democratic principles of social justice and solidarity by the market laws. The result of "globalization" has been a profound crisis – financial, ethical, nutritional, environmental – which can be seen as an opportunity as well as a shifting for a "new beginning" (Kotite 2012).

Peace education stems from the Charter of the United Nations, which was established in 1945 in order to spare successive generations from the ravages of war and to foster respect for fundamental human rights, justice, and other fundamental freedoms (UNESCO 1945a,b). The Charter of the United Nations promotes understanding, tolerance, and friendship among all nations and all racial and religious groups. Although the world has not achieved this ideal, the preamble of the Charter of the United Nations is still a point of departure for local and global peace. The preamble declared as follows: (i) to save succeeding generations from the scourge of war, which twice in our lifetime has brought untold sorrow to mankind; (ii) to reaffirm faith in fundamental human rights, in the dignity and worth of the human person, in the equal rights of men and women and of nations large and small; (iii) to establish conditions under which justice and respect for the obligations arising from treaties and other sources of international law can be maintained; (iv) to promote social progress and better standards of life in larger freedom, and for these ends, to practice tolerance and live together in peace with one another as good neighbors; (v) to unite our strength to maintain international peace and security; (vii) to ensure, by the acceptance of principles and the institution of methods, that armed force shall not be used, save in the common interest, and (viii) to employ international machinery for the promotion of the economic and social advancement of all peoples (UNESCO 1945a,b).

In other perspective, especially in teaching learning in higher education, peace education is deeply concerned with human life and human well-being, and students and teachers should become peace activists. Peace education aims to foster both achieved and experiential knowledge, skills, and attitudes (Kartadinata et al. 2015). These are needed to achieve a sustainable global culture of peace. What is really vital in peace education is that the attitudes of teachers and students can be transformed and the process of changing attitudes among teachers and students is evidence of peace education itself. As two thirds of school children in the world do not have enough schooling opportunities beyond the fifth and sixth grade and nearly 125 million school aged children are out of school, it is urgently necessary to guarantee a peaceful school climate with peace education materials for this age group.

As comparison, the following is a table concerning the basic skills, knowledge, and attitudes that should be possessed by the younger generation in order to apply the spirit of peace education (Fountain 1999).

Table 1. The basic skill, knowledge, and attitudes in peace education.

Skill	Knowledge	Attitude
□ Critical thinking	□ Self-awareness	□ Self-respect
□ Problem solving	□ Peace and conflict	□ Honesty
□ Self-solving	□ Justice and power	□ Open-mindedness
□ Self-awareness	□ Human rights	□ Fair play
□ Assertiveness	□ Globalization	□ Obedience
□ Reading	□ Duties and rights of citizens	□ Caring
□ Orderliness	□ Environment/ecology	□ Empathy
□ Perseverance	□ Social justice and power	□ Tolerance
□ Cooperation	□ Nonviolence	□ Adaptation to change
□ Cheerfulness	□ Conflict resolution and transformation	□ Sense of solidarity
□ Self-control	□ Culture and race	□ Respect for differences
□ Self-reliance	□ Gender and religion	□ Gender equity
□ Sensitivity	□ Health care and AIDS	□ Sense of justice
□ Compassion	□ Arms proliferation and drug trade	□ Sense of equality
□ Active listening		□ Reconciliation
□ Patience		□ Bias awareness
□ Mediation		□ Appreciation
□ Negotiation		□ Transparency
□ Conflict resolution		

As explained previously, peace education is an important component of educational work. The goal of peace education is essentially the acquisition of the knowledge, skills, attitudes, and values that are necessary for the behavior of learners, whether children, adolescents, or adults, to always avoid the occurrence of conflict and "violence" in the environment, and then able to create peaceful conflicts and conditions conducive to peace, whether at the intra-personal, interpersonal, intergroup, national, or international level (Fountain 1999, Kartadinata et al. 2015, UNESCO 2015). In addition, Table 2 shows a summary of the purpose of peace education is the achievement of knowledge, skills, attitudes, and value values needed for the change of behavior of learners.

Table 2. The objectives of peace education among countries (Fountain 1999).

No	Knowledge	Countries
1.	Awareness of own, self-awareness	Yugoslavia1966, Rwanda
2.	Understanding nature of conflict and peace	Liberia 1993
3.	Ability to identify conflict and nonviolence	Burundi 1994
4.	Conflicts analyses	Sri Lanka
5.	Maintaining peace condition	Tanzania 1997
6.	Mediation process	Liberia 1993, Sri Lanka
7.	Understanding rights and obligation	Burundi 1994, Lebanon
8.	Understand the meaning of freedom	Lebanon 1993
9.	Awareness of cultural heritage	Lebanon 1993
10.	Recognition of prejudice	Burundi 1994
No	Skills	Countries
1.	Communication, active listening,	Burundi 1994, Croatia
2.	Assertiveness	Egypt 1995, Sri Lanka
3.	Cooperation	Croatia 1997, Egypt
4.	Affirmation	Croatia 1997, Sri Lanka
5.	Critical thinking	Egypt 1995
6.	Ability to think critically about a thing	Burundi 1994, Tanzania
7.	Ability to understand *stereotype*	Tanzania 1997
8.	Manage emotion	Rwanda 1997
9.	Problem solving	Liberia 1993
10.	Ability to generate alternative solution	Sri Lanka
11.	Construct conflict solution	Croatia 1997, Egypt
12.	Avoiding conflict	Yugoslavia 1996
No	Attitude	Countries
1.	Self-respect, positive self-image	Burundi, 1994, Egypt
2.	Tolerance and respect for differences	Yugoslavia, Lebanon 1993
3.	Bias awareness	Croatia 1997
4.	Gender equity	Egypt, 1995
5.	Empathy	Egypt 1995, Lebanon 1993
6.	Reconciliation	Croatia 1997, Liberia 1993
7.	Solidarity	Burundi 1994, Lebanon
8.	Social responsibility	Yugoslavia 1993, Lebanon
9.	Sense of justice and equality	Burundi 1994
10.	Joy of living	Burundi, 1994

3 RESEARCH METHODS

This research uses research and development method (R & D). Gall et al. (1996) describes the R & D of education as "a process used to develop and validate educational product." The result of development research is not only the development of an existing product but also to find knowledge or to answer practical problems. Furthermore, Gall et al. (1996) describes the main features in R & D, namely, studying research findings and conducting a study or preliminary research related to the product to be developed, and developing the product based on the research findings.

4 RESULTS AND DISCUSSION

4.1 *Condition of learning process occurred in the context of peace education*

Based on data obtained, peace education in primary schools in Indonesia is learned integratively in many subjects in terms of cognitive, affective, and psychomotor elements. In terms of cognitive, students study the cultural, social, religious, ethnic, and racial conditions that exist in Indonesia. Students study various conflicts that have occurred in Indonesia. In addition, students are also required to look for various conflict analysis, such as using 5W + H (where, what, who, when, why, and how) or using SWOT analysis (strength, weakness, opportunities and treatment). Apart from the cultural aspect, students also study the various political differences that exist in Indonesia. It becomes important to learn because the conflict is not only caused by cultural differences, but on differences in political views. In terms of history, students study the various backgrounds that resulted in differences in political views that have occurred in Indonesia, as well as studying the legal consequences of differences in political views. In addition, students study the boundaries of political differences that are still considered reasonable and political views that have deviated from the Pancasila of the 1945 Constitution.

Galtung (1986) has explained that peace is the absence of violence, not just personal and direct violence but also structural and indirect violence. The forms of structural violence are the absence of the distribution of wealth and resources, as well as the absence of power distribution over decisions on the distribution of resources. Galtung (1986) and UNESCO (2015) preferred to formulate peace as: absence of violence and the presence of social justice. Galtung (1986) calls the former a positive meaning, and he calls it a negative peace; the latter is defined as a positive condition (equal distribution of power and resources), and it is a positive peace.

The fourth stage is the students must present the mini research results to be discussed in the class, so that each student can know the components related to the flower-petals model in fostering a culture of peace in society.

The procedures for implementing peace education can be measured by the indicator of a peaceful atmosphere. There are three major frameworks of how the peace lies, namely, social norms, state structures or political stability, and environmental characteristics. These three frameworks can be translated into some important issues as outlined below (Reardon 1997, Roche 2003).

- Social norms: (i) The growth of peace education, which includes cooperation and conflict resolution through dialogue, negotiation, and nonviolent relationships among the citizens; (ii) Respect for women and their activities or gender justice; (iii) Growing understanding, tolerance, solidarity, and the same obligation to achieve better social cohesion and reduce the growth of hostility.
- The construction of state structures and political stability: (i) The growing participation of a more democratic society capable of fighting for the needs of its citizens; (ii) Growing open communication characterized by the principle of transparency and accountability; (iii) Guarantees of human rights follow with real recognition and inclusion of the various groups.

- Environmental characteristics: (i) The growth of social security both locally and internationally rather than fueling power struggles and arms competition; (ii) Strengthening sustainable development that prioritizes harmony with the environment.

Implementation of a culture of peace at the University of Education of Indonesia can be viewed in terms of peaceful negative and positive peace. Negative peace is implemented by creating a state of direct conflict with the role of the school to create a culture of peace in the life of the school, in the sense that the condition of the students currently does not reflect the peaceful condition so it is necessary for the role of the school in creating a culture of peace to change. Seen in terms of positive peace, has created welfare in the components associated with school life, so as to bring a culture of peace by itself into the life of the school. The implementation of peace education at the University of Education of Indonesia aims to provide awareness to students to always cultivate a culture of peace. Implementation conducted by sampled schools is a continuous and sustainable activity to foster a culture of peace in school life.

In the implementation of peace education it is necessary to foster the character of peace. Roche (2003) shows some values that need to be taught to learners to foster the character of peace: (i) Affirmation is the recognition and appreciation that is open to the various strengths and potentials that exist in each person or group; (ii) Communication is the ability to not only convey ideas to others orally or written, but includes also the skills to listen; (iii) Cooperation is working together to achieve the same goals, share insights and findings and move together to reduce the climate of competition and hierarchy in social relations; and (iv) Conflict resolution is the solution to dispute in the community through a peaceful way, not violence.

In addition, the implementation of peace education needs to look at the sustainability of the culture of peace. Reardon (2000), Schreiber and Siege (2016), and UNESCO (2015) in the Declaration of a Culture of Peace mentioned that a culture of peace is an attitude, action, tradition, and model of behavior and way of life based on the following: (i) Appreciate life, end violence, and promote nonviolent action through education, dialogue, and cooperation; (ii) Full appreciation of the principles of sovereignty, territorial integration, political independence of the state, and the absence of intervention on the internal issues of a country relating to the UN Charter and international law; (iii) Full appreciation of and promoting respect for all basic human rights and freedoms; (iv) Commitment to peaceful resolution of the conflict; (v) Efforts to find development and environmental needs not only now but also for future generations; (vi) Respect and promote development rights; (vii) Respect and promote equality of rights and opportunities for men and women; (viii)Respect and promote the rights of everyone to freedom of expression, opinion and information; (ix) Follow the principles of freedom, justice, democracy, tolerance, solidarity, cooperation, respect for diversity, cultural differences, dialogue, and understanding at every level of society and nation.

In addition, the implementation of peace needs to be supported from every component of nation and state life, as efforts to create peace are a humanitarian task throughout history that also colored civilization. Creating peace is a concrete step to eliminate conflict and violence. Nevertheless it should be pointed out that conflict and violence have existed since humans existed and will continue to color human life if there is no attempt to uphold peace. Thus, commitment to peace must be more important than racial and ethnic ties. In pluralistic country life such as in Indonesia, intercultural and inter-religious dialogue becomes a central activity that can nurture a culture of peace and maintain that religion is not used to legalize violence including in the form of conflict or war between countries, via mass murder, trampling on the human rights of others through terrorism and organized violence.

5 CONCLUSION

The learning process of peace education was implemented in order to develop students" cognitive, affective, and psychomotor abilities. It can be applied both in the classroom and outside the classroom. The models of peace education are able to foster a culture of peace among students, characterized by harmony with nature; human rights and responsibility; respect for cultural differences and solidarity; justice in society; and the awareness in each individual of the need to create a culture of peace. The implementation of the peace education at sampled schools is done gradually, integrated into many subjects. It is also done in extra curricularly as in Pramuka (Scouts) (Kartadinata et al. 2015, Wahyudin 2017, Wahyudin 2018).

REFERENCES

Alimba, N. C. 2013. Peace education: Transformation of higher education and youths empowerment for peace in Africa. *International Journal of Scientific and Technological Research* 2(12): 338–347.

Darling-Hammond, L., & Bransford, J. (eds.) 2007. *Preparing Teachers for a Changing World: What Teachers Should Learn and Be Able to Do*. Hoboken, New Jersey: John Wiley & Sons.

Fountain, S. 1999. *Peace Education in UNICEF*. New York: UNICEF Publication.

Gall, M. D., Borg, W. R., & Gall, J. P. 1996. *Educational Research: An Introduction*. London: Longman Publishing.

Galtung, J. 1986. On the anthropology of the United Nations system. In D. Pitt & T. G. Weiss (eds.), *The Nature of United Nations Bureaucracies*: 1–22. London: Croom Helm.

Harris, I. 2009. Peace education: Definition, approaches, and future directions. *Peace, Literature and Art* 1: 77–96.

Hunkins, F.P. & Ornstein, A.C. 2016. *Curriculum: Foundations, Principles, and Issues*. London: Pearson Education.

Kartadinata, S., Affandi, I., Wahyudin, D., & Ruyadi, Y. 2015. *Pendidikan kedamaian*. Bandung: Remaja Rosdakarya.

Kotite, P. 2012. Education for conflict prevention and peacebuilding. In *Meeting the Global Challenges of the 21st Century*. Paris: International Institute for Educational Planning.

Oliva, P. F. 1988. *Developing Curriculum: A Guide to Problems, Principles, and Process*. New York: Harper & Publisher.

Reardon, B. A. 1997. Human rights as education for peace. In G. J. Andrepoulos & R. P. Claude (eds.), *Human Rights Education for the Twenty-First Century*: 21–34. Philadelphia, PA: University of Pennsylvania Press.

Reardon, B. A. 2000. Peace education: A review and projection. In B. Moon, S. Brown & M. Ben-Peretz (eds.), *Routledge International Companion to Education*. London, NY: Routledge.

Roche, D. 2003. *The Human Right to Peace*. Toronto, ON: Novalis.

Schreiber, J. R., & Siege, H. 2016. Curriculum framework education for sustainable development. In *On Behalf of: Standing Conference of the Ministers of Education and Cultural, Affairs (KMK), German Federal Ministry of Economic Cooperation and Development (BMZ), Engagement Global gGmbH* (2nd updated and extended edition). Berlin/Bonn: Cornelsen.

UNESCO. 1945a. *Charter of the United Nations and the Statue of the International Court of Justice*. New York, NY: United Nations Department of Public Information.

UNESCO. 1945b. *Constitution of the United Nations Educational, Scientific and Cultural Organization*. London, UK: UNESCO. Adopted on November 16, 1945.

UNESCO. 1980. *World Congress on Disarmament Education: Final Document and Report*. Paris, France: UNESCO.

UNESCO. 2012. *Education Sustainable Development*. Paris, France: UNESCO.

UNESCO. 2015. *Rethinking Education towards a Global Common Good?* Paris: UNESCO.

Wahyudin, D. 2016. A view on teaching philosophy in curriculum implementation at the Indonesia University of Education. *SOSIOHUMANIKA* 9(2): 235–248.

Wahyudin, D. 2017. *Curriculum Development and Teaching Philosophy*. Saarbrucken, Germany: Lambert Academic Publishing.

Wahyudin, D. 2018. Peace education curriculum in the context of Education Sustainable Development (ESD). *Journal of Sustainable Development Education and Research* 2(1): 21–32.

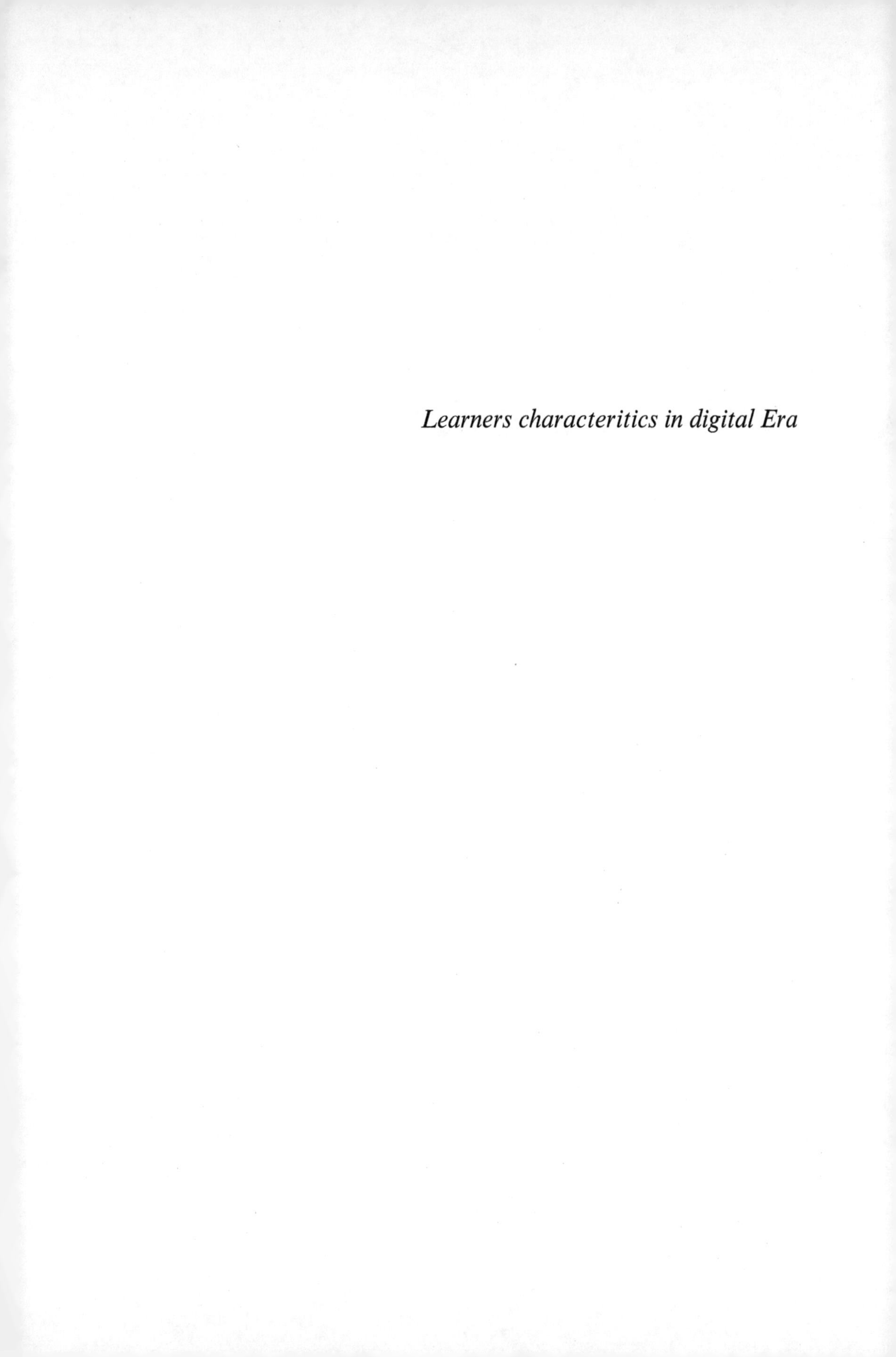

Learners characteritics in digital Era

Borderless Education as a Challenge in the 5.0 Society – Abdullah, Adriany & Abdullah (eds)
 ISBN 978-0-367-61960-2

Conceptions and awareness of information literacy among professionals and students of Vietnam's upper secondary schools

H.T. Ngo
University of Social Sciences and Humanities, Vietnam National University, Ho Chi Minh City, Vietnam

ABSTRACT: This study focuses on investigating the conceptions and awareness of Information Literacy (IL) among professionals and students in Vietnam's upper secondary schools in order to provide an appropriate suggestion regarding this issue in the implementation of an IL teaching model in the country. The investigation, aimed at achieving the above purpose, was conducted in three upper secondary schools in the country using an open-ended questionnaire and interviews. The questionnaire aims to explore students' conceptions and their awareness of the IL concept. Meanwhile, the interviews were used for professionals. This study contributes to existing knowledge by providing an understanding of professionals and students' conceptions and their awareness of IL in Vietnam's high schools, which is an under-researched context. The findings reveal that there is a real need to work towards enhancing professionals and students' understanding and their awareness of the role of IL to learning success.

1 INTRODUCTION

Vietnam is in the process of reforming its education system to improve the quality of education. Therefore, factors affecting the improvement of the education system should be considered comprehensively. Lifelong learning has begun to appear as one of the fundamental educational objectives in the country. This raises the question as to what schools in the country should do to prepare students to become lifelong learners. There is evidence that information literacy (IL) is an essential tool to promote lifelong learning (American Library Association [ALA] 1989). Thus, educators and students should gain an in-depth understanding of IL and its impact on their teaching and learning in order to achieve educational goals successfully.

There is a wide range of definitions of IL and this term has its own development history. A number of studies were conducted to explore the conceptions and awareness of IL in different contexts and populations. Studies introduced different results. Previous research as a whole indicate that there is a limited understanding of the different elements of IL among students. However, the literature shows that conceptions and awareness of IL in Vietnam remain under-represented. This research was conducted to redress this gap by exploring Vietnamese professionals and high school students' understanding of the IL concept and their awareness in this area. This study aims to (1) Explore whether professionals and students know the term 'information literacy'; (2) Determine how they define the term; (3) Explore their awareness of the concept. This study follows the view of Walton & Cleland (2013) who indicate that the development of IL capability ensues from a broader social setting and that IL comprises three spheres, finding, evaluating and using information, and each scope activates its own set of behavioral, cognitive, metacognitive and affective elements. This view was used as a framework to examine participants' IL conceptions and their awareness.

2 LITERATURE REVIEW

2.1 *Information literacy definitions*

In 1974, the phrase "information literacy" was first used by Paul Zurkowski. In his own definition, IL is considered as 'techniques and skills' in relation to the work environment and, specifically, problem-solving. The most influential definition to date is from ALA which states IL as a set of abilities allowing individuals to "recognize when information is needed and have the ability to locate, evaluate, and use effectively the needed information" (ALA 1989). ALA also stresses that "information literate people are those who have learned how to learn...They are people prepared for lifelong learning, because they can always find the information needed for any task or decision at hand" (ALA 1989). This process indicates the practice of searching for, selecting and evaluating information, as well as allows the user to create new ideas to interconnect to other people using a range of technological devices. This definition also regards IL as a set of skills and puts the emphasis on a lifelong learning context, showing that IL is required to promote independent learning (Andretta 2005), which equips people with the necessary capacities to enable them to learn during the course of life.

In a somewhat similar use of the term, the United Kingdom's Chartered Institute of Library and Information Professionals (CILIP) views IL as "knowing when and why you need information, where to find it, and how to evaluate, use and communicate it in an ethical manner" (CILIP 2013). One well-known IL definition was developed, in 2011, by the Society of College, National and University Libraries (SCONUL). This body provides a description of how an information literate individual might reveal "an awareness of how they gather, use, manage, synthesize and create information and data in an ethical manner and will have the information skills to do so effectively" (SCONUL 2011). It can be seen that the IL definitions provided by ALA (1989), CILIP (2013), and SCONUL (2011) have a common coverage of the stages of being information literate corresponding with explanations from other authors/organizations, but the difference is that these definitions highlight an ethical dimension. Ethical elements are a question of great interest, because publishing and using information nowadays is becoming easier than ever. Therefore, users should pay more attention to ethical issues when interacting with information to avoid derogations, for example, plagiarism.

Bruce provides a different perspective to researching and defining IL. She proposes that IL should be based on the under-standing of the users through their information experiences (Bruce 1997). She suggests that IL is a compound of diverse manners of engaging with information rather than being a set of skills, competencies and features (Bruce et al. 2006).

Bruce et al. (2012) divided the nuances in IL into two categories: "(1) the skills associated with using information in an ever-expanding range of contexts, representing a functional view of information and information literacy, and (2) the process of using information to learn, including communicating and creating in these contexts, representing transformative interpretations of information and information literacy."

Virkus (2013) believes that the skills-based approach should make room for an understanding based on information practice. Lloyd (2012) by examining how people connect with the information landscape that forms their set-tings or practices, sees IL "as a socially enacted practice". Instead of identifying IL as a set of skills, this point of view recognizes IL in its relationship with the information setting, which has a strong influence on how people understand information.

There are many different definitions of IL given by various institutions and individuals based on different benchmarks. A useful summary of the breadth of conceptualization of IL is given by Kerr who states that "Definitions of information literacy range from being equipped with discrete generic skills, constructing knowledge, critical thinking, enabling lifelong learning, a process of knowing, a process of acquiring new meaning and understanding, enabling the effective utilization of information for a purpose, and a complex of ways of experiencing information use" (Kerr 2010). Even if IL is approached from different viewpoints, the principal point of IL is being competent and confident in an ever-changing information environment (Institute of Development Studies & Information Training and Outreach Centre for Africa [IDS & ITOCA], 2010).

2.1.1 *The importance of information literacy*

The emergence of IL as an issue at a global level, along with the increase in the number of international conferences, publications and research projects around the world (Walton & Hepworth 2013), indicates its significance. We are witnessing the information explosion, especially the growth of digital information that has resulted from the development of digital technologies (Andretta 2005). According to Breivik (1998), by 2020, human knowledge will increase two-fold every 73 days. However, "more information is not always better" (Mai 2016). A large amount of information can bring many challenges to individuals when they engage with the information environment. IL can be considered a tool to help individuals know which information they need, where and how they can find it, and how to use it. Furthermore, lifelong learning is emerging in the goals of educational institutions and professional organizations. Students are required to have the ability to "make informed decisions and act effectively and responsibly" (Farmer 2013). Searching for meaning actively or exploring information freely creates favorable conditions for the development of critical inquiry. It is needed to equip students with a high level of IL to help them become effective information seekers and knowledge constructors (Chu 2012). Studies point out that this can be achieved by promoting the development of IL (Secker & Coonan 2011). Additionally, recent research points out that IL is an indispensable component for enhancing professional skills in the workplace as well as encouraging an informed citizenry and governance in a democratic society (Jinadu & Kaur 2014). Walton & Hepworth (2013) add weight to the significance of the concept by indicating that IL expresses the freedom of people, in other words, human rights. Thus, equipping students with a high level of IL is necessary to help them be successful in the school, the workplace and personal lives (Chu 2012). For example, the body of knowledge that students receive from their universities will soon be outdated, so knowing how to handle information after leaving university is essential to help them engage with an ever-changing information environment.

In the educational environment, schools are striving to prepare students to meet the changes of the information environment (Smith & Hepworth 2007). At the same time, librarians are expected to hold "a lead role in the development of students' information literacy skills" (CILIP 2014). Despite the important role of IL being widely acknowledged and a call for help from librarians, there is still a lack of IL programs implemented in educational institutions in general, and the school setting in particular (Shenton et al. 2014). For that reason, to date, learners have not been equipped with expected IL capability (Chu et al. 2011). This problem might result from the lack of awareness of the complex nature of the information engagement process from educators and learners (Pickard & Dixon 2004).

2.1.2 *Translation of the information literacy concept into Vietnamese*

It is not easy to find a Vietnamese term that matches with the term IL because there is no a concise equivalent term which expresses comprehensively the connotation of the 'literacy' concept in Vietnamese. Literacy is not simply the ability to read and write. It is "able to understand, interpret and assess texts, to evaluate statements, and to be able to take a standpoint when faced with flows of contradictory messages via various media and different types of sources. . .Literacy therefore extends from a mechanical skill to the ability to think critically and challenge dominant ideologies" (Limberg et al. 2012).

The term IL is translated into different terms in Vietnamese, such as **"kiến thức thông tin"** (kiến thức: knowledge/understanding; thông tin: information), **"kỹ năng thông tin"** (kỹ năng: skill; thông tin: information), and **"năng lực thông tin"** (năng lực: competence; thông tin: information). **"Kiến thức thông tin"** is the most commonly used term by Vietnamese information researchers and practitioners. It is also used in the IL conferences in Vietnam, for instance, "Kiến thức thông tin – Information literacy" held in 2006 by the University of Social Sciences and Humanities – Hanoi and "Vai trò kiến thức thông tin phục vụ học tập và giảng dạy trong trường đại học" (the role of IL in supporting teaching and learning in universities) held in 2012 by the Vietnamese Library Association of Southern Academic Libraries (VILASAL); and information professional education programs, for example, the program offered by the University of Social Sciences and Humanities – Ho Chi Minh City (HCMC). IL is also translated as **"kỹ năng thông tin"**, for example, the conference "Nội dung và phủóng pháp thực hiện các khóa huấn luyện kỹ năng thông tin cho độc giả" (content and methods for implementing IL training courses for readers) held in 2011 by Central Library – Vietnam National University – HCMC. Meanwhile, IL is considered **"năng lực thông tin"** by several researchers. Although IL is translated into different ways, until now there is no official discussion regarding the translation of IL into Vietnamese.

The term "**Kỹ năng thông tin**" (information skills) does not provide a comprehensive understanding of the IL concept because IL is the overarching concept and information skills are within that. Similarly, the more popular term "**Kiến thức thông tin**" (information knowledge/understanding) does not express comprehensively the connotation of IL. IL is the combination of many elements, including information skills. However, the term **"kiến thức thông tin"** does not represent this element because 'skill' is the ability to perform an action based on knowledge/experience to produce the desired results. Furthermore, based on the researcher's experience when conducting this study, the term **"kiến thức thông tin"** easily lead to confusion, especially for those who hear the term in the first time. They might regard it as an understanding of information around them that is published in newspapers and journals. Although the term "**Năng lực thông tin**" (information competence) is less commonly used, it provides a more comprehensive understanding of the IL concept. Competence can be described as the combination of knowledge, skills, and experience necessary to perform successfully one own mission (Reitz, 2004) rather than just focusing on knowledge/understanding as "**kiến thức thông tin**". Therefore, it is suggested that "**năng lực thông tin**" (information competence) should be used as an equivalent term to IL in Vietnamese.

3 METHODOLOGY

Under the control of the Ministry of Education and Training, Vietnam's high schools are divided into two groups: public schools and non-public schools (private schools, international schools). Public schools are established and managed by the state agencies. The state funds infrastructure construction and recurrent expenditure. A nominal fee can be contributed by students' families. Non-public schools are established and managed by social organizations, social professional organizations, economic organizations and individuals under the permission of the state agencies. Funding for infrastructure construction and recurrent expenditure is from tuition fees, organizations or individuals, not from the state.

The study employed a qualitative case study approach. Three upper secondary schools in Vietnam (labelled School A - an international school, School B - a public school and School C - a private school), based on their willingness, were purposively selected to participate in the study. There were four groups of participants from each institution, including school librarians, working as key informants to invite students (aged 15-18), teachers and administrators in their schools to take part in the study. The research used a combination of a questionnaire and interviews. A questionnaire was sent to students to determine whether they know the term IL by asking "have you heard or read about IL?", along with a supplement question "If Yes, what does IL mean to you?" to explore further students' perception of the concept. As there was no equivalent term for IL in Vietnamese, the researcher used the original term "information literacy" besides the translation such as "kỹ năng thông tin" and "năng lực thông tin" rather than using a single term. Two hundred students at level 10, 11 and 12 in the three schools were randomly selected to complete the questionnaire. It was assumed that the concept of IL was new to the students. There was a concern that a lack of understanding of IL might affect the students' awareness of the importance of IL. However, providing an explanation of the IL concept prior to the survey might have an impact on their perception of IL. Their answers to the question "what does IL mean to you?" might resemble the IL explanation that the researcher came up with in the question list. Therefore, the student questionnaire survey was divided into two occasions to avoid providing inaccurate information concerning their awareness. In the first questionnaire, the first two questions examined their perception of the IL concept were provided. In the second questionnaire aiming at exploring students' awareness of the IL concept, the students were provided with an explanation and examples of the IL concept in the questionnaire to provide them an understanding of the concept.

Fifteen professionals (three librarians, nine teachers and three administrators) from the three schools were also invited to take part in the interviews in order to explore their understanding and awareness of the concept. Participants' names were coded using two capital letters and an order number, for example AS1, to remain anonymous in which:

- The first capital letter is school name: A, B, C;
- The second capital letter is participant name: S, L, T, M for students, librarians, teachers and managers perspective.

By employing Nvivo, thematic analysis technique was used to analyze qualitative data. Two main themes associated with the research objectives, such as conception and awareness, were established prior to the coding process. This aimed to arrange and organize sub-themes more easily. The answers were then reviewed and categorized based on their content and the two main themes using the following process: (1) Read through transcripts/answers twice to identify uninteresting data; (2) Read through data again and make notes about important sentences; (3) Conduct a line-by-line analysis of the transcripts/answers and create codes under each main theme; (4) Review codes to identify the relationship between them; (5) General ideas about data; (6) Translate from Vietnamese to English only those portions that could be used in the paper.

4 RESULTS AND DISCUSSION

4.1 *Students' conceptions of information literacy*

This research highlights that students displayed a lack of comprehensive understanding of the IL concept. The survey found that 88% of the students had not heard or read about the term IL and its translation before taking part in the study. For those who had heard or read about the term (12%), they still could not provide a complete explanation of the IL concept. It was found that students had different conceptions of IL. There was a student who considered IL as the ability to use information, "*IL is the ability to use information reasonably and correctly*" (BS5), while another student believed that information literate individuals needed to know methods or techniques to search for information, "*IL is the way we find information*" (BS16). There were some students who understood that information literate individuals could understand the meaning of information through the evaluation of the content of information, for example, "*IL is the ability to understand correctly information provided*" (BS17), "*IL is the ability to understand the meaning of information*" (AS5 & CS57), "*IL is the ability to understand information and news that we update daily through the internet, books, journals, etc.*" (BS41). Interestingly, the idea "IL is knowledge" was agreed by the remaining students (16 students), for instance, "*IL is knowledge that I need to know*" (BS34). Respondents' explanation of the term was compared with Walton & Cleland's (2013) viewpoint. It can be seen that their IL explanation is extremely different in comparison with Walton & Cleland's (2013) perspective. IL elements, as provided by individuals and organizations such as ALA (1989), CILIP (2013), and SCONUL (2011), we're not fully mentioned in the students' IL explanation.

4.2 *Students' awareness of the role of information literacy*

In the second questionnaire, students were provided an explanation of the IL concept to help them understand what IL means. The study then explored students' awareness of IL regarding the importance of IL in general and the role of IL to learning success in particular. There was a difference among students concerning their awareness of the importance of IL and its role to learning success. There were students who had a positive viewpoint about IL, for example, *"IL is very necessary because of the information explosion, online libraries and the large amount of information on the internet. There is too much information, so we face many challenges in selecting appropriate information for use. There is information that we can find effortlessly, but it may be untrue or unreliable. Therefore, we need to have knowledge, we have to learn IL in order to search, evaluate and then use the information effectively"* (CS51), *"IL is very necessary. When I do literature essays, I need to use citations. If I know how to find appropriate citations, my essays will be more concise and achieve high scores. Or, when I do a presentation, if I can provide essential and appropriate information, listeners will be interested in it"* (BS74), *"I think that IL helps us understand and know much more, and broaden our minds…It is rather important"* (CS91).

However, there are a large number of students (67%) who did not highly value the importance of IL, *"IL is also unnecessary because it does not affect my life too much"* (AS28), *"IL may be important, but not now"* (BS10), *"I do not think it is that important for us to pay much attention to it, because we have been taught by teachers what we should learn"* (CS28). Thus, they were not motivated to develop their IL skills. They were not willing to attend or take part in activities that could develop their IL, as said by one of the librarians, *"As from this year, my school library will no longer organize a library introduction class because students refuse to go to the class. Students who want to read books can borrow books in the library"* (CL). One of the students states that: *"I think that if you provide an IL course to students, students will give time for private classes rather than the IL course"* (BS74). As a result, the schools in general and the libraries in particular, were not motivated to deliver activities that might be beneficial for the development of students' IL. Research indicates that school libraries create an environment that allows students to link what they learn from lessons to a broader world outside of the school (Mardis & Dickinson 2009). Students will face many challenges in conducting research projects if they are not familiar with the library and resources (Smith & Hepworth 2007). This research is supported by many studies which reveal that students infrequently use the library and librarian-related services for their course-related research assignment (Sokoloff 2012). In the context of this study, students rarely used the libraries for their learning. Several reasons that result in the above problem were found. Particularly, students think library research skills are not important in professional development (Novotny & Cahoy 2006). In addition, libraries are less convenient and more time-consume in locating information, as believed by students (Smith & Hepworth 2007). Also, reasons, such as fear of library staff, an affective sense of incompetence, feeling uncomfortable in the library, lack of knowledge about the library and discomfort using library equipment, significantly result in library anxiety (Onwuegbuzie et al. 2004). This research adds to the literature by indicating that lack of faith in librarians among students could result in not using the libraries. This is demonstrated in the following statement: *"I am not sure whether librarians can teach us IL or not"* (BS10 and BS55).

Cognitive elements are pervasive in IL models (Hepworth & Walton 2009). This demonstrates the significance of components related to awareness/cognition to the development of IL. However, Bundy (1999) indicates that students' information awareness has not been well developed by the time they move in to college and university education. As a result, there is a limited understanding of the different elements of IL among students (Smith & Hepworth 2007). This is demonstrated in the context of this study by indicating that IL elements were not fully mentioned in the students' IL explanation as indicated in the previous section. Therefore, IL programs need to concentrate on "fundamental task of shifting the youngsters' attitudes and changing their mindsets" (Pickard et al. 2014). From what has been discussed above, it is suggested that factors related to students' awareness/cognition should be considered in the implementation of an IL program in Vietnam's upper secondary schools.

4.3 *Professionals' conceptions of information literacy*

The IL perception of librarians and teachers was explored in a range of studies, such as Hepworth & Smith (2008), IDS & ITOCA (2010), Martin (2011), Smith (2013), and Tan et al. (2017). Most of them confirm the finding of this study by showing a limited understanding of the IL concept among librarians and teachers. Vietnamese professionals who took part in this research only knew about the term IL and its translation for the first time when they participated in the research, except the three librarians. For example, "*I had not heard about the term IL before*" (BT2), "*as for IL or IL skills, I have never used the terms before*" (CT3) and "*I know the term "information" or "IT", but I have never heard about IL*" (CM). It is not surprising to find that the IL concept was new to most of the professionals as the researcher predicted this result before conducting the study. It is assumed that the term IL can be understood and interpreted in different ways. Therefore, the professionals might explain IL in a different manner. However, the above finding shows that IL has not been widespread in the educational community in the schools under the common term "IL" and its translation in Vietnamese. The professionals' perception of the IL concept was then investigated, although most of them had not heard about the term.

The study found that the professionals could not provide a full explanation of the IL concept, including the three librarians who had heard about the notion before. Some of the professionals' explanations of the IL concept are presented as follows, "*IL is a method of finding information based on known information. Generally, IL is something relevant to information that individuals need to know based on known information*" (BT3), "*I think IL is retrieving information from the internet, and then identifying and selecting information that is appropriate to what we are looking for*" (AT1), and "*IL is gathering information to satisfy individuals' needs*" (CL). The above professionals indicated that IL mainly focuses on information engagement activities to retrieve appropriate information in order to satisfy individuals' information needs, in which finding information was clearly addressed.

Meanwhile, some of the professionals understood the term in a different way, as below, "*IL is information that individuals identify from the world around them*" (CT2) and "*IL is a noun that indicates a kind of knowledge*" (BT1). It can be seen that the above professionals thought IL is information or knowledge in general. It is interesting to find that the students and professionals understood IL in the same way (see Section 4.1 for an analysis of students' conception of IL). It was expected that the professionals might have a better understanding of IL than their students. Nevertheless, the study found that the conception of IL between the professionals and students had some similarities. For example, they both viewed IL as knowledge. The implementation of IL initiatives had not received much attention from professionals could result from a lack of understanding of IL. The literature indicates that teachers should be the person who is mainly responsible for delivering IL instructions (Bent 2013). Nonetheless, developing information literate students is not only the responsibility of librarians or teachers (Neely 2006), but also "all those who call themselves educators" (Bundy 1999). Many teachers consider IL as a separate subject rather than a way of learning and teaching (Williams & Wavell 2007). Bruce (1997) argues that "information literacy cannot be learned without engaging the discipline specific subject matter".

It is necessary to help academic staff and stakeholders have a common understanding of IL in order to integrate IL into the curriculum (Bent 2013). Lack of understanding of IL may result from the absence of IL training in librarianship courses (Weller 2006) and teacher training courses (Wilson 1997). Therefore, staff development is essential to enhance their understanding and competence of IL when implementing an IL program (UNESCO 2013).

4.4 *Professionals' awareness of the role of information literacy*

This study shows that administrators did not highly appreciate the importance of the school library and IL to teaching and learning activities. The librarians of the three schools said that: "*my school library does not receive much attention from the top-management panel*" (AL), "*my current rector does not pay much attention to the library. He thinks that the library is simply a place to store books. It does not need library management software or database, etc.*" (CL), "*we have not provided any library introduction class in this academic year, because the board of rectors has not given us permission to organize such classes. In practice, they think that the library is not important, although they do not say that. For example, they are always asking us to develop a digital library, but they do not know what a digital library is*" (BL). The librarians pointed out that their school administrators did not thoroughly appreciate the importance of the library to teaching and learning activities. Consequently, the library did not receive much support from administrators in the development of information sources and infrastructure or in organizing activities that could assist students in developing IL capability. It was found that, although teachers acknowledged the need for IL, they denied the significance of IL to school students' learning success, "*At upper secondary education level, students are mainly tested on what they have been taught. This means they can get high scores by memorizing what has been taught by teachers. IL is not helpful in this case, so we do not need to teach IL to students. However, it is needed at higher education level*" (BT1), and "*I do not think IL significantly affects students' learning. Students who take more effort to explore information can have a better understanding of issues than those who do not do that. However, it does not help students achieve good learning results*" (AT2). One of the teachers reinforced this finding when she indicated that teachers' awareness of IL could affect the implementation of IL initiatives, "*I only started to think about this issue while I was talking with you. Students actually need such instructions to be better in finding, evaluating and using information. They should not engage with information in a vague way anymore*" (BT2). The above statement shows that the teacher was not aware of the importance of IL to students' learning before she took part in the study. Therefore, she did not pay much attention to providing her students with IL instructions. In general, the professionals acknowledged the need for IL, because they believed that IL assisted students in obtaining a more in-depth understanding of issues. Nevertheless, they did not greatly appreciate the importance of IL to students' learning success, because they believe students could achieve high scores in learning without IL. This might result from the existence of the transmission approach in teaching and learning.

Although the teachers pointed out that IL did not have a significant impact on students' learning, they all agreed that IL was essential for their teaching activities, "*IL is definitely necessary to me. I can find information beyond the textbooks to pass on to students*" (CT2), and "*I mainly use it for my teaching activities, such as lesson preparation, presentation design, and so on*" (CT3).

It can be seen that, in the context of this research, Vietnam's high school professionals were not ready to deliver IL programs because of their lack of awareness of the importance of IL to students' learning success as well as the loose cooperation among the parties. The collaboration between librarians and faculty is required to integrate IL intervention into the curriculum (Anderson & May 2010). Librarians cannot assess the effectiveness of IL instructions without the support from the faculty (Anderson & May 2010). On the contrary, teachers can be supported by librarians in identifying the presence of skills in the curriculum (Pickard et al. 2011). Therefore, it is necessary to improve stakeholders' awareness of the importance of IL in order to impulse the success of an IL program.

5 CONCLUSION AND SUGGESTIONS

From what has been discussed above, it can be seen that some participants can grasp the concepts and role of IL or they are aware part of them, but others cannot. The number of people who do not have a comprehensive understanding of the IL concept and its importance to learning success is relatively large. This may affect the readiness to implement IL programs in educational institutions. It is suggested that factors related to participants' awareness/cognition should be involved in the IL teaching model in Vietnam's upper secondary schools. They need to have a thorough understanding of the IL concept in order to ensure the success of an IL program. Furthermore, there is a need to expand the study by exploring the reasons resulted in the issue.

It should be noted that IL is differently translated in Vietnamese, the use of the equivalent term of IL in Vietnamese "Năng lực thông tin" (information competence) is suggested to make it easy for team members. Providing an explanation of the IL concept is also necessary to help other team members have the same understanding of the term before embarking in implementing an IL program.

REFERENCES

American Library Association (ALA). 1989. *American library association presidential committee on information literacy*. [Online]. Retrieved from: http://www.ala.org/ala/acrl/acrlpubs/whitepapers/presidential.htm [Accessed on: 2019/7/20].

Anderson, K. & May, F.A. 2010. Does the method of instruction matter? An experimental examination of information literacy instruction in the online, blended, and face-to-face classrooms. *The Journal of Academic Librarianship* 36(6): 495–500.

Andretta, S. 2005. *Information literacy: A practitioner's guide*. Amsterdam, Netherlands: Elsevier.

Bent, M. 2013. Strand three: developing academic literacies. In J. Secker & E. Coonan, (eds.), *Rethinking Information Literacy: A Practical Framework for Teaching*: 27–40. United Kingdom, UK: Newcastle University.

Breivik, P.S. 1998. *Student learning in the information age*. Phoenix, AZ: Oryx Press.

Bruce, C. 1997. *The seven faces of information literacy*. Adelaide: Auslib Press.

Bruce, C., Edwards, S. & Lupton, M. 2006. Six frames for information literacy Education: a conceptual framework for interpreting the relationships between theory and practice. *Innovation in Teaching and Learning in Information and Computer Sciences* 5(1): 1–18.

Bruce, C., Hughes, H. & Somerville, M.M. 2012. Supporting informed learners in the twenty-first century. *Library Trends* 60(3): 522–545.

Bundy, A. 1999. Information literacy: the 21st century educational smartcard. *Australian Academic & Research Libraries* 30(4): 233–250.

Chartered Institute of Library and Information Professionals (CILIP). 2013. *Information literacy: Definition*. [Online]. Retrieved from: http://www.cilip.org.uk/cilip/advocacy-campaigns-awards/advocacy-campaigns/information-literacy/information-literacy [Accessed on: 2019/7/20].

Chartered Institute of Library and Information Professionals (CILIP). 2014. *The CILIP guidelines for secondary school libraries*. 3rd (eds.). London, UK: Facet Publishing.

Chu, S.K.W. 2012. Assessing information literacy: A case study of primary 5 students in Hong Kong. *School Library Research* 15: 1–24.

Chu, S.K.W., Tse, S.K. & Chow, K. 2011. Using collaborative teaching and inquiry project-based learning to help primary school students develop information literacy and information skills. *Library & Information Science Research* 33(2): 132–143.

Farmer, L.S. 2013. How AASL learning standards inform ACRL information literacy standards. *Communications in Information Literacy* 7(2): 13.

Hepworth, M. & Smith, M. 2008. Workplace information literacy for administrative staff in higher education. *The Australian Library Journal* 57(3): 212–236.

Hepworth, M. & Walton, G. 2009. *Teaching information literacy for inquiry-based learning*. Amsterdam, Netherlands: Elsevier.

Institute of Development Studies & Information Training and Outreach Centre for Africa (IDS & ITOCA). 2010. Strengthening information literacy intervention: creative approaches to teaching and learning. In *SCECSAL pre-conference seminar report*. [Online]. Retrieved from: http://blds.ids.ac.uk/files/dmfile/BotswanaCompressed5.pdf [Accessed on: 2019/7/20].

Jinadu, I. & Kaur, K. 2014. Information literacy at the workplace: A suggested model for a developing country. *Libri* 64(1): 61–74.

Kerr, P.A. 2010. *Conceptions and practice of information literacy in academic libraries: Espoused theories and theories-in-use*. Doctoral dissertation, New Brunswick, NJ: Rutgers University-Graduate School.

Limberg, L., Sundin, O. & Talja, S. 2012. Three theoretical perspectives on information literacy. *Human IT: Journal for Information Technology Studies as a Human Science* 11(2): 93–130.

Lloyd, A. 2012. Information literacy as a socially enacted practice. *Journal of Documentation* 68(6): 772–783.

Mai, J.E. 2016. *Looking for information: A survey of research on information seeking, needs, and behavior*. Bingley, UK: Emerald Group Publishing.

Mardis, M.A. & Dickinson, G.K. 2009. Far away, so close: Preservices school library media specialists' perceptions of AASL's "standards for the 21st-century learner". *School Library Media Research* 12: 1–16.

Martin, V. 2011. Perceptions of school library media specialists regarding their practice of instructional leadership. In D. Williams & J. Golden, (eds.), *Advances in Library Administration and Organization, Advances in Library Administration and Organization*: 207–287. Bingley, UK: Emerald Group Publishing Limited.

Neely, T. 2006. Beyond the standards: what now? In T. Neely, (ed.), *Information Literacy Assessment: Standard-Based Tools and Assignments*: 136–152. Chicago: American Library Association.

Novotny, E. & Cahoy, E.S. 2006. If we teach, do they learn? The impact of instruction on online catalog search strategies. *Portal: Libraries and the Academy* 6(2): 155–167.

Onwuegbuzie, A.J., Jiao, Q.G. & Bostick, S.L. 2004. *Library anxiety: Theory, research, and applications*. Lanham, Maryland, US: Scarecrow Press.

Pickard, A.J. & Dixon, P. 2004. Measuring electronic information resource use: towards a transferable quality framework for measuring value. *VINE* 34(3): 126–131.

Pickard, A., Gannon-Leary, P. & Coventry, L. 2011. The onus on us? Stage one in developing an i-Trust model for our users. *Library and Information Research* 35(111): 87–104.

Pickard, A.J., Shenton, A.K. & Johnson, A. 2014. Young people and the evaluation of information on the World Wide Web: Principles, practice and beliefs. *Journal of Librarianship and Information Science* 46(1): 3–20.

Reitz, J.M. 2004. *Dictionary for library and information science*. UK: Libraries Unlimited.

Secker, J. & Coonan, E. 2011. *A new curriculum for information literacy*. [Online]. Retrieved from: http://eprints.lse.ac.uk/37679/1/ANCIL_final.pdf [Accessed on: 2019/7/20].

Shenton, A.K., Pickard, A.J. & Johnson, A. 2014. Information evaluation and the individual's cognitive state: Some insights from a study of British teenaged users. *IFLA journal* 40(4): 307–316.

Smith, J.K. 2013. Secondary teachers and information literacy (IL): Teacher understanding and perceptions of IL in the classroom. *Library & Information Science Research* 35(3): 216–222.

Smith, M. & Hepworth, M. 2007. An investigation of factors that may demotivate secondary school students undertaking project work: Implications for learning information literacy. *Journal of Librarianship and Information Science* 39(1): 3–15.

Society of College, National and University Libraries (SCONUL). 2011. *The SCONUL Seven Pillars of Information Literacy: core model for higher education*. [Online]. Retrieved from: http://www.sconul.ac.uk/sites/default/files/documents/coremodel.pdf [Accessed on: 2019/7/20].

Sokoloff, J. 2012. Information literacy in the workplace: Employer expectations. *Journal of Business & Finance Librarianship* 17(1): 1–17.

Tan, S.M., Kiran, K. & Diljit, S. 2017. Examining school librarians' readiness for information literacy education implementation. *Malaysian Journal of Library & Information Science* 20(1): 79–97.

UNESCO. 2013. *Global media and information literacy assessment framework: country readiness and competencies*. Paris: UNESCO.

Virkus, S. 2013. Information literacy in Europe: Ten years later. In S. Kurbanoğlu, E. Grassian, D. Mizrachi, R. Catts & S. Špiranec, (eds.), *European Conference on Information Literacy*: 250–257. Berlin, Germany: Springer.

Walton, G. & Cleland, J. 2013. Becoming an independent learner. In J. Secker & E. Coonan, (eds.), *Rethinking Information Literacy: A Practical Framework for Teaching*: 13–26. United Kingdom, UK: Newcastle University.

Walton, G. & Hepworth, M. 2013. Using assignment data to analyse a blended information literacy intervention: A quantitative approach. *Journal of Librarianship and Information Science* 45(1): 53–63.

Weller, K. 2006. Putting our own house in order: CILIP-accredited LIS courses. *Library+ Information Update* 5(1-2): 30–32.

Williams, D.A. & Wavell, C. 2007. Secondary school teachers' conceptions of student information literacy. *Journal of Librarianship and Information Science* 39(4): 199–212.

Wilson, K. 1997. Information skills: the reflection and perceptions of student teachers and related professionals. In L. Lighthall & K. Haycock, (eds.), *Information rich but knowledge poor? Emerging issues for schools and libraries worldwide. Research and professional papers presented at the Annual Conference of the International Association of School Librarianship held in conjunction with the Association for Teacher-Librarianship in Canada, 26th, Vancouver, British Columbia, Canada.* US: Education Resources Information Center (ERIC).

Borderless Education as a Challenge in the 5.0 Society – Abdullah, Adriany & Abdullah (eds)
 ISBN 978-0-367-61960-2

Pre-service teachers' difficulty employing critical thinking to solve mathematical problems

I.P. Luritawaty & S. Prabawanto
Department of Mathematics Education, Universitas Pendidikan Indonesia, Bandung, Indonesia

ABSTRACT: Critical thinking is needed by teachers in dealing with major changes in the disruption era. With critical thinking, teachers can be adaptable, selective, and careful, and solve various problems that arise in the disruption era correctly and responsibly. Various attempts were made to improve critical thinking skills of teachers, but unfortunately not many people pay attention to the critical thinking skills of students, in this case, pre-service or future teachers. From previous studies, it is known that students' critical thinking skills are still low. Therefore this research was conducted to analyze the difficulties of students on employing critical thinking to solve mathematical problems. The method used in this research is descriptive qualitative. The subject of this research were 15 third-semester students in mathematics education research programs. The instrument used was a four-item test of critical thinking. The results showed that there were some difficulties experienced by students on employing critical thinking to solve mathematical problems. These difficulties include students who can't prove the relationship between spaces diagonal and edge concepts, can't generalize the volume of the pyramid in a cube, and can't consider alternative answers to find common conditions that must be met if many prismatic edges can be the same as many pyramid edges.

1 INTRODUCTION

The disruption era is the result of great revolutions in various fields in the 21st century. Disruption is a condition or period with many innovations that are not visible and not realized but can disrupt old habits or even destroy old systems (Kasali 2017, Prihanisetyo et al. 2018). Disruption occurs along with the development of technology from traditional to digital, the result of new ways of thinking as implemented by the millennial generation. This can lead to the damage or even the destruction of the old system and the emergence of a new system that is considered more effective. Humans must be vigilant and able to adjust to all changes that occur. Those who are unable to synergize with the changes will experience defeat. One area that is exposed to disruption is the education sector. The learning process that is an important part of education is starting to change. This is indicated by the abundance of information due to the ease of accessing learning resources and high flexibility in learning. Students can easily access a variety of knowledge references, not limited to knowledge from the source book in school or information provided by the teacher. This seems to be a good development in the world of education. However, seen from the other side, it can also be a distraction in the learning process. For example the problem of learning resources that are not valid resulting in a wrong understanding, or the appeal of technology that causes students to shift focus from the knowledge that should be learned.

The world of education requires a transformation to face major changes in the era of disruption. One important factor that must be considered is teacher competency. Shoop (2015) stated that teachers in the era of disruption were no longer focused solely on the cognitive world, but were required to be able to surf in cyberspace and the digital world. The teachers' task will extend to confirming the truth about the various students' findings from various sources they get. For this reason, teachers must be adaptive, that is, be able to respond to all developments that occur, but remain selective, that is, able to think carefully and be thoughtful to produce truth, a process called critical thinking. Lipman argues that critical thinking is a tool against thoughts and actions that are not considered (Daniel & Auriac 2011). Critical, creative, and innovative thinking was developed in America to develop an understanding of disruptive technology and the historical context of success and failure (Shoop 2015). Critical thinking is the ability to discuss all sides, consider all facts, decide what is relevant and irrelevant, and make wise decisions (Facione 2011). Critical thinking implies the ability to prove, generalize, consider possible alternatives, and solve problems (Daniel & Auriac 2011, Ennis 2011, Sanders 2016). In the critical thinking process, teachers must be able to prove through the integration of insights. It aims to make teachers more selective, careful, and not readily decide the truth of information before it is proven true. Teachers must also be able to generalize and consider alternative answers, especially as the open-access allows students to find information that provides variations of answers. Besides, teachers must also be able tp solve problems responsibly and guarantee the truth.

Critical thinking is closely related to various fields of learning that are essential. One of them is as a basis for learning and applying mathematics (Sanders 2016). Mathematics is the basis for national prosperity: it acts as a tool to understand knowledge, techniques, technology, and economics, while critical thinking enables students to work mathematically and become effective problem solvers (Chukwuyenum 2013, Sanders 2016). Therefore, various attempts were made to develop critical thinking skills in mathematics teachers through training and competency tests. But unfortunately many people don't pay attention to the critical thinking of the students who are to be the next generation of mathematics teachers and who are expected to develop education in a better direction. What we mean by students in this research are the pre-service teacher or future teacher generations. Based on research that Safrida et al. (2018) conducted with undergraduate mathematics education students in one of the university in Jember, only 23.3% of students had begun to think critically. The same thing is also known from Suparni's (2015) research in a mathematics research program at one of the state universities in Indonesia showing that only 17.4% of students can think critically. Therefore the research was conducted to analyze the difficulties of employing critical thinking to solve mathematical problems.

2 LITERATURE REVIEW

Critical thinking has become one of the tools used in everyday life to solve several problems involving logical thinking, interpreting, analyzing, and evaluating information to enable one to make reliable and valid decisions in the revolutionary science phase (Chukwuyenum 2013, Shoop 2015). Critical thinking is reflective thinking that makes sense by focusing on what must be believed or achieved through the integration of insights with certain characteristics (Daniel & Auriac 2011, Ennis 1993). Critical thinking is the ability to discuss all sides, consider all facts, decide what is relevant and irrelevant, and make wise decisions (Ennis 1993).

Critical thinking ability is considered fundamental for learning and applying mathematics (Sanders 2016). The development of critical thinking processes enables students to work mathematically and become effective problem solvers. In the problem-solving process, students think mathematically when generating and evaluating knowledge, finding various possible strategies, and proving and assessing the chosen strategy. Critical thinking ability can also improve students' understanding of mathematical concepts because it helps students' to be able to interpret, analyze, evaluate, and present results logically and systematically (Chukwuyenum 2013).

Indicators of critical thinking are stated by several experts. According to Sanders (2016), indicators of critical thinking skills include communication, reasoning, problem-solving, understanding, and fluency. Interpretation, analysis, evaluation, conclusions, explanations, and self-regulation can also be used as indicators of critical thinking (Facione 2011). Lipman also proposed indicators of critical thinking, namely conceptualization, generalization, reasoning, and evaluation (Daniel & Auriac 2011). The indicators of critical thinking ability used in this research are proving, generalizing, considering alternative answers, and problem-solving (Daniel & Auriac 2011, Ennis 2011, Sanders 2016).

3 RESEARCH METHODS

This type of research is descriptive qualitative. The research was conducted at a tertiary university in Garut regency, with 15 third-semester students in mathematics education research programs, academic year 2019/2020 with a Grade Point Average (GPA) above 3.00. The instrument used in this research was a four-item test of critical thinking skills, with an allocation of 90 minutes. The topic used in the instrument was solid figures. The research was conducted based on the guidelines for scoring critical thinking skills and indicators of critical thinking skills used on the instrument. Before being used, the instrument was tested for validity (i.e., face validity or legibility and content validity or conformity with indicators). The validation was carried out by an expert in the field of mathematics education, namely a lecturer at a university in Bandung. After being validated, the instrument was used to test students' critical thinking skills in solving mathematical problems. Data obtained from the results of tests of critical thinking skills were then analyzed. The first analysis compared the students' answers with the scoring guidelines. The ideal score for each item is 5, so, if students can answer all questions correctly, the ideal score is 20. The next analysis was based on the indicators of each item, namely proving, generalizing, considering alternative answers, and solving the problem. Each indicator was used to analyze students' difficulties on employing critical thinking to solve mathematical problem.

4 RESULTS AND DISCUSSION

This research provided analysis results of critical thinking skills test of students. The analysis consists of two parts. First, an analysis of the results of students' answers compared with scoring guidelines for critical thinking skills. Second, the analysis of the results of students' answers based on indicators of critical thinking skills, namely proof, generalization, considering alternative answers, and problem-solving.

Analysis of the results students' answers compared with the scoring guidelines for critical thinking skills is examined in two parts. The first part is comparing the number of students with the correct or incorrect answers. The results are described in Table 1, as follows.

Based on Table 1, item number 3 has the lowest results, no correct answers. Students can count many edges from prisms and pyramids, although it is still limited and not entirely accurate. Students also can see the relationship edges between prisms and pyramids by marking on many of the same edges. However, students' understanding is limited to the many prisms and pyramids edges they make. Students can't find general conditions that must be met related to the relationship of many edges of prisms and pyramids. The second part is comparing the score of each student's item based on the ideal score in the scoring guidelines. The results are described in Table 2, as follows.

Table 1. Analysis of the number of students with correct or incorrect answers based on scoring guidelines.

Item Number Question	Number of students with the right answer	Number of students with incorrect answers	Description
1	2	13	Students have difficulty analyzing the relationship between edge and space diagonal concepts
2	6	9	Students have difficulty calculating the volume of a pyramid in a cube
3	0	15	Students can not find general conditions that must be met related to the edge prism and pyramid
4	9	6	Students can calculate the volume of the geometric structure correctly, make a relationship, but are still wrong in solving problems and drawing conclusions

Table 2. Analysis of the score of each students' item based on the ideal score in the scoring guidelines.

	Item			
Student	1	2	3	4
M1	3	3	2	5
M2	5	5	2	5
M3	3	5	2	5
M4	3	5	2	3
M5	3	0	0	1
M6	2	5	0	2
M7	5	5	0	5
M8	2	1	0	2
M9	2	1	1	5
M10	0	5	2	2
M11	2	3	0	5
M12	1	3	1	5
M13	0	2	1	5
M14	2	2	2	1
M15	1	3	0	5

Based on Table 2, it is known that almost all students get a score on each item. In item 1, students generally get a score of 2 or more. According to the scoring guidelines, this shows that students can find and detect important things, write them down, and sketch pictures if needed along with the size they know. However, some students are wrong in determining the size of the image, especially analyzing the relationship between the concept of space diagonal and edge. In item 2, generally, students also get a score of 2 or more, and six students even obtained an ideal score. This shows that students have been able to drawing certain solid figure, even to the stage of calculating the volume of the pyramid properly. However, some students are still wrong in sketching the correct drawings. Some students also had difficulty in calculating the volume of the pyramid in the cube. This is partly due to the students' lack of analytical power. The difference in score achievement occurs in item 3. There are no students who get score above 2, and six students get a score of 0. Most students make mistakes in determining the type of prism and pyramid so that they can't see the linkage between edges of prisms and pyramids in general. Finally, in item 4, nine students received the ideal score. This shows that in general students' can make connections between cuboids and cubes to solve problems. The difficulty several students found in item 4 is connecting the comparison of the cuboid and cube. This makes some students's unable to solve the problems well.

Next research data analysis is based on indicators of critical thinking ability. The analysis is performed on each indicator consisting of proof, generalization, considering alternative answers, and problem-solving. The first analysis starts with proof. Consider the item 1 in the following Figure 1.

In item 1, generally, students still make mistakes in proving. Some of them can draw without analyzing the evidence.

In Figure 2, it appears the student was only able to draw a cube, but did not include the known diagonal size in the problem. Students can't analyze the relationship between space diagonal and edge concepts. This is different from the student's answers in Figure 3.

Prove that if a cube has a diagonal length$2p\sqrt{3}$ then the cube's rib

Figure 1. Critical thinking question with proof indicator.

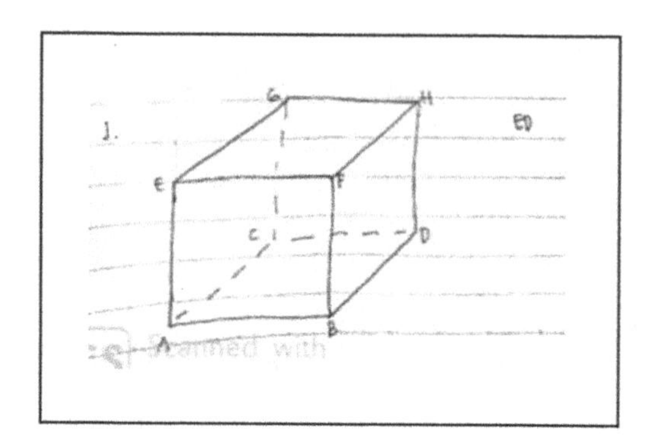

Figure 2. Results of student who is unable to prove.

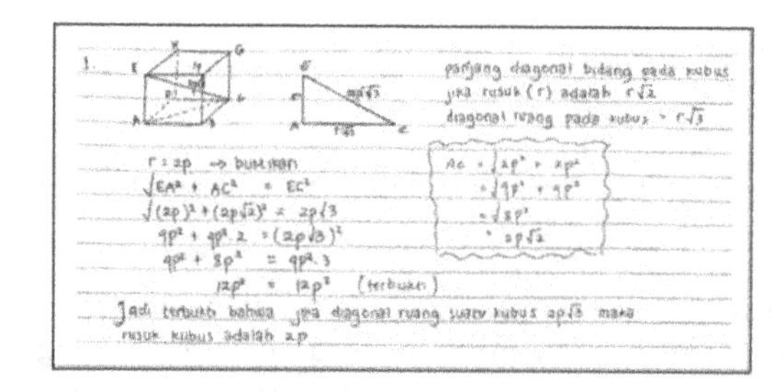

Figure 3. Results of student who can substitute for proof.

In Figure 3, it appears the student can do image analysis. The student calculated the diagonal value of the field first, then substituted the value of the known length of the edge on the problem to prove the relationship of the value of the existing size through the Pythagorean theorem. However, the evidence is quite complicated. Students seem not to have shown critical thinking about the relationship between the concept of diagonal space and the length of the rib.edge. Look at the answers of another student in Figure 4.

In Figure 4, it appears the student did not sketch pictures. Students try to think critically by linking the concept of diagonal space and edge length. The results of his work showed his mastery of the concept of diagonal space in a cube and succeeded in proving the truth that was asked about the problem well.

The next analysis is about generalization. Consider item 2 in Figure 5.

In item 2, some students seemed to have no difficulty in analyzing the images and making generalizations. This can be seen from the results of student work in Figure 6 below.

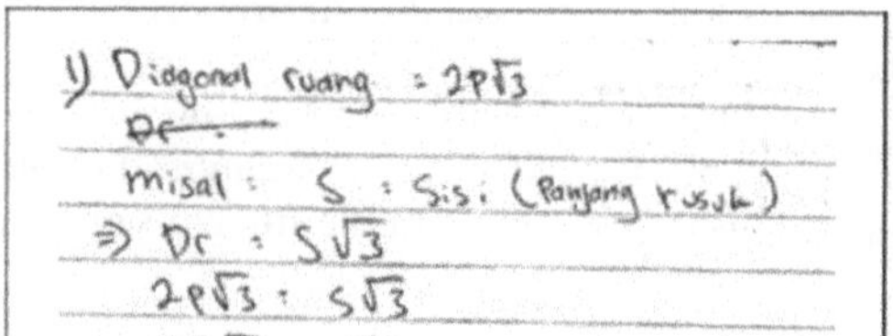

1) Diagonal ruang = 2p√3
misal : S : Sisi (Panjang rusuk)
⇒ Dr = S√3
2p√3 = S√3
2p√3 / √3 = S
S = 2p

Figure 4. The results of student who can proof.

Inside a cube, four lines are drawn that are diagonal of space so that they intersect at a point and form a space shape. What is the name called? Is it true that in general it can be said that the volume of one of these shapes is $\frac{1}{6}$ from the volume of the cube? Explain!

Figure 5. Critical thinking question with generalization indicator.

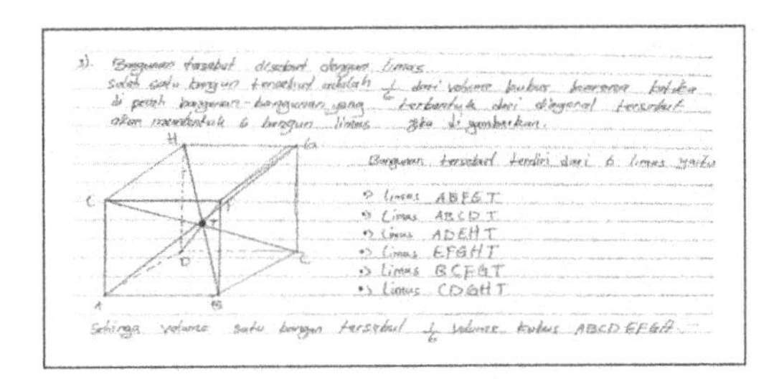

Figure 6. The results of student who can make generalizations.

In Figure 6, it appears the student can analyze the questions and draw well. Students can also find new shapes formed from four diagonals that are drawn and intersect at one point. Critical thinking can be seen from the results of the analysis of the formation of six pyramids in a cube so it is possible to generalize that the volume of the pyramid is 1/6 of the volume of the cube. However, not all students can think up to this stage. Look at Figure 7 below

In Figure 7, the students can sketch pictures correctly. However, students are still wrong in analyzing images when the images are opened into nets. The student wrongly made a pyramid in cube nets so that only two pyramids appear. Critical thinking of students has not yet appeared here. The student is trying to make generalizations but not following the analysis done in the previous stage.

The next analysis is about considering alternative answers. Consider item 3 in Figure 8 below.

In item 3, in general, students seem to have difficulty in finding alternative answers. This can be seen from Figure 9.

In Figure 9, the students seems to be only able to think that the prism and the pyramid in question are triangular grounded. But if you think about it critically, the base of the prism and pyramid form varies, for example, triangles, rectangles, pentagons, hexagons, until the

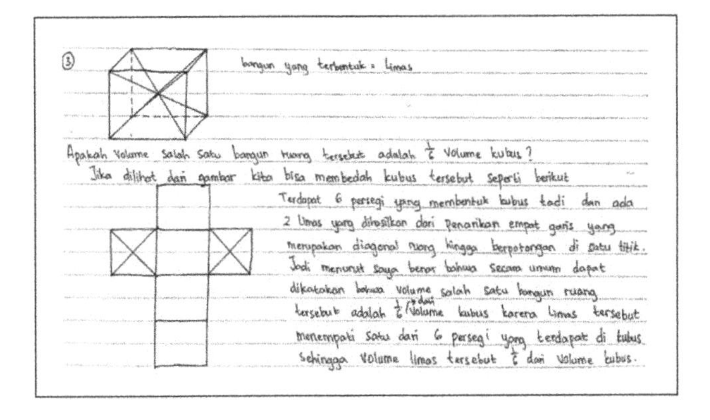

Figure 7. Results of student to analyze a generalization.

Can the number of ribs in a prism be the same as the number of ribs in a pyramid? If you can, in general, what conditions must be met?

Figure 8. Critical thinking question with considering alternative answers indicator.

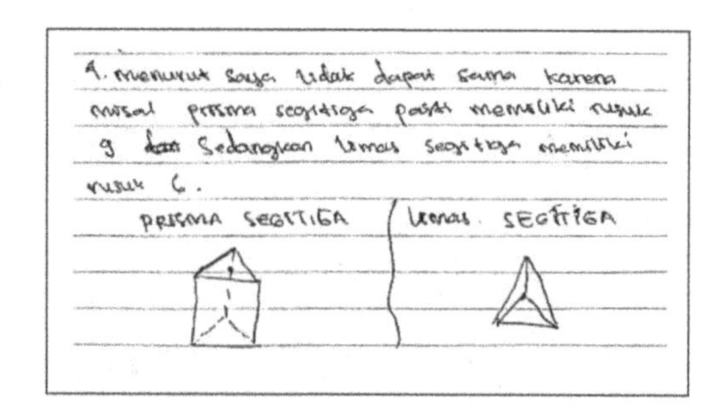

Figure 9. Results of student who can analyze to consider alternative answers.

n-facet. Many choices of answers that can be used by students to analyze finally finding a general condition that must be met if many prismatic edge can be the same as many pyramid edge. A different answer is shown by a student who seems to have begun to think critically as in the following Figure 10.

In Figure 10, the student is beginning to think critically about a variety of types of prisms and pyramids. The student is also starting to count the same number of edges on the prism and the pyramid and then try to find a connection from the results. Analysis conducted by the student has almost shown the consideration of alternative answers, it's just not reached the expected stage. Students can analyze the relationship between the aspects of the base of the pyramid with many prismatic ribs, but can't arrive at a general condition that must be met so that the edge of a prism can be the same as that of the pyramid edge.

The next analysis is about problem-solving. Consider the item 4 in Figure 11 below.

In item 4, in general, students can solve problems well. For example in Figure 12 below.

In Figure 12, the student seemed to do detailed problem-solving with the correct stages. Besides calculating correctly, students can solve problems and conclude correctly. Some students made mistakes, but generally only in the calculation or pattern errors.

Based on the results previously described, it was found that there are some difficulties experienced by students in applying critical thinking to solve mathematical problems. These

4) banyaknya rusuk suatu prisma dapat sama dengan banyaknya rusuk suatu limas bisa saja terjadi dengan kondisi:
- Alas dari prisma tersebut harus bersegi genap
- Jumlah rusuk pada prisma harus dapat di bagi 2
- Segi dari alas limas tersebut berjumlah setengah dari banyak rusuk pada prisma

Contoh bangun prisma dan bangun limas dengan jumlah banyak rusuk yang sama

Prisma	Limas	banyak rusuk
Segi 4	Segi 6	12
Segi 6	Segi 9	18
Segi 8	Segi 12	24
Segi 10	Segi 15	30

Figure 10. Results of student beginning to show alternative answers.

A box of instant noodles in the form of blocks with a length of 24 cm, width of 5 cm, and height of 2 cm will be filled with 2 boxes of chocolate cubes measuring 35 cm. In the opinion of trader A, the box can hold all the existing chocolate boxes. Meanwhile, according to trader B, the box is too small to contain all the chocolate. Which opinion from the merchant is correct?

Figure 11. Critical thinking question with problem-solving indicator.

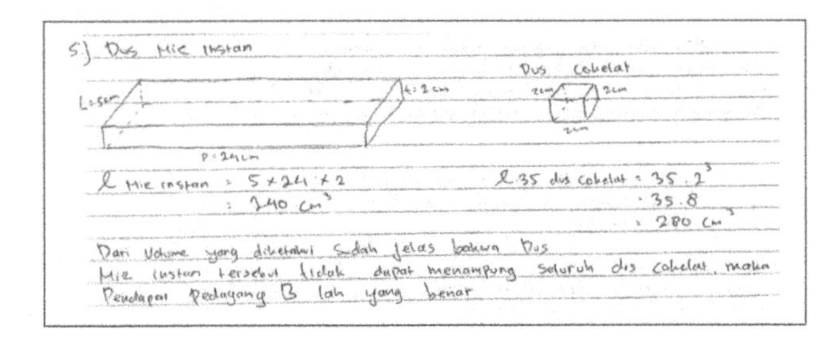

Figure 12. Results of student who can solve problems correctly.

difficulties include analyzing the interrelationship of concepts, the interrelationships between concepts, drawings, proving, generalizing, and providing alternative answers. There are many factors that affect these results. From the observations of researchers based on casual chat with research subjects, certain things are thought to be the cause of the emergence of these difficulties, including learning patterns. The average subject is only learning with the material provided in class. When at home, students are preoccupied with learning assignments of the same type as those studied in class so that it is not necessary to look for other material related to the assignment. In addition, the type of problem given can be solved with the same steps as the solution to the example problem so that it does not motivate the subject to look for solutions. The subjects also felt that the questions given by the researchers were quite difficult because usually the questions were limited to simple problems that did not require deep understanding. The learning pattern is predicted to influence the subject's motivation in learning. Lack of stimulus in the learning process makes the subject comfortable and not challenged to seek more knowledge, so that motivation to learn tends not to develop. However, further research is needed to ensure these predictions through in-depth analysis of the subject's learning motivation.

5 CONCLUSION

The conclusion of the research is that there are some difficulties experienced by students on employing critical thinking to solve mathematical problems. These difficulties are found through analysis of the results of critical thinking test answers, both compared with the guidelines for scoring critical thinking skills or based on indicators of critical thinking skills – proofing, generalizing, considering alternative answers, and problem-solving. Based on the scoring guidelines, it is known that students can find and detect important things from problems, write them down, and sketch pictures if needed. But at the next level of thinking, students still seem to make a lot of mistakes. For example, determining the relationship between the diagonal concept of the plane and the edge, calculating the volume of the pyramid in the cube based on the drawing, determining the different types of prism and pyramid and looking at the relationship between the edge in the prism and the pyramid in general, and connecting the comparison of the cuboid and cube volume in the problem. Based on the indicators of critical thinking skills, students appear to be less critical in proving. This is evident from the error of not yet being able to prove the relationship between the concept of space diagonals and edges. In terms of generalization, in general, students can analyze the questions and find new shapes formed from the given conditions. Critical thinking also appears from the ability to generalize that the volume of the pyramid is 1/6 of the volume of the cube. However, not all students can think up to this stage. Students also have difficulty in considering alternative answers. This is evident from the limitations of thought that prisms and pyramids can only be triangular. Students can't also analyze to find a general condition that must be met if many prismatic ribs can be the same as many pyramid ribs. However, students can do detailed problem-solving with the correct stages in some cases. Analysis of students' difficulty on employing critical thinking as has been described can be used as a basis for further research on how to grow and develop critical thinking skills in solving mathematical problems.

REFERENCES

Chukwuyenum, A. N. 2013. Impact of critical thinking on performance in mathematics among senior secondary school students in Lagos State. *IOSR Journal of Research & Method in Education* 3(5): 18–25.

Daniel, M. F., & Auriac, E. 2011. Philosophy, critical thinking and philosophy for children. *Educational Philosophy and Theory* 43(5): 415–435.

Ennis, R. H. 1993. Critical thinking assessment. *Theory into Practice* 32(3): 179–186.

Ennis, R. H. 2011. The nature of critical thinking: An outline of critical thinking dispositions and abilities. In *Sixth International Conference on Thinking, Cambridge, MA*: 1–8. Champaign, IL: University of Illinois.

Facione, P.A. 2011. Critical thinking: What it is and why it counts. *Insight Assessment* 2007(1): 1–23.

Kasali, R. 2017. *Disruption: Tak ada yang bisa diubah sebelum dihadapi, motivasi saja tidak cukup.* Jakarta: Gramedia Pustaka Utama.

Prihanisetyo, A., Pebrianto, D., & Fitriasari, P. 2018. Era disruption sebuah tantangan atau bencana sebuah telaah literatur. *Jurnal MEBIS (Manajemen dan Bisnis)* 3(1): 10–21.

Safrida, L. N., Ambarwati, R., Adawiyah, R., & Albirri, E. R. 2018. Analisis kemampuan berpikir kritis mahasiswa program studi pendidikan matematika. *EDU-MAT: Jurnal Pendidikan Matematika* 6(1): 10–16.

Sanders, S. 2016. Critical and creative thinkers in mathematics classrooms. *Journal of Student Engagement: Education Matters* 6(1): 19–27.

Shoop, B.L. 2015. Developing critical thinking, creativity and innovation skills of undergraduate students. *Proc. of SPIE* 928904(1): 1–6.

Suparni, S. 2015. Pengembangan bahan ajar berbasis integrasi interkoneksi untuk memfasilitasi peningkatan kemampuan berpikir kritis mahasiswa. *Jurnal Derivat: Jurnal Matematika dan Pendidikan Matematika* 2(2): 1–19.

Learning media development

Borderless Education as a Challenge in the 5.0 Society – Abdullah, Adriany & Abdullah (eds)
© 2021 Taylor & Francis Group, London, ISBN 978-0-367-61960-2

The development of pop-up book media on the concept of area and wide area of plat build

K. Karlimah, E.T. Amelia & L. Nur
Universitas Pendidikan Indonesia, Tasikmalaya, Indonesia

ABSTRACT: The purpose of this study is to develop, test the effectiveness and apply pop-up book media to the concept of a wide and flat area. Researchers developed a pop-up book media concept around and wide flat area for fourth grade elementary school students. The method used is Design Based Research (DBR) which goes through four stages including problem identification, prototype development, due diligence, and reflection to produce the final product. This research shows that; 1) the development of pop-up book media is made according to the curriculum as well as from predetermined indicators, 2) pop-up books are made using the Canva application and printed on 260 gsm art paper A3 size, 3) the use of pop-up book media the circumference and area of the flat build area are suitable for use in the learning process, in terms of the results of validation and testing. Product trials were conducted at two schools, which were reviewed from students' responses and observations during learning. In the learning process students are very enthusiastic about learning using pop-up book media, this is indicated by the positive response to the pop-up book media with the results achieved 96.54% and 99.78%, 4) the final product of this study is in the form media pop-up book concept around and wide flat area that is suitable for use. From the results of the study pop-up book media around and wide flat area for fourth grade elementary school students is suitable for use in learning mathematics.

1 INTRODUCTION

One of the mathematics goals contained in the learning manual of mathematics and education of physical, sports and health curriculum 2013 is to understand the concept and apply it in daily life. But these goals are not easily achieved if mathematics is always considered as a difficult subject by many students (Indiyani & Listiara 2006, Lado et al. 2016, Supriyanto 2014). The current condition of mathematics learning does not involve students in the learning process, students can only copy and record the solutions given by the teacher and if students learn mathematics separately from their daily experiences, students cannot apply mathematics (Fuadi et al. 2016, Sundayana 2015). To facilitate students in understanding the concept of material in mathematics, it is necessary to use media in learning. Media is a components that have been contained into the curriculum and has a very important role in learning process (Masturah et al. 2018, Pramesti 2015, Simatupang 2016, Vebrianto & Osman 2011). The media can influence the success of student learning that is able to actively involved in seeing, touching and experiencing it themselves through the media used so that learning seems more meaningful for students (Safri et al. 2017).

Based on the results of an interview on September 14th, 2018 with a grade IV elementary school teacher stated that the existing mathematics learning media are not specific to the concept of circumference and area of the plane figure. Media that are usually used during learning are objects contained in class that resemble the shape of a square and rectangular flat shape or ask students to understand the concept of the material contained in the book used. However, the results of learning activities show students are still lacking in depth to understand the concept of the circumference and area of the plane figure, this is seen when doing exercises in determining the circumference and area of the plane figure students still find it difficult in the way it works. Therefore the media is needed to improve students' understanding of concepts in explaining and determining the circumference and area of the plane figure.

There are several types of media that can be used in learning, one of them is pop-up book media. A pop-up book is a book in which pages have folded paper that is cut and shaped and appears to form three-dimensional layers when opened and two dimensions when closed and has movable parts (Iizuka et al 2011, Khoirotun et al. 2014, Ruiz Jr et al. 2014, Sholikhah 2017). Pop-up book media can be attract student attention and give the impression of learning that is more fun because it looks packaged with a more attractive structure of three dimensions and has a part that can be moved, shifted, even changed shape (Hanifah 2014, Safri et al. 2017, Simatupang 2016, Ukhtinasari et al. 2017).

Some research on the development of pop-up media has been carried out by Mulianti (2017) on mathematics subjects for grade II elementary schools the effectiveness of its use reaches 92%. In accordance with the results of Resmaniti (2019) in mathematics for first grade elementary schools with an average student response of 100% and shows that the pop-up book media is suitable for use in learning.

Pop-up book media in mathematics learning is not enough, but there have been several developments that have been carried out namely in the material addition of whole numbers, the concept of fraction numbers, as well as building cubes and blocks. The novelty of this development is that by developing different types of material content, a pop-up book media for the concept of circumference and area of the plane figure is made to facilitate students in learning mathematics. Another development is to add and subtract some use of paper types and pop-up book techniques that are adjusted to the material to be presented.

Based on the above background, it is necessary to do research with the title of developing the concept of pop-up book media on circumference and area of the plane figure for students in grade IV elementary school. The problem formulation is as follows: how to develop and the feasibility of pop-up book media for the concept of circumference and area of the plane figure for grade IV elementary school students. The purpose of this study was to develop and find out the feasibility of pop-up books as a medium for elementary school mathematics learning in the concept of circumference and area of the plane figure.

2 METHODS

This research method adopts the steps of Design Based Research (DBR) by using the Reeves (2006) model (Figure 1).

The steps are explained as follows:

- Problem identification and problem analysis through preliminary studies to elementary schools by means of observation and documentation study of the focus of research.

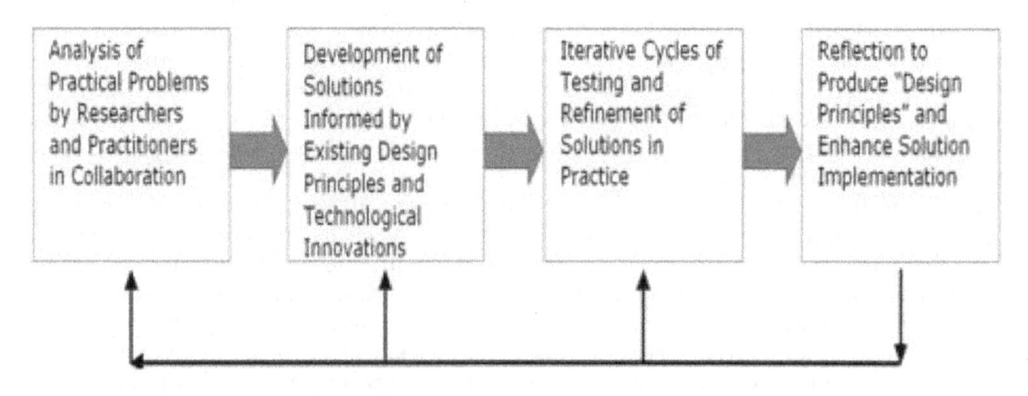

Figure 1. Model Design Based Research (DBR) by Reeves (2006).

- This stage is carried out after obtaining information about the problem to be studied, researchers develop solutions to these problems.
- At this stage a research product trial is carried out to find out the practicality of what was developed based on the results of problem identification and analysis. After knowing the shortcomings of the products that have been tested, then a product revision is carried out and validation is carried out by the validator or expert practitioners.
- At this stage, the final design of the research product is produced after several trials and validation by expert practitioners.

Design Based Research is defined as "a series of approaches used with the intent to produce new theories, artifacts, and practical models in explaining and potentially impacting learning in natural conditions (naturalistic)" (Barab & Squire 2004). Therefore, "design-based research integrates the development of solutions to practical problems in the learning environment with the identification of reusable design principles" (Herrington et al. 2007). "The research step with development is carried out as a prefix to carry out complex and innovative activities by carrying out several validation principles in compiling then supporting a design and development activity" (van den Akker 1999).

3 RESULTS AND DISCUSSION

3.1 *Problem identification and analysis by researchers and practitioners collaboratively*

In this stage the researcher conducted a preliminary study through direct interviews with class IV teachers and documentation studies. Interviews were conducted to obtain information about the use of media in the activities of the mathematics learning process. While the documentation that the researchers studied is the pop-up book media with the content of mathematics and elementary/MI grade IV elementary school mathematics books. The results of this analysis are used as a reference by researchers to develop pop-up book media about the concept of circumference and area of the plane figure for grade IV elementary school students. Pop-up book media that are designed are based on the theories and techniques that should be, so that the products produced are in accordance with predetermined indicators and are suitable for use.

3.2 *Design for making pop-up book media*

The design of making pop-up media is developed based on the indicators that have been set. Following is the design of making pop-up book media developed:

- The design is created using the Canva application.
- The type of letters used are laila bold, adigiana toybox, alegreya sans black, bree serif, barrio, montserrat, bryndan write, atma bold.
- Background uses a light blue base.
- The type of paper used is 260 gsm art paper, duplex paper and the cover is coated with glossy lamination (Ives 2009).
- Paper size uses A3 with a size of 42 x 29.7 cm (Ives 2009).
- Pop-up techniques used are waterfall, corousel, flaps, full tab, v-fold, step-fold, maze and slider techniques (Siregar & Rahmah 2016).

3.3 *The feasibility of pop-up book media the concept of the circumference and area of the plane figure for grade IV elementary students*

3.3.1 *Validation*

This product validation is carried out to determine the feasibility of the product that was designed before use in conducting a trial. The product validation was carried out by a validator or a team of four experts, namely two lecturers from UPI Campus in Tasikmalaya, one teacher from SDN Sukamenak Indah and another teacher from SDN Karsanagara. The validator or the expert team provides an assessment of the pop-up book media product that has been designed by the researcher, this assessment is based on the instrument in the form of a validation sheet. This validation sheet consists of an expert validation sheet for the media and material containing indicators for the content of the pop-up book media. At this stage of validation, the researcher has more discussion with the validator so the validator does not only provide limited to the assessment but providing advice and input that can be used as material for improvement, so that the pop-up book media products are already suitable for use in the trial phase.

3.3.1.1 TRIAL PHASE 1

From the results of student responses, students gave positive responses to pop-up media as the concept of circumference and area of the plane figure for grade IV elementary school students. The percentage of student responses obtained to the pop-up book media obtained the following results: 1) the pop-up book media 97.22%; 2) 95.37% objectives; 3) 97.04% benefit. The average percentage of use of pop-up book media reached 96.54%.

3.3.1.2 TRIAL PHASE 2

From the results of the students' responses gave a positive response to the pop-up book media about the concept of circumference and area of the plane figure for grade IV elementary school students. The percentage of student responses obtained to the media pop-up book obtained the following results: 1) the media pop-up book 100.00%; 2) 100.00% goal; 3) benefits of 99.35%. The average percentage of use of pop-up book media reaches 99.78%.

3.3.1.3 PRODUCT REFLECTION

After going through several stages of development, starting from problem identification and analysis, product design development, product validation and testing, the reflection of the development of this learning media product is to produce a final product in the form of a pop-up book media about the concept of circumference and area of the plane figure for class students IV elementary school which is packaged in book form.

After the product is made and tested on learning activities, there are several advantages and disadvantages of the pop-up book media about the concept of circumference and area of the plane figure for grade IV students, including the following:

(1) Strengths

- Designed based on the 2013 mathematics curriculum reference.
- Can be used in classroom learning or independently.
- Providing hands-on experience.
- Has a three-dimensional display, so that objects contained in the media pop-up book can move and play.
- Provide a visualization of the circumference and area of the plane figure that can be more easily understood.
- Provides interesting surprises on each page.
- Can form a more careful attitude in using books, and increase patience and accuracy in using it.

(2) Weaknesses

- Making a product requires a long time.
- Requires quite expensive costs.

- It takes extra patience, meticulous and careful in its manufacture.
- In the learning process, each group is limited to a maximum of 5 people, this is so that learning is more effective.
- High skills are needed when guiding the learning process using pop-up book media.

4 CONCLUSION

The development of a pop-up book media about the concept of circumference and area of the plane figure is based on two aspects, including aspects of conformity with the curriculum and aspects of the mechanism of making pop-up books. The development of pop-up book media is adjusted to the 2013 fourth grade elementary school mathematics curriculum which is based on basic competencies and predetermined indicators. In addition to conformity with the curriculum, the development of pop-up book media was developed based on the manufacturing mechanismpop-up by paying attention to the paper type, paper texture, paper size, paper weight and thickness, the pop-up book method, the pop-up book technique and the tools and materials used during the manufacturing process.

Development of pop-up book media about the concept of circumference and area of the plane figure for grade IV elementary school students is designed based on predetermined indicators, while the design is made namely 1) product design using the *canva* application; 2) the color used in the background uses a light blue color with the addition of a transposed square shape; 3) the type of letters used are *laila bold, adigiana toybox, alegreya sans black, bree serif, barrio, montserrat, bryndan write, atma bold*; 4) Paper type uses 260 gsm mate and glossy textured art paper with A3 size; 5) the techniques used include watrerfall, v-fold, flaps, carousel, full tab, maze, slider and step-fold.

The feasibility of the pop-up book media about the concept of circumference and area of the plane figure through two stages, namely the validation stage by experts and the pilot phase in elementary schools. The assessment is carried out based on the validation instrument made, after which an improvement or revision is made on the product that was developed according to the advice of an expert to produce a media that is feasible to try out. This trial was conducted in two different schools, the data obtained at the testing stage were student responses and the results of observations on the use of pop-up book media during learning activities. The trial results showed positive results even though in the first stage there were some improvements made to the product. The results of the responses of students' responses and the results of observations on the use of pop-up book media on the concept of circumference and area of the plane figure shows that the pop-up book media about the concept of circumference and area of the plane figure for grade IV elementary school students is suitable for use in the learning process.

Reflection of product development is to produce a final product in the form of a pop-up book media about the concept of circumference and area of the plane figure for grade IV elementary school students. This product is packaged with a three-dimensional structure that can be played and moved by book users. Presentation of material was based on the 2013 fourth grade elementary school mathematics curriculum.

REFERENCES

Barab, S. & Squire, K. 2004. Design-based research: Putting a stake in the ground. *The Journal of the Learning Sciences* 13(1): 1–14.

Fuadi, R., Johar, R. & Munzir, S. 2016. Peningkatkan kemampuan pemahaman dan penalaran matematis melalui pendekatan kontekstual. *Jurnal Didaktik Matematika* 3(1): 47–54.

Hanifah, T.U. 2014. Pemanfaatan media pop-up book berbasis tematik untuk meningkatkan kecerdasan verbal-linguistik anak usia 4-5 tahun (Studi eksperimen di TK Negeri Pembina Bulu Temanggung). *BELIA: Early Childhood Education Papers* 3(2): 46–54.

Herrington, J., McKenney, S., Reeves, T. & Oliver, R. 2007. Design-based research and doctoral students: Guidelines for preparing a dissertation proposal. In *EdMedia+ Innovate Learning*: 4089–4097. Waynesville, USA: Association for the Advancement of Computing in Education (AACE).

Iizuka, S., Endo, Y., Mitani, J., Kanamori, Y. & Fukui, Y. 2011. An interactive design system for pop-up cards with a physical simulation. *The Visual Computer* 27(6-8): 605–612.

Indiyani, N.E. & Listiara, A. 2006. Efektivitas metode pembelajaran gotong royong (cooperative learning) untuk menurunkan kecemasan siswa dalam menghadapi pelajaran matematika (suatu studi eksperimental pada siswa di SMP 26 Semarang). *Jurnal Psikologi* 3(1): 10–28.

Ives, R. 2009. *Paper engineering and Pop-ups for dummies*. Indianapolis: John Wiley & Sons.

Khoirotun, A., Fianto, A.Y.A. & Riqqoh, A.K. 2014. Perancangan buku pop-up museum Sangiran sebagai media pembelajaran tentang peninggalan sejarah. *Jurnal Art Nouveau* 2(1): 134–141.

Lado, H., Muhsetyo, G. & Sisworo, S. 2016. Penggunaan media bungkus rokok untuk memahamkan konsep barisan dan deret melalui pendekatan RME. *Junal Pembelajaran Matematika* 3(1): 1–9.

Masturah, E.D., Mahadewi, L.P.P. & Simamora, A.H. 2018. Pengembangan media pembelajaran Pop-up Book pada mata pelajaran IPA kelas III Sekolah Dasar. *Jurnal EDUTECH Undiksha* 6(2): 212–221.

Mulianti, E.S. 2017. *Pengembangan media pembelajaran Pop-up Book pembelajaran matematika kelas II MI Ma'arif Bego Maguwoharjo Sleman Yogyakarta*. Tesis, Yogyakarta: Sekolah Pascasarjana, Universitas Islam Negeri Sunan Kalijaga.

Pramesti, J. 2015. Pengembangan media pop-up book tema peristiwa untuk kelas III SD. *Jurnal Pendidikan Guru dan Sekolah Dasar* 4(16): 1–11.

Reeves, T.C. 2006. Design research from a technology perspective. In J. Van den Akker, K. Gravemeijer, S. McKenney & N. Nieveen, (eds), *Educational Design Research*: 52–66. London: Routledge.

Resmaniti, D.M. 2019. Rancangan media Pop-up Book tentang konsep operasi hitung penjumlahan bilangan cacah. *Indonesian Journal of Primary Education* 3(1): 1–8.

Ruiz Jr, C.R., Le, S.N., Yu, J. & Low, K.L. 2014. Multi-style paper pop-up designs from 3D models. *Computer Graphics Forum* 33(2): 487–496.

Safri, M., Sari, S.A. & Marlina, M. 2017. Pengembangan media belajar Pop-up Book pada materi minyak bumi. *Jurnal Pendidikan Sains Indonesia (Indonesian Journal of Science Education)* 5(1): 107–113.

Sholikhah, A. 2017. Pengembangan media Pop-up Book untuk meningkatkan kemampuan menulis kreatif pada mata pelajaran Bahasa Indonesia materi menulis karangan kelas V SDN Rowoharjo tahun ajaran 2016/2017. *Jurnal Simki Pedagogia* 1(8): 1–12.

Simatupang, H.A. 2016. Pengembangan media Pop-up pada materi organisasi kehidupan untuk meningkatkan motivasi dan hasil belajar peserta didik SMP kelas VII. *Jurnal Pendidikan Matematika dan Sains* 5(1).

Siregar, A. & Rahmah, E. 2016. Model pop-up book keluarga untuk mempercepat kemampuan membaca anak kelas rendah sekolah dasar. *Ilmu Informasi Perpustakaan dan Kearsipan* 5(1): 10–21.

Sundayana, R. 2015. *Media dan alat peraga dalam pembelajaran matematika*. Bandung: Alfabeta.

Supriyanto, B. 2014. Penerapan discovery learning untuk meningkatkan hasil belajar siswa kelas VI B mata pelajaran matematika pokok bahasan keliling dan luas lingkaran di SDN Tanggul Wetan 02 Kecamatan Tanggul Kabupaten Jember. *Pancaran Pendidikan* 3(2): 165–174.

Ukhtinasari, F., Mosik, M. & Sugiyanto, S. 2017. Pop-up sebagai media pembelajaran fisika materi alat-alat optik untuk siswa sekolah menengah atas. *Unnes Physics Education Journal* 6(2): 1–6.

Van den Akker, J. 1999. Principles and methods of development research. In *Design Approaches and Tools in Education and Training*: 1–14. Dordrecht: Springer.

Vebrianto, R. & Osman, K. 2011. The effect of multiple media instruction in improving students' science process skill and achievement. *Procedia-Social and Behavioral Sciences* 15: 346–350.

Borderless Education as a Challenge in the 5.0 Society – Abdullah, Adriany & Abdullah (eds)
© 2021 Taylor & Francis Group, London, ISBN 978-0-367-61960-2

Developing learning media based on magnetic white board on motion dynamics

B.R. Kurniawan, W. Winarto, M.I. Shodiqin, D.E. Saputri, S. Kusairi & S. Sulur
Physics Department, Faculty of Mathematics and Sciences, Universitas Negeri Malang, Malang, Indonesia

ABSTRACT: The interaction between teacher and student is important to provide students' understanding of learning materials. The fact shows that teachers only focus on providing information for students, so interaction with students is still not optimal. Therefore, we have to decide the right media to increase interaction to improve students' understanding of physics learning material. A good understanding of physics learning material is an investment required by students in solving physics problems. One of the important learning materials in physics is motion dynamics. Therefore, this research and development target is to develop media that can increase interactions between teachers and students on motion dynamics. The research was conducted by adopting six of the ten steps from Sugiyono's method. The validation results show that the media is categorized as very feasible with a score of 89% and the results of limited trials indicate that the media is categorized as very practical with a percentage of 85%. Learning media based on the magnetic whiteboard is expected to improve activities and interactions between teachers and students in learning material of motion dynamics at schools

1 INTRODUCTION

The teacher has an important role in the learning process. The teacher's role influences student learning outcomes (Dasem et al. 2018). Generally, the center of the learning process is the teacher, but students must participate actively (Antara et al. 2014). Teachers and students must be actively involved in the learning process. In reality, the teacher is more active in the learning process while students tend to be passive. Teachers tend to be providers of information (Nugroho et al. 2013). There is almost no interaction between the teacher and the students. The lack of teacher-student interaction results in students' lacking understanding of the material.

One of the learning materials that less understood by students is Vector. Vector is a basic learning material in physics and has an important role in many learning materials in physics (Bollen et al. 2017, Kurniawan et al. 2019). Yet the fact shows that students still have difficulty in drawing vector (Nguyen & Meltzer 2003, Shaffer & McDermott 2005). The difficulty of students in describing vectors is found in the learning material of equilibrium, rotational dynamics, mechanics, and electrostatics (Rosengrant et al. 2009, Sarkity et al. 2016). Moreover, the lack of understanding in vector also impacts to dynamics of motion (Flores et al. 2004, Flores-García et al. 2008, Jim nez-Valladares & Perales-Palacios 2001, Knight 1995). This was confirmed by the findings of Astuti & Syafitri that showed a mutual relationship between the understanding of vector and motion dynamics learning materials (Astuti & Syafitri, unpubl). The study of Handhika et al. (2016) to 94 students of the Physics Department in IKIP PGRI Madiun found that physics students lack understanding in learning material of dynamics of motion.

Students lacking understanding related to the fact that they only receive the theory without constructing the understanding of their knowledge (Potimbang 2014, Tsegaye et al. 2010). Activities to build knowledge of students can be done by providing practical learning

(Nugroho et al. 2013). Providing practical learning to students can be done by using instructional media. The use of instructional media makes students actively interact and unleash the potential they have (Falahudin 2014).

Learning media could be in the form of audio, graphics, animation, and/or game. The selection of instructional media must consider the characteristics of the presented learning material (Ningrum et al. 2018). Appropriate learning media can motivate students to be actively involved in the learning process. Active students are expected to able to construct their understanding of physical learning material.

The efforts to improve understanding of motion dynamics learning material through the use of media have been carried out by Primanda et al. (2015) who developed book supplements. The development of book supplements by Primanda et al. (2015) received a positive response from students. The efforts to develop media to improve interactions between teachers and students were conducted by Ningrum et al. (2018) using a movable free body diagram Charta for equilibrium and rotational dynamics. The developed chart is in the form of cardboard pieces of separate images to visualize the free body diagram. The use of moveable images can attract the students' attention and active role. Ayesh et al. (2010) states that the use of free body diagram re-presentation has clear impact on the student performance.

Based on the description above, this study aims to develop learning media based on a magnetic whiteboard on motion dynamics. The developed media helps the teacher to display force vector images that acting on free objects based on the motion dynamics learning material. The development of this media is expected to improve the interaction of teachers and students in the learning process of motion dynamics.

2 RESEARCH METHODS

Research and development that has been done in the development of learning media are based on a magnetic whiteboard on motion dynamics. The research was conducted using six out of ten development methods of Sugiyono (2007). Development methods used include (1) potential and problem analysis, (2) information gathering, (3) media design, (4) media validation testing, (5) media revision, and (6) media testing. In detail, the research method is presented in Figure 1.

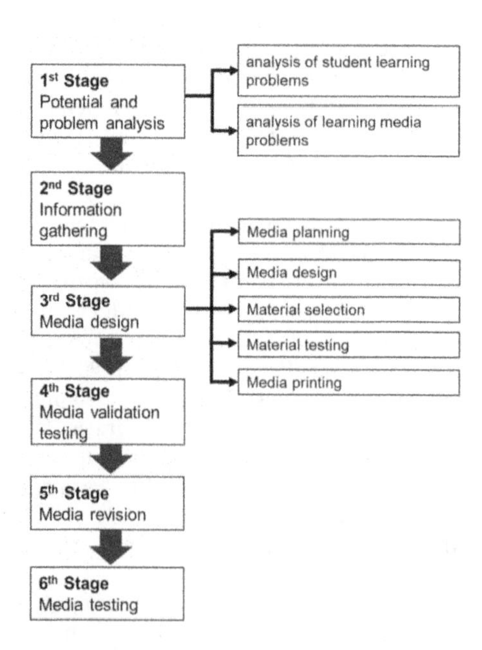

Figure 1. Research and development methods.

This research and development begin with an analysis of student learning difficulties on the motion dynamics learning material and an analysis of learning media problems. Based on the analysis of the problems, the information collecting was conducted to resolve the difficulties. The yielded result was used to design learning media. Media design starts with planning, designing, material selection, media testing, and media printing. After that, the printed media are validated by experts based on its conformity with the learning material and the media appearance. The validation test results and expert insights are used as references for media revision. The revised media were subsequently trialed at the Lustrum XIII education exhibition in Universitas Negeri Malang. Limited trials were conducted to determine the practicality of media use. The practicality test of the media has been done by evaluating the aspects of ease, attractiveness, usefulness, and motivation in media use. The test subjects were visitors to the exhibition such as high school students, teachers, and college students.

The data obtained in this study are quantitative and qualitative. Quantitative data were obtained through expert validation tests in the form of media prototype assessment scores and results of media trials at the Lustrum XIII UM educational exhibition. The percentage of the validation test results was used to determine the category of media validity using Arikunto (2006) criteria presented in Table 1.

On the other hand, the percentage of media trial results is used to determine the practicality criteria of the media using the criteria from Yamasari (2010) presented in Table 2.

The qualitative data in this research includes the validator's suggestion and the trial subject's suggestion. The obtained qualitative data was used as material for consideration in conducting media revisions.

3 RESULTS AND DISCUSSION

The stages of this research and development begin with an analysis of student learning problems on the learning material of motion dynamics, and an analysis of learning media problems. One of the fundamental difficulties of students in learning motion dynamics material is to determine the direction of the force vector (Sarkity et al. 2016). This situation happened due to the teacher's tendency to be an information provider (Nugroho et al. 2013) so that students only receive material without constructing their understanding (Potimbang 2014). The

Table 1. Eligibility criteria.

No.	Score (%)	Eligibility Criteria
1.	< 21 %	Very infeasible
2.	21 – 40 %	Infeasible
3.	41 – 60 %	Feasible enough
4.	61 – 80 %	Feasible
5.	81 – 100 %	Very feasible

Table 2. Practicality criteria.

Percentage	Criteria	Information
$1 \leq P < 25$	Impractical	Can't be used
$25 \leq P < 50$	Less practical	Can be used with many revisions
$50 \leq P < 75$	Practical	Can be used with less revision
$75 \leq P \leq 100$	Very practical	Can be used without revision

use of instructional media on dynamics of motion is dominated by worksheets (Ningrum et al. 2018).

According to the analysis of the problem and information, learning media was designed to improve activities and interactions between the teacher and students on learning the motion dynamic material. The media was designed using two acrylics that were glued together. The top acrylic was made of clear acrylic which was printed according to the desired design, while the bottom acrylic was made of white acrylic. Magnets were planted in the acrylic thus the media could be used on the resulting magnetic whiteboards. Material testing was obtained by testing the strength of the magnet and the appearance of the material. This media requires a strong magnet to support learning motion dynamics in complex problems.

The products of learning media based on the magnetic whiteboard are in the form of blocks, vector directions, force symbols (F), weight symbols (W), normal forces symbols (N), and angular symbols. The media was printed with color variations for better use. The printed media was tested for validation by experts. The expert validation test results are 82% for its compatibility with the learning material and 96% for media appearance as presented in Table 3.

Based on these results, overall learning media based on the magnetic whiteboard is 89% so that it could be categorized as very feasible. The suggestion from the validator is needed to add more media components so that the media can be used in the pulley problem.

The stages of media trials have been done by displaying media on the Lustrum XIII education exhibition Universitas Negeri Malang. The visitor of the exhibition were high school students, teachers, and college students that were allowed to try and give an assessment of the media through a provided questionnaire as shown in Figure 2.

60 exhibition visitors including 22 high school students, 1 teacher, and 37 students who had tried and provided an assessment of the learning media based on the magnetic whiteboard. The results of visitor ratings are presented in Table 4.

Based on the trial results, learning media based on the magnetic whiteboard can be categorized as very practical up to 85%. These results indicate that the media received a good response from visitors. Even from the questionnaire given, 100% of visitors supported the use of learning media based on the magnetic whiteboard at schools. Visitors also hope that this media could be developed for other learning materials.

Table 3. Validation test result.

No.	Aspect	Score	Information
1.	Compatibility with learning material	82 %	Very feasible
2.	Media Appearance	96%	Very feasible
	Average	89 %	Very feasible

Figure 2. Media testing.

Table 4. The results of visitor ratings.

No.	Aspect	Score	Criteria
1.	Ease	89 %	Very practical
2.	Attractiveness	85 %	Very practical
3.	Usefulness	85 %	Very practical
4.	Motivation	82 %	Very practical
Average		85 %	Very practical

4 CONCLUSION

Based on the analysis of the results of research and development that has been conducted, it can be concluded that the learning media based on the magnetic whiteboard on motion dynamics is categorized as very feasible and very practical. The media developed are in the form of blocks, vector directions, force symbols (F), weight symbols (W), normal forces symbols (N), and angular symbols. Learning media based on the magnetic whiteboard is expected to improve activities and interactions between teachers and students in learning material of motion dynamics at schools. To meet this goal, we need collaborative effort with all parties, including researchers, teachers and educational institutions.

ACKNOWLEDGMENT

This research and development were carried out with the financial support from the PNBP Faculty of Mathematics and Science 2019, Universitas Negeri Malang.

REFERENCES

Antara, I.N.R., Haris, I.A. & Nuridja, I.M. 2014. Pengaruh kesiapan dan transfer belajar terhadap hasil belajar ekonomi di SMA Negeri 1 Ubud. *Jurnal Pendidikan Ekonomi Undiksha* 4(1): 1–12.

Arikunto, S. 2006. *Prosedur penelitian suatu pendekatan*. Jakarta: Rineka Cipta.

Astuti, D. & Syafitri, E. (n.d.) Pengaruh pemahaman teori vektor terhadap pemecahan masalah kinematika dan dinamika teknik. Unpublished.

Ayesh, A., Qamhieh, N., Tit, N. & Abdelfattah, F. 2010. The effect of student use of the free-body diagram representation on their performance. *Educational Research* 1(10): 505–511.

Bollen, L., van Kampen, P., Baily, C., Kelly, M. & De Cock, M. 2017. Student difficulties regarding symbolic and graphical representations of vector fields. *Physical Review Physics Education Research* 13(2): 020109.

Dasem, A.A., Laka, B.M. & Niwele, A. 2018. Peranan guru dalam proses pembelajaran Bahasa Indonesia di SD Inpres Komboi Kabupaten Biak Numfor. *Wacana Akademika* 2(2): 126–136.

Falahudin, I. 2014. Pemanfaatan media dalam pembelajaran. *Lingkar Widyawiswara* 1(December): 104–117.

Flores, S., Kanim, S.E. & Kautz, C.H. 2004. Student use of vectors in introductory mechanics. *American Journal of Physics* 72(4): 460–468.

Flores-García, S., Alfaro-Avena, L.L., Dena-Ornelas, O. & González-Quezada, M.D. 2008. Students' understanding of vectors in the context of forces. *Revista Mexicana de Fisica E* 54(1): 7–14.

Handhika, J., Cari, C., Soeparmi, A. & Sunarno, W. 2016. Student conception and perception of Newton's law. *AIP Conference Proceedings* 1708(1): 070005.

Jim nez-Valladares, J.D.D. & Perales-Palacios, F.J. 2001. Graphic representation of force in secondary education: Analysis and alternative educational proposals. *Physics Education* 36(3): 227–235.

Knight, R.D. 1995. The vector knowledge of beginning physics students. *The Physics Teacher* 33(2): 74–77.

Kurniawan, B.R., Saputri, D.E. & Shoiqin, M.I. 2019. Analisis pemahaman konsep mahasiswa pada topik vektor. *Efektor* 6(2): 107–114.

Nguyen, N.-L. & Meltzer, D.E. 2003. Initial understanding of vector concepts among students in introductory physics courses. *American Journal of Physics* 71(6): 630–638.

Ningrum, A.S., Susanto, H. & Mindyarto, B.N. 2018. Pengembangan media charta Free Body Diagram (FBD) yang moveable untuk meningkatkan kemampuan multirepresentasi siswa pada materi kesetimbangan dan dinamika rotasi. *Physics Education Journal* 7(3): 43–50.
Nugroho, A., Raharjo, T. & Wahyuningsih, D. 2013. Pengembangan media pembelajaran fisika menggunakan permainan ular tangga ditinjau dari motivasi belajar siswa kelas VIII materi gaya. *Jurnal Pendidikan Fisika* 1(1): 11–18.
Potimbang, K. 2014. Meningkatkan hasil belajar siswa pada materi gaya terhadap gerak benda melalui penerapan model pembelajaran kooperatif tipe jigsaw kelas IV SD Inpres 2 Slametharjo. *Jurnal Kreatif Tadulako* 4(12): 181–188.
Primanda, A., Maharta, N. & Sesunan, F. 2015. Pengembangan Suplemen buku siswa materi dinamika gerak dengan pendekatan scientific. *Jurnal Pembelajaran Fisika* 3(3): 13–23.
Rosengrant, D., Van Heuvelen, A. & Etkina, E. 2009. Do students use and understand free-body diagrams? *Physical Review Special Topics-Physics Education Research* 5(1): 010108.
Sarkity, D., Yuliati, L. & Hidayat, A. 2016. Kesulitan siswa SMA dalam Memecahkan masalah kesetimbangan dan dinamika rotasi. In *Prosiding Semnas Pendidikan IPA Pascasarjana UM*: 166–173. Malang: Pascasarjana Universitas Negeri Malang.
Shaffer, P.S. & McDermott, L.C. 2005. A research-based approach to improving student understanding of the vector nature of kinematical concepts. *American Journal of Physics* 73(10): 921–931.
Sugiyono. 2007. *Metode penelitian kuantitatif kualitatif dan R&D*. Bandung: CV. Pustaka Setia.
Tsegaye, K., Baylie, D. & Dejne, S. 2010. Computer based teaching aid for basic vector operations in higher institution physics. *Latin-American Journal of Physics Education* 4(1): 3–6.
Yamasari, Y. 2010. Pengembangan media pembelajaran matematika berbasis ICT yang berkualitas. In *Seminar Nasional Pascasarjana X-ITS, Surabaya*, 12 August 2009. Surabaya: FPMIPA Universitas Negeri Surabaya.

Borderless Education as a Challenge in the 5.0 Society – Abdullah, Adriany & Abdullah (eds)
 ISBN 978-0-367-61960-2

Developing computer-based assessment to support the basic skills of teaching course

B.R. Kurniawan, A. Suyudi, K. Nurhidayah, S. Fawaiz & E. Purwaningsih
Physics Department, Faculty of Mathematics and Sciences, Universitas Negeri Malang, Malang, Indonesia

ABSTRACT: The quality of teachers plays an important role in determining the success of students. Prospective teachers must have a good understanding of the basic skills of teaching. Therefore, it is necessary to conduct research and development computer-based assessment which is not only able to assess the prospective teachers but also can assist the prospective teachers to understand the basic skills of teaching better. Computer-based assessment consisting of a true-false quiz with three levels, the learning material, and video simulation. Each level consists of five questions about the basic skills of teaching with different duration of work. Research and development were done by using the model research of ADDIE. The results of test validation by experts showed that the media developed is very decent with a percentage of 87.75 %. The results of limited testing show that the media is very practical with a percentage of 78.29%. This shows that the media ratings that have been developed can be used without revision. The selection of Microsoft Power-Point in developing computer-based assessments is expected to motivate the prospective teachers to be able to develop the computer-based assessment independently.

1 INTRODUCTION

The quality of a teacher determines the students' success rate. The teacher's quality in teaching and learning process affects the achievement of competence, feeling, involvement, and academic performance (LoCasale-Crouch et al. 2018), attitude and students' habits (Blazar & Kraft 2017). There is a positive correlation between the teacher's quality towards learning achievement (Adnot et al. 2017, Canales & Maldonado 2018) and students' success (Baumert et al. 2010). Besides, Abell (2008), Demirbağ & Kingir (2017), and Valdmann et al. (2017) confirm that the teacher is the most important factor in learning.

Entering the Era of 4.0 Industry and the 21st century, the teacher as a determinant factor of educational advance in a country, having many demands of expertise. Therefore, the education of prospective teachers needs more attention to prepare prospective teachers to be professional teachers. The importance of a teacher's role is evident from the educational course that presents the Basic Skill of Teaching course since the beginning as preparation for prospective teachers to become a professional teacher. Koehler & Mishra (2011) explain that a professional teacher needs three basic knowledge such as pedagogical knowledge, technology, content and the integration of this knowledge.

To guarantee the course achievement, then it needs assessment implementation. The observation results in the field showed that paper-based assessment is still frequently used. The same condition also mentioned by Setyoko (2018). Paper-based assessment enables the students to cheat (Mastuti 2016). The research results by Istiyono & Subroto (2017) showed that paper-based assessment is less accurate in measuring students with medium to high ability. Besides that, the process of giving the feedback in the paper assessment cannot be done directly because in correcting the answers, the lecturer needs quite a long time (Haigh 2010, Khalanyane & Hala-hala 2014, Pratomo & Mantala 2016).

In line with the advance of computer technology usage in all aspects of life, the assessment activity nowadays starts to shift to technology use. Various advantages are obtained by using computer technology and becomes one of the reasons for using a computer-based assessment. In the computer-based assessment, the students can do discussion with the educators outside the class, the educator does not need a certain classroom for assessment implementation (Fuentes et al. 2014, Haigh 2010, Setyoko 2018). The computer-based assessment also enables the existence of direct feedback which can be accepted by the students and the assessment results are more accurate and reliable (Fuentes et al. 2014, Haigh 2010, Shute & Rahimi 2017). Besides, computer-based assessment can assist in achieving excellent learning and assessment achievement (Serradell-López et al. 2010), and improving the learning motivation (Nikou & Economides 2016).

Based on the advantages obtained through computer-based assessment, a computer-based assessment is developed to support the basic skill of teaching courses. This research and development aim to develop and test the content validity along with the construct validity of the computer-based assessment until it is feasible to be used in the course as students' guidance in understanding the basic skill of teaching. The media is expected to be able to support the course and generate competent and skillful prospective teachers in teaching.

2 METHODS

The development of a computer-based assessment was conducted to support the course of Basic Skill of Teaching for Physics Prospective Teachers. The product of this research and development is in the form of a computer-based assessment. The research model used in this research and development was ADDIE which consisted of five stages such as (1) the analysis stage, (2) the designing stage, (3) the development stage, (4) the implementation stage, and (5) the evaluation stage.

The analysis stage is done by conducting a need analysis and a field analysis. The need analysis was conducted through a literature study presented in the journal. The condition analysis in the field is conducted through interviews. The designing stage was conducted by designing the storyboard which would be developed. Based on the storyboard design, the development stage was conducted by arranging the assessment in Microsoft Power-Point. The assessment is arranged by utilizing the hyperlink feature in the program of Microsoft Power-Point. After that, the computer-based assessment then was validated by the experts.

Expert validation was conducted to know the feasibility of a computer-based assessment. The computer-based assessment is then revised by the suggestion from the validator. The product of computer-based assessment which was revised then tested on a limited number of prospective teachers to know the practicality of product usage.

The types of data collected in this research were quantitative data and qualitative data. Quantitative data were in the form of assessment scores of the product, while the qualitative data were in the form of suggestions related to computer-based assessment. The technique of data analysis used to know the feasibility of the product is the average calculation technique. The results obtained then determined based on the criteria of feasibility by using criteria by Arikunto (2006) presented in Table 1.

The test of product practicality of computer-based assessment is conducted covering the aspect of easiness, attractiveness, advantages, and motivation. The percentage of the practicality test result is used as the basis of determining the usage practicality of the computer-based assessment with the criteria by Yamasari (2010) presented in Table 2.

Table 1. Eligibility criteria.

No.	Score (%)	Eligibility Criteria
1.	< 21 %	Very infeasible
2.	21 – 40 %	Infeasible
3.	41 – 60 %	Feasible enough
4.	61 – 80 %	Feasible
5.	81 – 100 %	Very feasible

Table 2. Practicality criteria.

Percentage	Criteria	Information
$1 \leq P < 25$	Impractical	Can't be used
$25 \leq P < 50$	Less practical	Can be used with many revisions
$50 \leq P < 75$	Practical	Can be used with less revision
$75 \leq P \leq 100$	Very practical	Can be used without revision

3 RESULTS AND DISCUSSION

The research and development of a computer-based assessment are started by conducting the condition analysis in the field. The result of the condition analysis and the need analysis in the field shows that there are still prospective teachers who cannot deliver the learning materials well during the teaching practice. Even a finding conducted by Adji et al. (2019) shows that the prospective teacher forgets the learning materials that they would like to deliver. This condition was caused by the fact that they did not acquire enough learning materials for the basic skill of teaching (Adji et al. 2019). Besides, it was also found that the assessment media to support the course of basic skill of teaching was still less provided. Thus, the solution which has been used to strengthen the skill of basic teaching for prospective teachers was by using lesson study (Lestari 2018, Rizki 2014, and Zunaidah 2016). The lack of assessment media in the basic skill of teaching courses causes the prospective teachers we're not ready in the teaching practice. Therefore, the assessment media which can give direct feedback is needed to give strengthening towards the basic skill of teaching owned by the students.

The results of the condition analysis and the need analysis in the field show that media assessment with computer-based is needed to support the course of basic skill of teaching. The selection of computer as the media of assessment was caused by the fact that the computer skill can give direct feedback, more accurate, and more reliable (Fuentes et al. 2014, Haigh 2010, Shute & Rahimi 2017). Therefore, a computer-based assessment is arranged to support the course of the basic skill of teaching. This computer-based assessment is designed to provide three features that can be chosen by the users in acquiring the basic skill of teaching. The quiz feature consists of three levels of questions that were differed based on the time given to work on it. The type of questions presented in the form of isomorphic questions with right or wrong format. At the end of the quiz, the users can get feedback which shows their understanding of the basic skill of teaching. If the users do not understand the concept well, then the users are directed to the material features. Beside material features, there is also a simulation feature which is provided to strengthen the understanding of basic teaching skill. The results of this design are the basis for the development of computer-based assessment.

The development stages of computer-based assessment are started by arranging all features and content in Microsoft Power-Point which has been developed then validated by the experts to know the feasibility of the assessment media covering the content aspect and the media aspect. The result of product validation shows that the product is categorized into very feasible with a percentage of 87.5%. The suggestion given by validators are as follow: (1) the less right words need to revised, (2) add the home button on the quiz materials. The validated product then is revised based on the suggestions from the expert.

The product of computer-based assessment which has been revised consists of: (1) the opening page, (2) the menu page, (3) the material page, (4) the simulation page, (5) the quiz page, and (6) the feedback page. The following is the picture of the assessment media product which had been developed. At the beginning of the media use, the users will be presented on the opening page (see Figure 1).

After the opening page, then the users will go to the menu page which helps them to go to the instruction page, material, simulation, and quiz. The material page (see Figure 2) consists of eight chapters of the basic skill of teaching. The users are given the freedom to choose any materials they want to learn.

Figure 1. Opening page.

During learning the materials, the users can read and also listen to music at once which has been provided. The users can also shift to other materials by using the menu button at the right bottom (see Figure 3). The next and previous buttons are provided to ease the users to access the material page. Besides, it also provides a home button to go back to the menu page.

At the simulation page (see Figure 4), a video simulation about the basic skill of teaching is presented. The simulation video can be accessed by the users through the internet network.

In the quiz page, the isomorphic questions with right or wrong types are provided. The quiz consist of three levels with each level consists of five questions and have different time allocation. The users are free to choose any level of questions they want to work on. At the end of the quiz, the users can get the feedback based on the results of the quiz they worked on. The users who can answer all questions correctly will obtain feedback after understanding the materials. While the users who cannot answers all questions correctly yet also will obtain the feedback that they do not understand the whole materials yet.

Figure 2. Material page.

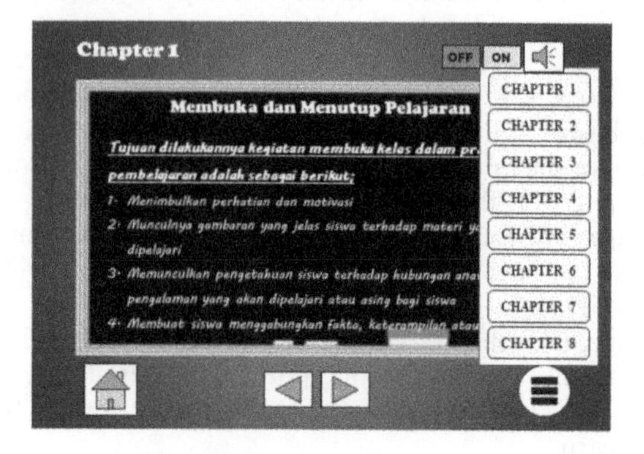

Figure 3. Menu button.

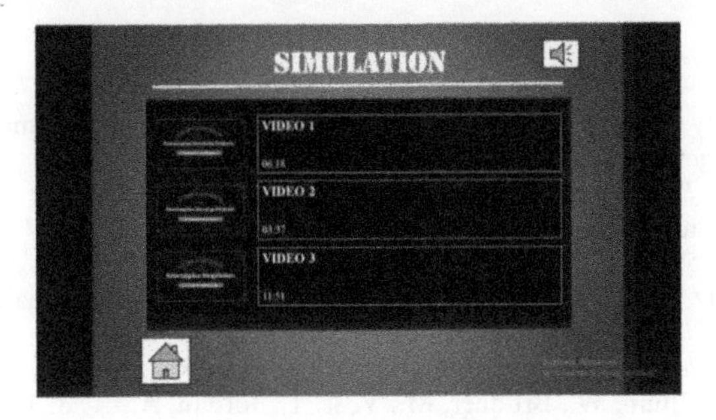

Figure 4. Simulation page.

Table 3. Limited-trial result.

No.	Indicator	Percentage	Criteria
1.	Ease of use	79.20 %	Very practical
2.	Interest	76.59 %	Very practical
3.	Benefits	78.98 %	Very practical
4.	Motivation	78.38 %	Very practical
	Average	78.29 %	Very practical

After conducting the development of the computer-based assessment, then limited try out was conducted on the product towards 75 students. The selection of students as the subjects of the try-out is because the Physics Education students are viewed as capable of representing the point of view of the students and the teachers. The results of limited try-out show that the product of computer-based assessment is categorized into practical with percentage 78.29%. The results of the limited try out are showed in Table 3. Based on the results of the limited try-out, then it can be concluded that the product of computer-based assessment is categorized into very practical.

The development results of the computer-based assessment with the criteria of very feasible and very practical shows that the development of computer-based assessment could be conducted using simple software such as Microsoft Power-Point. Microsoft Power-Point is a well-known software for most people, especially prospective teachers. Thus, this work result is expected to be able to motivate the prospective teacher to develop computer-based assessments independently using Microsoft Power-Point software. This is very important because teachers nowadays are necessary to have the ability to assess students using technology such as computers or Android.

4 CONCLUSION

The developed product of computer-based assessment media consists of (1) the opening page, (2) the menu page, (3) the material page, (4) the simulation page, (5) the quiz page, and (6) the feedback page. The results of expert validation and the limited try out show that the computer-based assessment is categorized very feasible and very practical. The selection of Microsoft Power-Point in developing computer-based assessments is expected to motivate the prospective teacher to be able to develop a computer-based assessment independently.

ACKNOWLEDGMENT

This research and development were carried out with the financial support from the PNBP Faculty of Mathematics and Science 2019, Universitas Negeri Malang.

REFERENCES

Abell, S.K. 2008. Twenty years later: Does pedagogical content knowledge remain a useful idea?" *International Journal of Science Education* 30(10): 1405–16.

Adji, V.P., Masykur, R. & Sodiq, A. 2019. Pengaruh matakuliah matematika dasar dan strategi belajar mengajar terhadap matakuliah micro teaching. *Prosiding Seminar Nasional Matematika dan Pendidikan Matematika* 2(1): 59–63.

Adnot, M., Dee, T., Katz, V. & Wyckoff, J. 2017. Teacher turnover, teacher quality, and student achievement in DCPS. *Educational Evaluation and Policy Analysis* 39(1): 54–76.

Arikunto, S. 2006. *Prosedur penelitian suatu pendekatan*. Jakarta: Rineka Cipta.

Baumert, J., Kunter, M., Blum, W., Brunner, M., Voss, T., Jordan, A., ... & Tsai, Y.M. 2010. Teachers' mathematical knowledge, cognitive activation in the classroom, and student progress. *American Educational Research Journal* 47(1): 133–180.

Blazar, D. & Kraft, M.A. 2017. Teacher and teaching effects on students' attitudes and behaviors. *Educational Evaluation and Policy Analysis* 39(1): 146–170.

Canales, A. & Maldonado, L. 2018. Teacher quality and student achievement in Chile: Linking teachers' contribution and observable characteristics. *International Journal of Educational Development* 60: 33–50.

Demirbağ, M. & Kingir, S. 2017. Promoting pre-service science teachers' conceptual understanding about boiling by dialogic teaching. *Journal of Baltic Science Education* 16(4): 459–71.

Fuentes, J.M., García, A.I., Ramírez-Gómez, Á. & Ayuga, F. 2014. Computer-based tools for the assessment of learning processes in higher education: A comparative analysis. In *8th International Technology, Education and Development Conference, Valencia, Spain, 10-12 March 2014*. Valencia, Spain: IATED Academy.

Haigh, M. 2010. Why use computer-based assessment in education. A literature review. *Research Matters* 10(6): 33–40.

Istiyono, E. & Subroto, S. 2017. Pengembangan instrumen asesmen pengetahuan fisika berbasis komputer untuk meningkatkan kesiapan peserta didik dalam menghadapi ujian nasional berbasis komputer. *Jurnal Pendidikan Matematika Dan Sains* 5(1): 89–97.

Khalanyane, T. & Hala-hala, M. 2014. Traditional assessment as a subjectification tool in schools in Lesotho. *Educational Research and Reviews* 9(17): 587–93.

Koehler, M.J. & Mishra, P. 2011. Introducing TPCK. In *Handbook of Technological Pedagogical Content Knowledge (TPCK) for Educators*: 13–40. NY: Routledge.

Lestari, R. 2018. Pengaruh model lesson study terhadap kemampuan dasar mengajar mahasiswa biologi Universitas Pasir Pengaraian. *Bio-Lectura* 5(1): 103–110.

LoCasale-Crouch, J., Jamil, F., Pianta, R.C., Rudasill, K.M. & DeCoster, J. 2018. Observed quality and consistency of fifth graders' teacher–student interactions: Associations with feelings, engagement, and performance in school. *Sage Open* 8(3): 2158244018794774.

Mastuti, E. 2016. Pemanfaatan teknologi dalam menyusun evaluasi hasil belajar. *Jurnal Penelitian Psikologi* 7(1): 10–19.

Nikou, S.A. & Economides, A.A. 2016. The impact of paper-based, computer-based and mobile-based self-assessment on students' science motivation and achievement. *Computers in Human Behavior* 55: 1241–1248.

Pratomo, A. & Mantala, R. 2016. Pengembangan aplikasi ujian berbasis komputer beserta analisis uji guna sistem perangkat lunaknya menggunakan metode SUMI (Software Usability Measurement Inventory). *POSITIF: Jurnal Sistem dan Teknologi Informasi* 2(1): 1–11.

Rizki, S. 2014. Efek lesson study terhadap peningkatan kompetensi pedagogik calon guru. *AKSIOMA: Jurnal Program Studi Pendidikan Matematika* 3(1): 17–27.

Serradell-López, E., Lara, P., Castillo, D. & González, I. 2010. Developing professional knowledge and confidence in higher education. *International Journal of Knowledge Society Research (IJKSR)* 1(4): 32–41.

Setyoko, S. 2018. Implementasi pembelajaran blended learning berbasis media google classrom terhadap hasil belajar mahasiswa pendidikan fisika. *Jurnal Pendidikan Fisika dan Sains* 1(02): 5–10.

Shute, V.J. & Rahimi, S. 2017. Review of computer-based assessment for learning in elementary and secondary education. *Journal of Computer Assisted Learning* 33(1): 1–19.

Valdmann, A., Holbrook, J. & Rannikmae, M. 2017. Determining the effectiveness of a design–based, continuous professional development programme for science teachers. *Journal of Baltic Science Education, 16*(4): 576–91.

Yamasari, Y. 2010. Pengembangan media pembelajaran matematika berbasis ICT yang berkualitas. In *Seminar Nasional Pascasarjana X-ITS, Surabaya*, 12 August 2009. Surabaya: FPMIPA Universitas Negeri Surabaya.

Zunaidah, F.N. 2016. Meningkatkan kompetensi calon guru melalui kegiatan microteaching berbasis Lesson Study (LS) mahasiswa pendidikan biologi. *Efektor* 3(2): 21–24.

Learning model development

Borderless Education as a Challenge in the 5.0 Society – Abdullah, Adriany & Abdullah (eds)
 ISBN 978-0-367-61960-2

Developing online learning communities

R.C. Johan, M.R. Sutisna, G. Rullyana & A. Ardiansah
Department of Education and Technology, Study Program Library and Information Science, Universitas Pendidikan Indonesia, Bandung, Indonesia

ABSTRACT: Learning innovation is an effort to improve the quality of learning. Improving the quality of learning can be done in various ways; such as changes in learning strategies, the provision of instructional media and quality information resources, and the use of technology that supports teaching and learning. Through this study, an analysis of needs was used to direct the form of learning innovation designed for online, technology-based learning with internet connection, and the form of digital learning materials. The research method used is action research, with a data collection tool in the form of an assessment guide to the development of learning needs, involving three similar study programs, namely library and information science study programs at the Indonesian Education University, Malang State University, and Padang State University. The expected research results of curriculum studies from three similar study programs and the availability of an agreed learning plan will be developed in the form of online learning, a subject that is designed to be studied together. This initial research is expected to provide a basis for implementing forms of collaborative action in online learning, assisting the implementation of the Ministry of Research and Higher Education program of the Republic of Indonesia to refer to meeting the needs of the educational community in the era of society 5.0.

1 INTRODUCTION

Education is the process of conveying an experience in the form of both knowledge and skills to others, from older people or people who are more experienced to achieve certain goals. In addition, education is also a human right that must be obtained for every human being. This is confirmed by the 1945 Constitution article 31 paragraph (1) which states that "Every citizen has the right to education", and paragraph (2) "Every citizen is required to attend basic education and the government is obliged to finance it."

Advances in Information and Communication Technology (ICT) now bring about various changes in human life. The ICT is increasingly felt in various sectors, especially in the field of education where it is expected to improve the quality of education. Online learning technologies are becoming an integral part of the learning experience at the university (Ellis & Bliuc 2019). Improving the quality of education is a priority given the awareness that the success of a nation in the future is highly dependent on the quality of education and the lecture process, and given it takes a learning pattern that integrates technology so that it allows students to communicate and obtain information from various sources (Hasmunarti et al. 2019).

Access to education using network technology and communication technology is an appropriate means of learning without limits of place and time. Universities may often behave quite conservatively and be slow to implement the structural and organizational changes required for the use of online technologies (Larionova et al. 2018). This learning is very efficient, because the same learning resources can be used by thousands of people at the same time. Online learning is becoming increasingly affordable, accessible, and suitable for today's students, who can enjoy the convenience of learning while staying connected (Luo et al. 2018). The primary goal in online courses is the creation of communities of learners, where learners find opportunities to be leaders and teachers (Maddix 2013). These results suggest that environmental design impacts the learning environment, but the relation between personality and online performance is still unclear (Stone 2018). Resistance to change and fear of different things is not something new to human beings, and fear of technology still holds a grasp on people even in this fast-paced mobile-device-obsessed time (Vivolo 2016). The benefits for student learning and performance of online learning have also been investigated. The results so far are equivocal (Upton & Adams 2006).

Learning will be interesting especially for students, if the information presented is easy to understand, enjoyable, thought-provoking, and less expensive (Suwardiyanto & Yuliandoko 2017). Online learning media built using the Moodle application provide features that can accommodate online learning needs. Some of the features provided by Moodle include: assignments, examinations, communication, collaboration, as well as the main features that can upload a variety of learning material formats. Complete, clear content, fostering interest in learning, will be increasingly favored for the growth of an intelligent, knowledge-rich society able to develop their knowledge through experiments, research, and studies that will ultimately be empowered by developing their competencies.

To enrich the content of this learning resource requires the involvement of universities in the field of ITC and knowledge in their fields. The development of learning resources based on digital technology is accessable by many users, cheap, and dynamic so that the learning content is to develop digital assets. Learning materials used in the learning process are expected to help students to understand the material presented, and can be used by students to study individually or in groups (Annajmi & Isharyadi 2019).

Learning innovation is one of the efforts to improve the quality of learning. Improving the quality of learning can be done in various ways; such as changes in learning methods, the provision of learning tools and quality books, and the use of technology that supports teaching and learning. Many technologies have been developed to support the teaching and learning process. One of them is online learning. The use of technology in various fields is needed especially in the current modern era. The education sector is one of the important aspects, because education can improve and develop the quality of the next generation of a nation (Priwantoro et al. 2018).

Online learning is an internet-based learning technology that can be used to carry out distance learning activities (online) or also be used to complement face-to-face learning methods (blended learning). Online learning is an ICT-based learning model. The carrying capacity of this program is ICT facilities in the form of web LMS, monitoring programs, and modular and multimedia supplements (Dewi 2017). Thus students can access learning wherever they are and whenever they want to learn. The term online learning model (OLM) was originally used to describe learning systems that utilize computer-based learning (CBL) technology (Kuntarto 2017). Online learning can contain teaching materials in the form of document files, audio, and video, and can also be used to improve the interaction of lecturers and students because there are chat facilities, discussions, forums, etc. Likewise, lecturers can provide quizzes or tests. Online learning is a new way of teaching and learning in various education institutions to overcome the limitations of space, time, and energy. In previous research, the online learning model that was tested on a small scale was very valid for users, as well as for students (Adhe 2018).

Online learning is the basis and logical consequence of the development of ICT. With online learning, participants (learners or students) do not need to sit sweetly in the classroom to listen to every remark from a lecturer directly. Student activities are supported by a competitive learning environment, case studies that challenge and encourage learning, the formation of scientific discussion forums, the creation of research topics, and an assessment system that motivates students to learn. Students also get direct feedback in the form of comments about activities, and explanations from the lecturer (Yuhdi & Amalia 2018). Online learning can also shorten the target schedule of learning time, and of course save costs that must be incurred by a study program. This platform can be used to carry out distance learning activities (online) or to complement face-to-face learning methods (blended learning) (Rahayu 2019).

Online learning (SPADA) will be carried out at three universities, namely the Indonesian University of Education, Padang State University, and Malang University, Library and Information Science Study Program in Web Design and Information Literacy courses. The program is based on the spirit of sharing open educational resources between UPI, UNP, and UM, so that quality material, lecturers, and facilities can be accessed and enjoyed by all universities in Indonesia by utilizing ICT. The SPADA program consists of (1) Open Material, which provides course material that is presented online in various media forms so that it can be accessed by students and lecturers anytime and anywhere. (2) Open Courses, which is an online learning system of one whole course that allows it to be used by related lecturers as online courses. (3) Online Courses, that is, courses in the form of complete online learning, which are ready to be offered by one of the organizing universities to other PTs (partner PTs) to be followed by partner PT students as a vehicle for credit transfer from PT the organizer to the PT where they are registered.

1.1 *Needs analysis*

Needs assessment is the process of determining the priority of educational needs. Needs are basically discrepancies between what is available and what is expected, and needs assessment is the process of gathering information about gaps and determining priorities of gaps to be solved (Sanjaya 2008).

Needs analysis as a formal process to determine the distance or gap between actual outputs and impacts with the desired outputs and impacts, then place this sequence of gaps on a scale of priorities, then choose how to solve the problem (Warsita 2011). Then the needs analysis is a tool or method for identifying problems in order to determine the right action or solution.

There are several things attached to the understanding of needs assessment, as stated by McNeil and Glasgow. First, it is a process, which means that there is a series of activities in the implementation of needs assessment, and not an outcome but a certain activity in an effort to make certain decisions. Second, the need itself is essentially a gap between expectations and reality. Thus, the needs assessment is an activity to gather information about the gap between what all students should have and what they do have.

The following is a function of Morrison's learning needs analysis (Warsita 2011):

- Identify needs that are relevant to the current job or task, namely problems that affect learning outcomes.
- Identifying urgent needs related to financial, security, or other problems that interfere with the work or educational environment.
- Present a priority scale for choosing the right action in overcoming learning problems.
- Provide a database to analyze the effectiveness of learning activities.

1.2 *Purpose of requirement analysis*

The following are the objectives of learning needs analysis (Warsita 2011):

1.2.1 *Inventory or identify learning problems*

Problem identification is the process of comparing the present situation with the expected situation. The results will show the gap between the two conditions. This gap is called necessity. If the gap between the two conditions is large, the needs need to be addressed or resolved. The large and determined needs to be addressed are called problems. Therefore, smaller needs, that is, needs that are not considered a problem may be temporarily or permanently ignored. The final result of problem identification is the formulation of general learning goals.

1.2.2 *Develop priority problem solving scale*

After you know the learning problems faced, you need to find alternative solutions to these problems by using the priority scale of problem solving. Some considerations to consider in assessing or determining the priority scale of problem solving include (a) the level of significance of its influence, (b) the extent of its scope, and (c) the importance of the role of the gap on the future of the institution or program.

1.2.3 *Formulate goals*

The results of the learning needs analysis activities are a list of knowledge, skills, and attitudes – in other words, competencies – that need to be but are not yet mastered by students.

This basic competence will be the basis for the next stage of reference, namely the formulation of general instructional objectives (TIU) or general learning objectives (TPU).

1.3 *Steps for conducting a needs analysis*

Following are the steps in analyzing Gosslow's needs in the learning system planning and design book:

1.3.1 *Information collection*

When designing learning for the first time, a designer needs to understand what students can do, who understands what, who will learn, what obstacles will be faced, and the influence of certain circumstances on student characteristics. Various information collected will be useful in determining the objectives to be achieved along with the scale of priorities in solving a problem.

1.3.2 *Identify gaps*

Gap identification determines the gap through the organizational elements model (OEM). The OEM model explains the existence of five interrelated elements. The first two elements, namely input and process, are how to use every potential and existing source, while the last elements, products, outputs, and outcomes, are the final results of a process. Input components include the conditions currently available, for example, finance, time, buildings, teachers, students, needs, problems, goals, curriculum materials, etc. Process components include the implementation of ongoing education that consists of staff formation patterns, and education that takes place in accordance with competencies, planning, methods, individual learning, and applicable curriculum. Product components include completion of education, skills, knowledge and attitudes possessed, as well as passing the competency test.

Output components include graduation certificates, prerequisite skills, and licenses. The outcome component includes the adequacy and contribution of individuals or groups at present and in the future. Outcome is the final result obtained. Through results analysis, designers can determine the extent to which the results obtained can contribute to the achievement of objectives. This is the process that essentially determines the gap between expectations and what happens. Based on this analysis, designers can describe the problems and needs of each component, namely inputs, processes, products, and outputs.

1.3.3 *Performance analysis*

The third stage in the needs assessment process is the stage of analyzing performance. Analyzing performance is done after the designer understands various information and identifies gaps. After finding gaps, then identify which gaps can be solved through learning planning and which need to be resolved in other ways, such as through new management policies, determining a better organizational structure, or perhaps through developing materials and tools. This requires understanding the factors that cause the gap, and this can be done when the needs assessment takes place. Performance analysis includes identification of teachers, identification of suggestions and completeness of student learning support, identification of school policies, and identification of social climate and psychological climate.

1.4 *Identification of obstacles*

The fourth step in a needs assessment is to identify various obstacles that arise, along with their sources. In implementing a program, various obstacles can arise that can affect the smooth running of a program. Various obstacles can include facility time, materials, grouping and composition, philosophy, personal factors, and organization. The sources of constraints can come from, first, people involved in a learning program, for example teachers, principals, and students themselves, and the philosophy or outlook on work, work motivation, and abilities each of these people has. Second, constraints can come from the existing facilities, including the availability and completeness of the facilities and the condition of the facilities. Third constraint can be related to the amount of funding and its arrangements.

1.5 *Identification of purpose*

Needs assessment is a process of identifying, documenting, and justifying the gap between what happens and what will be produced through determining the priority scale of each need. The definition stated is closely related to the objectives to be achieved, therefore, identifying the objectives to be achieved is one of the activities that must be carried out in the needs assessment process. Not all needs are objectives in instructional design. A designer needs to determine what needs are considered urgent, which essentially determines the scale of priority in needs assessment. There are several techniques in determining the priority scale of the data that has been collected, for example, ranking techniques such as Delphi technique, focus group discussion, q-sort, and storyboarding. These techniques are used to capture various objectives deemed necessary through the assessment of experts involved in the discussion. Thus, goal formulation is really the result of a study that is necessary and identifies needs to be solved.

1.6 *Formulate the problem*

The final stage in the problem analysis process is writing a problem statement as a guideline in the preparation of the instructional design process. A problem statement is basically a summary or essence of the problem specified. The problem statement must be written in a clear and concise form and usually does not exceed one or two paragraphs.

2 RESEARCH METHODS

This research was conducted from June to October at three universities, namely the Indonesian University of Education, Padang State University, and Malang University. This research uses a quantitative approach with a survey method, that is, a method for describing phenomena in the form of attitudes, opinions, behavior, etc.

3 RESULTS AND DISCUSSION

The rapidly developing information and communication technology has positive potential in the world of education, including online learning, where students can access education wherever and whenever. The purpose of implementing online learning is to expand access to education, increase the efficiency and effectiveness of learning through the use of ICT, and provide services to students who cannot attend face-to-face or regular learning. One form of online learning programs that is currently being developed is the online learning system (SPADA).

Indonesia has become part of a knowledge-based global society so demand to always be in touch with the world community will increase. Advances in technology, information, and communication provide a world community that is increasingly easy to interact with, such as the Free Trade Area, which requires people to be able to compete with other countries. Movement and integration of trade in the Asian region and the world is very fast so that automatic mobility of people between countries is also significantly increased.

The education sector has a very large role in the development of a nation. The level of education in Indonesia needs to be improved to be able to create and maintain a knowledge-based global society. Human resource development is required in order to compete in all fields with other countries, therefore Indonesia, as a country with a large population, must be able to adjust to current global conditions in order to anticipate the swift global competition in various aspects.

Indonesia is able to compete in all fields with other countries, increasing the role of tertiary education institutions to develop competent human resources. Higher education in Indonesia, as one of the drivers of the national development, is responsible for producing human resources or graduates who have knowledge, as well as extensive skills, nationally and internationally, and are able to compete with university graduates from other countries. Increasing the competitiveness of Indonesian tertiary institutions and universities with those of other countries must be taken seriously.

Improving the quality of tertiary institutions needs to be done; one way is by collaborating with other tertiary institutions that have a good reputation and high quality. Higher education cooperation is an agreement between universities in Indonesia with universities, the business world, or other parties, both at home and abroad. Collaboration with universities aims to increase the effectiveness, efficiency, productivity, creativity, innovation, quality, and relevance of higher education Tridharma, an Indonesian ancient expression meaning three main principles, to enhance the nation's competitiveness. Collaboration between universities is one of the efforts to develop human resources capable of creating and maintaining a global and knowledge-based society.

Higher education institutions can collaborate in the academic and/or non-academic fields with other tertiary institutions in the country and abroad. Collaboration organized by tertiary institutions is based on the strategic plan and statutes of each tertiary institution. One form of academic cooperation between universities is the transfer or acquisition of credit numbers. Academic credit transfer is the process of evaluating the qualification component to determine overall/equality with other qualifications by bringing together comparable credit for academic achievement and individual achievement. Transfer credit is a mechanism for acknowledging workload and achievement at an institution of higher education with other institutions of higher education. Credit transfer and acquisition is recognition of the results of the educational process expressed in semester credit units or other measures to achieve learning competency in accordance with the curriculum. Transfer of credit numbers and obtaining credit numbers can be done between the same or different study programs. Study program institutions are free to determine the courses that will be transferred to the study.

The collaboration that will be carried out between UPI through the Library and Information Science Study Program and Malang State University and Padang State University is academic cooperation through the transfer and/or acquisition of credit numbers. Credit transfer is one of the keys to student mobility and cooperation among higher education institutions in an effort to anticipate swift global competition. Collecting credit scores will be implemented through online learning under the ministry of research and higher education program, namely the online learning system (SPADA). So, before the academic cooperation program through obtaining credit numbers in the online learning system (SPADA) is implemented, program needs analysis was conducted, as follows:

3.1 *Study the curriculum of each study program*

The needs analysis was done by formulating various aspects of the requirements needed in the process of designing and developing this online lecture. The analysis process carried out by the research team determined the design and development phase in the curriculum field of each study program. Analysis of the curriculum of the library study program and information science programs at UPI, UM, and UNP, obtained courses that will be used as online learning are seen in Table 1:

3.2 *Analysis of online learning strategy*

The purpose of the semester learning plan analysis is to establish and define conditions in the learning process. The initial analysis aims to bring up and determine the basic problems faced in learning so that the development of online courses is needed. Initial analysis discusses the limitations when learning takes place, the time of student interaction with lecturers in the classroom, in addition to learning activities that are only limited to 2 credits, so that students do not have time to discuss the material related to the initial lecture.

Table 1. Courses that are used as online learning.

No	Course	Credit	Semester
1	Information literacy	2	2
2	Legal aspects of information	2	2
3	Digital library development	3	6

Concept analysis is carried out to identify the main concepts that will be taught to students and make details of relevant concepts. Supporting tools for conducting concept analysis are (1) analysis of learning outcomes aimed at determining teaching material; and (2) analysis of learning resources, namely collecting and identifying learning resources that will be used. Concept analysis is based on the Semester Learning Plan. Learning outcomes are adjusted to the KKNI curriculum policy, which includes the achievement of knowledge, attitudes, and skills. This learning achievement analysis is carried out by discussion with the research team to obtain agreement.

This task analysis aims to analyze the tasks or activities to be carried out by students in the learning process in the context of achievement. Analysis of the specifications of learning objectives is the formulation of learning objectives to be achieved in the learning process and the expectation students will master the material that has been taught.

The learning design is an online learning program plan for one course for one semester that is used as a guideline for the course implementation. The semester learning plan (RPS) or other terms contains at least;

- name of study program, course name and code, semester, credits, name of lecturer;
- graduate learning outcomes that are charged to the course (see Table 2);
- the final ability planned at each stage of learning to meet the learning outcomes of graduates;
- study material related to the ability to be achieved;
- learning methods;
- the time allotted to reach the ability at each stage of learning;
- student learning experience that is manifested in the description of the tasks that must be done by students for one semester;
- criteria, indicators, and rating weights; and
- reference list used.

Table 2. Online learning design.

Study Program	Code Course Info	Course Info	Semester	Credit	Learning Outcome
UPI – Library and Information Science	Information Literacy - LM106	This course provides knowledge and understanding of theory and practice about the role of librarians as information literacy agents who are able to meet the information needs of users. Includes identifying information needs, recognizing and finding suitable sources of information, accessing information from those sources effectively and efficiently and evaluating information obtained to be used ethically and legally. Providing web-based information search techniques and strategies and knowledge to design information literacy programs according to the type of library and library.	2	2	After attending this course, students are expected to understand the role and duties of librarians as information literacy agents and understand how to identify information and information sources, access information from information sources in various formats, evaluate information to be used as needed, and design information literacy programs.

(*Continued*)

Table 2. (*Continued*)

Study Program	Code Course Info	Course Info	Semester	Credit	Learning Outcome
UM – Library and Information Science	Legal Aspect Information - PUST6117	Competence in this course requires students to understand the legal aspects in the generation, dissemination, and use of information including censorship, copyright, and patents.	2	2	After this course, students understand the legal aspects of the generation, dissemination, and use of information including censorship, copyright, and patents.
UNP – Library and Information Science	Digital Library Development - PII1.62.6002	Competence in this course requires students to understand library software development techniques, database design, and simple website design.	3	6	After, this course, students understand library software development techniques, database design, and simple website design.

3.3 *Subject of online learning*

Subject of online learning can be seen in the Table 3.

Table 3. Subject of online learning.

Study Program	Code Course Info	Week	Indicators of Subject Learning Achievement	Study Materials
UPI – Library and Information Science	Information Literacy - LM106	1	After studying this, students can explain the scope of the definition and development of information literacy	Development and definition of LI and related terms
		2	After studying this, students can explain the role of LI as a lifelong learning supporter	LI and lifelong learning
UM – Library and Information Science	Legal Aspect Information - PUST6117	1	After studying this, students understand the legal aspects in the originating, dissemination and use of information including censorship, copyright and patents	Law information aspect
		2	After studying this, students understand copyright and patents	Copyright
UNP – Library and Information Science	Digital Library Development - PII1.62.6002	1	Rounding up the SD digitalization insight	Definition of digital libraries
		2	Rounding up ICT network insights	Definition

4 CONCLUSION

Cooperation that will be carried out between UPI through the Library and Information Science Study Program with Malang State University and Padang State University is an academic collaboration through the transfer and/or acquisition of credit numbers. Credit transfer is one of the keys to student mobility and cooperation among higher education institutions is a manifestation of an effort to anticipate swift global competition in various aspects. Collecting credit scores will be implemented through online learning under the ministry of research and higher education program, namely the online learning system (SPADA). So, before the academic cooperation program was implemented, the following needs analysis was carried out: (1) Curriculum review of each study program of the same type was carried out by formulating various aspects of the requirements needed in the process of designing and developing this online lecture. The analysis process carried out by the research team decided that the design and development stage in the curriculum field of each library study program and information science program at UPI, UM and UNP. (2) The objective of the semester learning plan analysis was to establish and define the conditions in the learning process. The initial analysis aimed to bring up and determine the basic problems faced in learning through online courses. Initial analysis considered the limitations of when learning takes place, the time of student interaction with lecturers in the classroom, in addition to learning activities that are only limited to 2 credits, so that students do not have time to discuss the material related to the initial lecture. (3) The analysis of this subject looked at tasks or activities to be carried out by students in the learning process in the context of achievement. The analysis of the subject matter specification formulated learning objectives to be achieved in the learning process and the expectation in this study that students are able to master the material that has been taught.

REFERENCES

Adhe, K. R. 2018. Pengembangan media pembelajaran daring matakuliah kajian PAUD di jurusan PG PAUD Fakultas Ilmu Pendidikan Universitas Negeri Surabaya. *JECCE (Journal of Early Childhood Care and Education)* 1(1): 26–31.

Annajmi, A., & Isharyadi, R. 2019. Analisis kebutuhan pengembangan bahan ajar mata kuliah kalkulus peubah banyak berbantuan software geogebra bagi mahasiswa pendidikan matematika universitas pasir pangaraian. *Mathline: Jurnal Matematika dan Pendidikan Matematika* 4(2): 85–97.

Dewi, L. 2017. Rancangan program pembelajaran daring di perguruan tinggi: Studi kasus pada mata kuliah kurikulum pembelajaran di Universitas Pendidikan Indonesia. *EDUTECH* 16(2): 205–221.

Ellis, R. A., & Bliuc, A. M. 2019. Exploring new elements of the student approaches to learning framework: The role of online learning technologies in student learning. *Active Learning in Higher Education* 20(1): 11–24.

Hasmunarti, H., Bahri, A., & Idris, I.S. 2019. Analisis kebutuhan pengembangan blended learning terintegrasi strategi PBLRQA (Problem-Based Learning and Reading, Questioning & Answering) pada pembelajaran biologi. *Biology Teaching and Learning* 1(2): 101–108.

Kuntarto, E. 2017. Keefektifan model pembelajaran daring dalam perkuliahan Bahasa Indonesia di perguruan tinggi. *Indonesian Language Education and Literature* 3(1): 99–110.

Larionova, V., Brown, K., Bystrova, T., & Sinitsyn, E. 2018. Russian perspectives of online learning technologies in higher education: An empirical study of a MOOC. *Research in Comparative and International Education* 13(1): 70–91.

Luo, Y., Pan, R., Choi, J. H., & Strobel, J. 2018. Effects of chronotypes on students' choice, participation, and performance in online learning. *Journal of Educational Computing Research* 55(8): 1069–1087.

Maddix, M. A. 2013. Developing Online Learning Communities. *Christian Education Journal* 10(1): 139–148.

Priwantoro, S. W., Fahmi, S. & Astuti, D. 2018. Analisis kebutuhan pengembangan multimedia. *AdMathEdu* 8(1): 49–58.

Rahayu, M. K. P. 2019. Peta penggunaan e-learning oleh dosen fakultas ekonomi dan bisnis pasca hibah spada. *Jurnal Manajemen Bisnis* 9(2): 175–192.

Sanjaya, W. 2008. *Perencanaan dan desain sistem pembelajaran*. Jakarta: Kencana Prenada Media Group.

Stone, N. J. 2018. Environmental design, personality, and online learning. In *Proceedings of the Human Factors and Ergonomics Society Annual Meeting* 62(1): 1171–1175. Sage CA: Los Angeles, CA: SAGE Publications.

Suwardiyanto, D., & Yuliandoko, H. 2017. Pemanfaatan teknologi sebagai media pembelajaran daring (on line) bagi guru dan siswa di SMK Nu Rogojampi. *J-Dinamika* 2(2): 96–100.

Upton, D., & Adams, S. 2006. Individual differences in online learning. *Psychology Learning & Teaching* 5(2): 141–145.

Vivolo, J. 2016. Understanding and combating resistance to online learning. *Science Progress* 99(4): 399–412.

Yuhdi, A., & Amalia, N. 2018. Desain media pembelajaran berbasis daring memanfaatkan PortalSchoology pada pembelajaran apresiasi sastra. *Basastra* 7(1): 14–22.

Warsita, B. 2011. *Pendidikan jarak jauh*. Bandung. PT Remaja Rosdakarya.

Management, supervision and assessment

Borderless Education as a Challenge in the 5.0 Society – Abdullah, Adriany & Abdullah (eds)
 ISBN 978-0-367-61960-2

Teacher's perception of orientation and mobility assessments of children with visual impairments

E. Heryati, B. Susetyo & E. Ratnengsih
Special Education Department, Universitas Pendidikan Indonesia, Bandung, Indonesia

ABSTRACT: Orientation and mobility skills for children with visual impairment will not only develop psychomotor and motion skills, but will also develop their cognitive and affective domains. Orientation and mobility skills assessment is a process of evaluating abilities, needs, and difficulties in orientation and mobility of children with visual impairment. The results of assessment process will be the basis for determining the learning program conducted by the teacher. This study examines teacher's perceptions about orientation and mobility assessments of children with visual impairment. This research uses a qualitative approach through a case study of three special education teachers who teach children with visual impairment. Findings in the form of similarities and differences in perception between each case will be described. Some practical implications related to orientation and mobility assessment will emerge from this research.

1 INTRODUCTION

Visual impairment is a general term used to indicate the state of vision loss in a wide variety or range. The understanding of children with visual impairments is legally based on the ability to use visual sensors in the learning process after maximum correction which is divided into two groups; blind and low vision. Children with blindness cannot use vision in the learning process but can still respond to light and some still have visual imagery. Children with low vision have difficulty completing visual tasks but they can still learn to use visual sensors with the help of special learning tools and techniques (Kirk et al. 2009).

Vision loss can have an impact on many aspects of an individual's daily life. Furthermore, it can significantly limit some important aspects. One very important aspect for students with visual impairments is the area of Orientation and Mobility (Malik et al. 2018).

The main limitations due to visual impairment experienced by children as stated by Lowenfeld (1973) include; (1) limitations in terms of variety and breadth of experience, (2) limitations in terms of mobility, and (3) limitations in terms of interactions with the environment. The same thing was stated by Kirk et al. (2009) that one of the main problems due to visual impairments or blindness is related to the ability of orientation towards the environment and the ability to mobilize in the environment.

Development of Orientation and Mobility (O&M) is needed to cover limitations as a direct result of visual impairments. The skill is needed to be able to access and interact with the environment. The O&M is a skill of readiness, and ease of moving from one position/place to another position/place which is done correctly, precisely, effectively, and safely. Orientation and Mobility skills will not only develop the psychomotor and movement domain, but will also develop the cognitive and affective domains of students with visual impairments.

O&M skill helps individuals achieve freedom to participate in all aspects of life in society. O&M learning promotes movements, which are very important for the development of concept, exploration, and environmental awareness in children. Beginning at an early age and continuing into adulthood, O&M learning facilitates the development of concepts, skills and knowledge needed for someone to achieve their life goals now and in the future (Kaiser et al. 2018).

O&M learning programs aim to enable children with visual impairments to enter both familiar/unfamiliar environments correctly, safely, effectively and efficiently without much help from others (Wall Emerson & Corn 2006). The determination of an appropriate O&M learning program is determined by an accurate assessment of the skills of students prior to the learning (Jacobson 2013). Thus, the assessment process becomes the initial step that needs to be done before the learning program begins.

Effective O&M assessments must be carried out by instructors who are skilled and equipped with adequate knowledge. However, due to the limited number of O&M instructors, it is the teachers who teach children with visual impairments which are required to have sufficient skills and knowledge about O&M.

Many important components need to be included in assessing O&M skill. These components are body image, motor skill, posture and gait, sensory and perception (tactile, auditory, and visual), as well as concept of body, space, time, and environment (Jacobson 2013). Importantly, before the O&M assessment begins, O&M teachers also need to know the age, maturity, as well as the condition and function of the students' vision at that time. It is important to determine the approach to be carried out during the assessment.

2 RESEARCH METHODS

This research objective is to explore how teachers perceive assessments of children with visual impairments, especially assessment of orientation and mobility. This research used descriptive qualitative research methods with a case study approach.

Participants involved in this research were three teachers who teach children with visual impairments. The three teachers teach in different extraordinary schools in the province of West Java, Indonesia. All three teachers have similarities and differences in characteristics related to the experience in teaching, experience as instructor for orientation and mobility, and academic qualifications.

The data collection technique was done through oral interviews and written questionnaires. The data that has been collected were compiled into a comprehensive description that includes all the main information used in case analysis and case studies. Each individual case study analysis began with a description of each teacher's perception. Eventually, those case studies were integrated, exploring the similarities and differences between the three teachers.

3 RESULTS AND DISCUSSION

3.1 *Result*

3.1.1 *Case 1*

The first participant, Mr. Fa, is a teacher who had taught students with visual impairments for about 5 years. He is qualified undergraduate special education, has attended O&M instructor training, and is certified. He taught O&M to 3 second grade students and 4 fourth grade students. The O&M assessment in his school is carried out at the beginning of the new academic year on an ongoing basis for both new and old students. The assessments are carried out individually in the early weeks of the new school year and the results will be used as a guide the O&M program for the coming year. Mr. Fa also stated about the importance of O&M assessments, "it is very important because O&M is an inseparable part of the lives of those students, from waking up back to going to sleep, every day ... and everything starts from the assessment process". He also added that there are many aspects that must be assessed when conducting O&M assessments, including O&M needs, body concepts, spatial concepts, independence, motor skills, and so on. He mentioned that those who must be involved in the assessment process are students, O&M teachers, class teachers, and parents. During the assessment process, several obstacles were found, such as inadequate assessment media and limited time, so Mr. Fa only used the available media and used the learning time in the early weeks of the new academic year.

3.1.2 *Case 2*

The second participant is Ms. Za. She is a teacher who had taught students with visual impairments for around 3 years, a bachelor of special education qualification, has attended an O&M training program, but is not certified yet. She taught O&M to 2 fourth grade students. The assessment of the O&M skill of the students at her school is conducted the first time the students enter the school. Ms. Za explained that "assessment is usually done before teaching O&M and it covers many aspects ... there are aspects of body, motor, cognitive, language, social and communication concepts, as well as what O&M techniques children had mastered beforehand ...". According to her, O&M assessments are important and all must be involved when assessments are conducted, including O&M teachers, class teachers, P.E. teachers, and parents. The obstacles faced when conducting an assessment was insufficient time. Ms. Za said "time is limited to teach O&M and to assess. Therefore, I often spontaneously assess when children are doing activities, but first I read the points that must be assessed ..." Other obstacles were concerning the media for conducting assessments, no existing instruments standard, and some assessments remain unclear.

3.1.3 *Case 3*

The third participant, Ms. An, is a teacher who had taught students with visual impairments for around 4 years, a bachelor of special education qualification, and had never attended an O&M training program. She taught O&M to 1 third grade student and 2 fifth grade students. She explained that the assessment of O&M skills for visually impaired students is done when they first enter the school. This assessment is important to be a benchmark of what should be taught to children. She added "aspects that need to be assessed are orientation skill and mobility skill. The orientation indicates that the child knows his position, whereas mobility means the ability to move around". She mentioned that people who should be involved in the assessment process are the O&M teacher and the class teacher. Her reasoning was that the O&M teacher will teach the students, but the class teacher can also help teaching process in class". Some constraints faced related to the assessment process were limited time, lack of media, and unclear instrument format.

3.2 *Discussion*

3.2.1 *The importance of O&M assessment for students with visual impairments*

Appropriate O&M learning programs for children with visual impairments will be determined by an accurate assessment of the skills the children have. All three respondents in this research agreed that orientation and mobility assessments are important, although there were differences in their implementation. Two respondents mentioned that the assessment was carried out the first time students entered school, while one respondent (Mr. Fa) explained that the assessment was carried out at the beginning of the new academic year for both new and old students.

3.2.2 *Aspects of O&M*

Orientation and mobility assessment in children with visual impairments is a comprehensive assessment of all aspects concerning to O&M. Based on the results of interviews with the three participants, it can be seen that there were similarities and differences in the perceptions of each related aspects that must be assessed when conducting an O&M assessment. In general, it was explained that the concepts of the body and motor are important aspects in orientation and mobility, but the third participant only stated that orientation abilities and mobility abilities are aspects that must be assessed. The first participant mentioned four aspects, namely the concept of the body, the concept of space, independence, and motoric. Meanwhile, according to the second participant, in addition to the four aspects above, it is necessary to add cognitive, language, social and communication aspects as well as O&M techniques that have been mastered by children.

3.2.3 *Parties involved in O&M assessment*

The implementation of O&M assessment involved several parties. The three participants explained that those who need to be involved are O&M teachers and class teachers, and specifically the first and second participants added that parents and P.E. teachers also need to be involved. The involvement of parents and family is important because they can provide information about medical records related to visual impairments that the children have. The assessment process involves interviewing people who interact with children such as parents, family members, caregivers, and teachers (Kaiser et al. 2018).

3.2.4 *Obstacles in the process of O&M assessment*

The process of implementing the O&M assessment faced some obstacles such as the allotted time and required media. All three participants agreed that the time for assessment was still limited because students only had two hours of study or about 60 minutes a week for O&M learning. Likewise, the availability of media or tools in schools was still limited. Specifically, the third participant also added another obstacle, which was the unavailability of a clear O&M instrument format.

4 CONCLUSION

The results of this research illustrate that teachers who teach children with visual impairments have the same perception about the importance of assessment of O&M skill as an initial step that must be carried out before starting an O&M learning program. Furthermore, accurate assessment results will help determine the appropriate learning program. Differences in perceptions regarding aspects of O&M and those who must be involved in conducting the assessment expose the need for O&M training for teachers to synchronize perceptions and at the same time improve support for students with visual impairments. In addition, school as a service point needs to provide serious responses related to time constraints and media limitations for the implementation of O&M assessments.

REFERENCES

Jacobson, W.H. 2013. *The art and science of teaching orientation and mobility to persons with visual impairments.* (2nd ed.). New York, NY: AFB Press.

Kaiser, J.T., Cmar, J.L., Rosen, S. & Anderson, D. 2018. *Scope of practice in orientation and mobility.* Association for Education and Rehabilitation of the Blind and Visually Impaired O&M Division IX. Alexandria, VA: Association for Education and Rehabilitation of the Blind and Visually Impaired.

Kirk, S., Gallagher, J.J., Coleman, M.R. & Anastasiow, N. 2009. *Educating exceptional children.* USA: Houghton Mifflin Harcourt Publishing Company.

Lowenfeld, B. 1973. *Our blind children: Growing and learning with them.* (3rd ed.) Springfield, Illinois: Charles C. Thomas.

Malik, S., Abdul Manaf, U.K., Ismail, M. & Ahmad, N.A. 2018. Conceptualising orientation and mobility practices within the expanded core curriculum. *Global Journal of Human-Social Science: G Linguistics & Education* 18(9): 1–7.

Wall Emerson, R., & Corn, A.L. 2006. Orientation and mobility instructional content for children and youths: A Delphi study. *Journal of Visual Impairment & Blindness* 100: 331–342.

Borderless Education as a Challenge in the 5.0 Society – Abdullah, Adriany & Abdullah (eds)
© 2021 Taylor & Francis Group, London, ISBN 978-0-367-61960-2

Teacher competency enhancement for 21st century educators and effective school management in the context of continuing professional development

S. Sururi
Faculty of Educational Sciences, Universitas Pendidikan Indonesia, Bandung, Indonesia

ABSTRACT: In order to achieve educational goals maximally, the role of the teacher is very important and it is hoped that the teacher has a good teaching model and is able to choose the right learning model and is in accordance with the concepts of the subject matter to be provided and adapted to the development of existing technology. Therefore, an effort is needed in order to develop teacher professionalism which is supported through training to improve teacher competence as a 21st century educator and effective school management in the context of sustainable professional development. Through this activity it is hoped that school principals and teachers will gain an understanding of the concepts of 21st century teachers and effective school management in the context of sustainable professional development. Community service activities organized by the Department of Education Administration in collaboration with UPTD Kertasari sub-district and PGRI Kertasari sub-district Bandung Regency is actually a continuous process of a product that is complete and easy to see the results, so that it can help that is being developed with the development of problems education that appears in schools. Therefore continued efforts were followed by a positive attitude to improve the ability of principals, teachers and supervisors in the management of internal education resources.

1 INTRODUCTION

The creation of quality learning processes and outcomes is influenced by many things including the competent teacher. However, if the teacher competency is low, it will have an impact on the quality of learning that is also low.

The success of implementing quality education is very closely linked to the success in increasing the competence and professionalism of educators and education personnel (Sururi & Kurniady 2014). The teacher is a crucial factor in the effort to create dynamic conditions in learning. Learning objectives will be achieved if the teacher has a sense of optimism during the learning process. The assumption underlying this argument is that the teacher is the main driver in learning. Success in learning lies with the teacher in carrying out his mission. Because the teacher is one of the supporting factors to obtain success in learning. In this connection the teacher must be able to encourage students to be active in learning. Thus it is probable that student interest and learning activities are increasing.

Various studies have found that the most dominant school factors in basic education are teachers and school management (Riswandi 2016). The school principal partially influences the effectiveness of the school and is an important figure in determining school success (Bolanle 2013, Jacobson et al. 2005, and Pandoyo & Wuradji 2015).

The principal as the leader of the school organization plays an important role in leading, managing, directing and fostering all activities related to the school organization (Sururi et al. 2016). The accuracy of the principal in carrying out its functions will greatly affect the achievement of overall organizational performance. In addition, the results of the school principal's performance appraisal will provide important information in the process of developing the school principal itself. Efforts are being made by principals as school leaders to improve the quality of their subordinates, especially teachers. Efforts made are involving teachers in cluster activities, PKG, seminars, workshops, and other training seminars (Sunaengsih et al. 2019).

Based on what has been described, an effort is needed in order to develop sustainable professionalism of teachers by increasing teacher competency through teacher learning activities and school governance. This is in line with the steps established by the Ministry of Education and Culture where as a step to actualize professional teachers through facilitation programs for teachers to carry out professional development activities to support Teacher Learner which is a teacher's self-development activity. This needs to be done with the hope of reducing the gap in knowledge, skills, social abilities, and personality among teachers, and ultimately can improve the quality of learning in the classroom. Increased competence has implications for recognition or rewards in the form of credit numbers which can then be used to improve his career.

Improving the quality of education services, the role of education stakeholders is needed, including universities to build quality education in the district of Bandung. Because of the vast area, the improvement of the quality of education in the district of Bandung where one of them is Kertasari sub-district, a professional learning community that is built based on a culture, as well as human values and professional communication that is strong and continuous.

2 METHODS

Community service activities using the Participatory Normative Method. A method that requires participants to have theoretical and practical insights about the concepts of 21st century educators and effective school management in the context of sustainable professional development through several stages of activities which include the presentation of information, discussion and simulations/exercises/mentoring.

The target audience for community service is the Principal and elementary school teachers (prospective principals) in the UPTD District of Paperari District, Bandung Regency. After this activity, it is hoped that there will be implications in the dissemination of the results of this community service to other teachers in other regions as implementing partners in education operations.

3 RESULTS AND DISCUSSION

Based on the program coordination meeting, a formulation of operational service programs for the community of the Department of Education Administration has been produced on training to improve teacher competency through teacher competency improvement activities as a 21st century educator and effective school management in the context of sustainable professional development for elementary school principals and elementary school teachers in the UPTD of Kertasari sub-district. Bandung district. This operational program is a manifestation and good coordination between the implementation team and the field parties by considering local conditions and situations or implementation physiability. Thus the resulting program formulation takes into consideration the interests of the target parties in overcoming the problem at hand.

In every activity, all parties would want success in realizing the program. Therefore the increasing understanding of the participants in responding to issues of increasing teacher competency through activities to increase teacher competency as a 21st century educator and effective school management in the context of sustainable professional development is an encouraging indicator.

Efforts to solve problems through the presentation of information, discussions and exercises have broadened the participants' insights both theoretically and practically. Based on the results of the evaluation of the activities of 70 people participating in community service activities, the reality shows in Table 1.

Table 1. Evaluation results of activities.

No	Statement	Response		
		Very Satisfying	Satisfying	Less satisfying
1	The suitability of the material with the needs of participants	90%	10%	0%
2	Clarity of material	30%	70%	0%
3	Conformity with field experience	12.5%	80%	7.5%
4	Use of the presentation method	12.5%	75%	12.5%
5	Use of language terms	10%	70%	20%
6	Usability level	90%	10%	0%
7	The seriousness of the facilitator	90%	6%	4%

This community service activity is actually a continuous process of a product that is complete and easy to see results, because it is a continuous development in line with the dynamic development of educational problems that arise in schools. Therefore the follow-up effort is a positive attitude to improve the ability of principals, teachers and supervisors in the management of internal education resources.

Some conditions that can be used as a driver for the success of community service, among others:

- The responsiveness of the field is an indicator that efforts to reform education have gained significant support.
- The strong enthusiasm of the participants is a very important capital to realize its role and contribution in realizing effective school management.
- Efforts to restructure the management system through new policies are a breath of fresh air for participants in managing the school.

In carrying out this activity it is not free from sharing obstacles, that is:

- The attitude of some school principals who still think that the efforts to reform education are mere dreams. This kind of attitude will give a negative impact on his peers even more so for the teachers they lead.
- Structuring a conventional management system will narrow efforts to develop education in schools.

4 CONCLUSION

Training on increasing teacher competency through teacher competency improvement activities as a 21st century educator and effective school management in the framework of continuing professional development has broadened the perspectives of school principals and teachers in the UPTD Paper District of Bandung District, so they are aware of the importance of continuing professional development in efforts to improve the quality of education in school. The material on improving teacher competency as a 21st century educator and effective school management in the context of sustainable professional development is considered a topic that interests them considering this topic is important in improving the professional quality of school personnel.

The attitude of the participants who were still unsatisfied with the situation they had faced so far and the participants had the desire to manage their schools professionally, especially in developing the quality management of education in the district of Kertasari, Bandung.

The intensive efforts to renew education in schools have had positive implications for improving the ability of school personnel. Therefore principals and teachers must take advantage of the opportunities available for the improvement and improvement of the quality of education through the process of learning towards the 21st century.

In the context of strengthening cooperation, community service programs are an effective alternative as an effort to professionalize education staff. However, more intensive attention and guidance from the parties concerned will provide higher added value. Therefore the Education Office is an institution that is also competent to develop coaching programs in a more planned and organized manner.

REFERENCES

Bolanle, A.O. 2013. Principals' leadership skills and school effectiveness: The case of South Western Nigeria. *World Journal of Education* 3(5): 26–33.

Jacobson, S.L., Day, C., Leithwood, K., Gurr, D., Drysdale, L. & Mulford, B. 2005. Successful principal leadership: Australian case studies. *Journal of educational administration* 43(6): 539–551.

Riswandi, R. 2016. Pelatihan manajemen sekolah sebagai upaya untuk menciptakan sekolah efektif pada Sekolah Dasar di Kabupaten Tanggamus. *Jurnal Tarbiyah* 22(1): 148–168.

Sunaengsih, C., Anggarani, M., Amalia, M., Nurfatmala, S. & Naelin, S.D. 2019. Principal leadership in the implementation of effective school management. *Mimbar Sekolah Dasar* 6(1): 79–91.

Sururi, S. & Kurniady, D.A. 2014. Pengembangan dan pemberdayaan pengawas sekolah dari sudut pandang manajemen sumber daya manusia dan kebijakan yang berlaku. *Jurnal ABMAS* 14(14).

Sururi, S., Sa'ud, U.S. & Suryana, A. 2016. Studi efektivitas penilaian kinerja kepala SMP Negeri. *Jurnal Administrasi Pendidikan* 23(2): 161–172.

Pandoyo, R. & Wuradji, W. 2015. Pengaruh kepemimpinan kepala sekolah, kinerja guru, komite sekolah terhadap keefektifan SDN Se-Kecamatan Mlati. *Jurnal Akuntabilitas Manajemen Pendidikan* 3(2): 250–263.

Borderless Education as a Challenge in the 5.0 Society – Abdullah, Adriany & Abdullah (eds)
© 2021 Taylor & Francis Group, London, ISBN 978-0-367-61960-2

The contribution of supervision and the role of the city government to the implementation of school-based management at junior high school in Padang Panjang City

U.B. Fitrillah
Educational Administration Department, Padang State University, Padang, Indonesia

A. Bentri, H. Hadiyanto & R. Rifma
Faculty of Education,Padang State University, Padang, Indonesia

ABSTRACT: The purpose of this study was to determine the contribution of supervision and the role of the district government to the implementation of school-based management in the high school in Padang Panjang City. This research uses quantitative methods with correlational type of research. The population in this study was 41 principals and vice-principals in junior high schools in the city of Padang Panjang. The sample of this research is 37 people, determined by simple random sampling. The results of this study indicate that there is a significant contribution of supervision of the implementation of school-based management of 18.7%, the role of the district government toward the implementation of school-based management of 45.1%, and the implementation of supervision and the role of the district government together toward the implementation of management-based schools of 46.6%.

1 INTRODUCTION

School-based management in Indonesia was implemented in 2003, namely in Article 51 Act of the Republic of Indonesia No. 20 Year 2003 on National Education System and Article 49 Government Regulation of the Republic of Indonesia No. 19 Year 2005 on National Education Standards. The legislative policy gives more authority to schools in order to provide stimulation to improve academic and nonacademic activities by utilizing all the potential possessed by its stakeholders. The policy was formulated in the form of a program called school-based management (SBM).

SBM, as Caldwell said, is an effort to decentralize authority from the central government to the school level (Bank 2007). Furthermore, through SBM, solving internal school problems – both concerning the learning process and supporting resources – is sufficiently discussed internally within the school and with the community, so there is no need to involve the city (regional) or central government. The task of the government (central and regional) is only to provide facilities and assistance to schools and communities when they find a dead end in solving a problem. These facilities may take the form of capacity building, technical assistance for learning or school management, subsidies for educational resource assistance, as well as national curriculum and quality control of education at both the regional and central levels (Gerungan 2006, Hasbullah 2006). SBM can improve the quality of educators, which in turn will increase student achievement (Rodriguez & Slate 2010).

In the SBM model, school principals and teachers have broad and autonomous freedom to manage schools without ignoring policies and priorities determined by the government (Sagala 2004). School management is the responsibility of the government. Several attempts have been made to resolve school-level problems by improving regulations at the central and regional levels (Hadiyanto & Komariah 2019). Therefore, in accordance with Government Regulation No. 17 Year 2010, SBM can be implemented well if supported by the second level local government as the organizer and manager of education at the district/city level. In addition, supervision is also needed, in the form of ongoing and continuous guidance with targeted and systematic programs for teachers and other education personnel in schools. Supervision activities carried out by supervisors and school principals in the implementation of SBM including assessment of teaching and learning activities in the classroom, straightening deviations, improvement of the situation, improvement of the program, and development of professional skills of teachers (Mulyasa 2002). Implementation of the curriculum is one of the principal components that are very important in education. The curriculum is an absolute requirement and is an integral part of education (Sukmadinata 2010). In addition, the implications of the decentralization of curriculum education can also be developed in accordance with local wisdom (Bentri 2017). Supervision is an improvement in teaching conducted by teachers with the help of supervisors through supervision activities, direction, guidance, examples, regulations, controls, efforts toward achieving learning objectives, and other forms of activity (Tim Dosen Administrasi Pendidikan 2009). In addition to improvement, supervision is also an effort to improve teacher competency in learning (Rifma et al. 2019a). Teacher competency development aims to implementation of quality learning (Rifma et al. 2019b). Therefore, supervision is very important in implementing SBM.

The purpose of this study was to determine the contribution of supervision and the role of the district government to the implementation of SBM in junior high schools in Padang Panjang City. The results of the observation carried out on February 18–19, 2019, showed that junior high schools in Padang Panjang City had run SBM. However, the reality discovered by researchers in the field shows that SBM is still not maximally implemented. This can be seen from the following conditions: (1) participation of all parties in the school program has not been maximized, (2) cooperation that has not been well coordinated, and (3) schools have not been independent in meeting their needs. In addition, according to the principal of SMP Negeri 1 Padang Panjang, the implementation of SBM is inseparable from the role of supervision, which plays an important role in the implementation of SBM; for example, every activity has a report as a form of accountability, and supervision of one teacher is related to the making of lesson plans that must be included in the design of literacy activities, thinking, skills, adiwiyata programs, and characters as a derivative of the school's vision and mission. Principal of SMP Negeri 1 Padang Panjang also explained that the physical construction of the school did not receive assistance from the government, so the school sought to optimize the role of alumni and the community.

This article reveals the contribution of supervision and the role of the city government to the implementation of SBM in junior high schools in the city of Padang Panjang.

2 RESEARCH METHODS

This study uses a quantitative method with the type of correlational research. Arikunto (2010) said "correlation research aims to find whether there is a relationship and if there is how close the relationship is and whether the relationship is meaningful or not."

This study uses two independent variables (X), namely: supervision (X1) and the role of the city government (X2), while the dependent variable is the implementation of SBM (Y). This study will reveal the contribution of each independent variable to the dependent variable both individually and collectively.

The population in this study was the headmasters (principals) and vice-principals in the junior high schools in Padang Panjang City, consisting of six state junior high schools and eight private junior high schools. Total population was 41 people. The sample of this research was 37 people, determined by simple random sampling.

Data collection techniques carried out by distributing questionnaires. Data analysis techniques using simple regression and correlation techniques. Data analysis was performed using SPSS application version 24.00.

3 RESEARCH RESULTS AND DISCUSSION

The first hypothesis tested in this study was the implementation of supervision (X1) contributing to the implementation of SBM (Y). To find out these contributions, correlation analysis and simple regression analysis were used. The results of the calculation of the correlation coefficient of supervision of the implementation of SBM was 0.432, with $p = 0.008 < 0.05$. This means a contribution to the implementation of supervision of the implementation of school-based management. The coefficient of determination (r^2) was 0.187. To determine the predictive relationship between the implementation of supervision and the implementation of school-based management, a simple regression analysis was performed.

Regression analysis results obtained by the regression equation $\dot{Y} = 198.003 + 0.280$. This equation was then tested for significance. Results of regression analysis of variables for supervision (X1) and SBM (Y) showed that Fcount = 8.035 with $p\ 0.008 < \alpha\ 0.05$.

Then, the significance of the regression coefficient test was performed. Based on the results of the regression coefficient test of supervision of the implementation of SBM it can be seen that the price (value) t coefficient was 12.942 and the significance level of 0.00. This means that the regression coefficient of 0.280 was significant and can be used to predict the implementation of school-based management. The regression equation $\dot{Y} = 198.003 + 0.280X1$ explains that each increase in the implementation of SBM by 1 scale will contribute to an increase in the implementation of SBM by 0.280 scale, while the score of implementing SBM already exists at 198.003 scale without supervision.

Based on the results of these tests it appears that all are significant. Then the hypothesis stating that the implementation of supervision contributes to the implementation of SBM can be accepted at the 95% level of confidence. Furthermore, it can be interpreted that the implementation of supervision has a significant predictive power to the implementation of school-based management. The contribution of supervision to the implementation of SBM in junior high schools in Padang Panjang City was 18.7%.

The second hypothesis tested in this study was the role of the city government (X2) and its contribution to the implementation of SBM (Y). To find out these contributions, correlation analysis and simple regression analysis were used. The results of the calculation of the correlation coefficient of the city government on the implementation of SBM was 0.671, with $p = 0.000 < 0.05$. This means that the role of the city government contributes to the implementation of school-based management. The coefficient of determination (r^2) is 0.451.

To determine the predictive relationship between the role of the city government and the implementation of school-based management, a simple regression analysis was performed. Regression results obtained by the regression equation showed $\dot{Y} = 153.078 + 0.752X2$. This equation was then tested for significance. The results of the regression analysis of the variables of the role of the city government (X2) and the implementation of SBM (Y). Fcount = 28.717 with $p\ 0.000 < \alpha\ 0.05$. This means the regression equation is of significance at the 95% confidence level and can be used to predict the implementation of school-based management.

Then, the significance of the regression coefficient test was performed. Regression coefficient test of the role of the city government on the implementation of SBM shows that the price (value) t coefficient is 9.277 and the significance level is 0.00. This means that the regression coefficient is 0.752 significant and can be used to predict the implementation of school-based management. The regression equation $\dot{Y} = 153.078 + 0.752X2$ explains that each increase in the role of government by 1 scale will contribute to an increase in the implementation of SBM by 0.752 scale, while the score of implementing SBM already exists at 153.078 scale without the role of the city government.

Based on the results of these tests, it appears that everything is very significant. Then the hypothesis stating that the role of the city government contributing to the implementation of SBM can be accepted at the level of 95% confidence. Furthermore, it can be interpreted that the role of the city government has a significant predictive power on the implementation of school-based management. The contribution of the role of the city government to the implementation of SBM in junior high schools in Padang Panjang City is 45.1%.

The third hypothesis tested in this study is the implementation of supervision (X1) and the role of the city government (X2) jointly contributing to the implementation of SBM (Y). To find out these contributions, multiple correlation analysis is used. The results of the calculation of the multiple correlation coefficient of supervision implementation and the role of the city government together contributing to the implementation of SBM is 0.683.

The results of the correlation analysis between the implementation of supervision variables (X1) and the role of the city government (X2) on the implementation of SBM (Y) can be seen from the calculation results that show the correlation coefficient (Ry1.2) 0.683 with $p = 0.000 < 0.05$. This means that there is a contribution to the implementation of supervision and the role of the city government together in the implementation of school-based management. The coefficient of determination (r^2) is 0.466.

To find out the predictive relationship between the implementation of supervision and the role of the city government on the implementation of school-based management, a multiple regression analysis was performed. Regression results obtained by the regression equation $\dot{Y} = 152.887 + (-0.126)X1 + 0.918X2$. This equation is then tested for significance.

The results of the multiple regression analysis between the implementation of supervision variables (X1) and the role of the City government (X2) toward the implementation of SBM (Y) showed that Fcount = 14.861 with $p = 0.000 < \alpha\ 0.05$. This means the regression equation is of significance at the 95% confidence level and can be used to predict the implementation of school-based management. Then, the significance of the regression coefficient test is performed.

Regression coefficient test results of the implementation of supervision and the role of the city government on the implementation of SBM test results show that the t coefficient price of 9.265 and the significance level of 0.000 means that the regression coefficients of –0.126 and 0.918 are significant and can be used to predict the implementation of school-based management. The regression equation $\dot{Y} = 152.887 - 0.126X1 + 0.918X2$ explains that the X1 direction coefficient of –0.126 and X2 coefficient of 0.918 means that any increase in supervision implementation of 1 scale will contribute to an increase in the implementation of SBM by –0.126 scale, and the role of government city of 0.918 for 1 scale will contribute to the implementation of SBM (Y), while the constant is 152.887 without the effect of the two predictors.

Based on the above test results that are all significant, the hypothesis stating there is a contribution to the implementation of supervision and the role of the city government in the implementation of SBM can be accepted with a level of confidence of 95%. Furthermore, it can be interpreted that the implementation of supervision and the role of the city government has a significant predictive power toward the implementation of school-based management. The contribution of supervision and the role of the city government to the implementation of SBM in junior high schools in Padang Panjang is 46.6%.

Based on the results of descriptive analysis, it appears that the implementation of supervision is good with an average score of 82% of the ideal score. Meanwhile, if observed from each of the research indicators, it can be seen that the indicator with the highest score in the implementation of assessment indicators is in a 'good' category (85% of the ideal score), while the lowest achievement indicator is the implementation of mentoring and training, which in the 'fair' category (74.73% of the ideal score). It turned out that of the four indicators of supervision that were analyzed, three indicators were in the 'good enough' category and one indicator was in the 'poor' category.

The results of this analysis indicate that the implementation of supervision in junior high schools in the city of Padang Panjang needs to be improved. The supervision can be improved by supervisors increasing the intensity of monitoring, coaching, evaluating, and providing guidance and training to schools on an ongoing basis. Increasing the implementation of supervision can improve the implementation of school-based management at the junior high schools in the city of Padang Panjang.

Based on the results of descriptive analysis, it can be seen that the role of the city government is good with an average score of 80% of the ideal score. Meanwhile, if observed from each of the research indicators, it can be seen that the indicators with the highest level of achievement in the city government policy indicators are in the 'good' category (81.89% of the ideal score), while the lowest performance indicators are the support and motivation in the category which is 'early good' (78.02%) of the ideal score.

It turns out that of the four indicators of the role of the city government analyzed, three indicators were in the 'early good' category and one was in the 'good' category. The results of this analysis indicate that the role of the city government in junior high schools in the city of Padang Panjang needs to be improved. Increased performance of the Padang Panjang city government, can improve the implementation of school-based management in the junior high school in the Padang Panjang city.

The results of the study found that the implementation of supervision (X1) and the city government (X2) contributed significantly to the implementation of SBM by 46.6%, while the remaining 53.4% was attributed to other variables not examined in this study.

Analysis results show that the implementation of SBM is influenced by the implementation of supervision and the role of the city government, both individually and collectively. The implementation of supervision and the role of the city government are two very important factors because they can influence the implementation of school-based management.

Based on the results of this study, it can be concluded that if the implementation of the supervision is good, and the role of the city government can also carry out its role and function properly, the implementation of school-based management can be carried out well too.

4 CONCLUSION

The implementation of supervision in this study is described as good. Supervision has contributed to the increase of implementation of school-based management in junior high schools in the city of Padang Panjang by 18.7%. The role of government in Padang Panjang city in the education sector is quite good. The government of Padang Panjang city has contributed to the implementation of school-based management in junior high schools in the city of Padang Panjang by 45.1%. While the supervision and role of government in Padang Panjang city together contributed significantly to improving the implementation of school-based management in the junior high schools of Padang Panjang city by 46.6%. This means that the better the implementation of supervision and the role of the city government, the better the implementation of school-based management in the junior high schools of Padang Panjang city.

REFERENCES

Arikunto, S. 2010. *Prosedur penelitian*. Jakarta: Rineka Cipta.

Bank, T. W. 2007. *What is school-based management?* Washington, DC: The International Bank for Reconstruction and Development.

Bentri, A. 2017. A model of local content disaster-based curriculum at elementary schools. *International Journal of GEOMATE* 13(40): 140–147.

Gerungan, R. A. 2006. *Otonomi pendidikan (Kebijakan Otonomi daerah dan implikasinya terhadap penyelenggaraan pendidikan)*. Jakarta: Divisi Buku Perguruan Tinggi PT. Raja Grafindo Persada.

Hadiyanto & Komariah, A. 2019. The fluctuation of school-based management implementation. *Revista De Ciencias Humanas y Sociales, Opción, Año* 35(21): 1012–1027.

Hasbullah, O. P. 2006. *Kebijakan Otonomi daerah dan implikasinya terhadap penyelenggaraan pendidikan*. Jakarta: PT Raja Grafindo Persada.

Mulyasa, E. 2002. *Manajemen berbasis sekolah: Konsep, strategi, dan implementasi*. Bandung: PT. Remaja Rosdakarya.

Rifma, Alkadri, H., & Ermita. 2019a. Supervision service practices by school principals and impacts towards the implementation of teacher duties in primary schools. *Advances in Social Science, Education and Humanities Research* 337: 46–49.

Rifma, R., Alkadri, H., Ermita, & Meizatri, R. 2019b. Teacher prototype for supervision services effectiveness. *Advances in Social Science, Education and Humanities Research* 382: 438–441.
Rodriguez, T. A., & Slate, J.R. 2010. *Site-based management: A review of the literature*. Kansas City, USA: University of Missouri.
Sukmadinata, N.S. 2010. *Pengembangan kurikulum teori dan prakteknya*. Bandung: PT Remaja Rosdakarya Offset.
Sagala, S. 2004. *Manajemen berbasis sekolah dan masyarakat*. Jakarta: Rakasta Samasta.
Tim Dosen Administrasi Pendidikan. 2009. *Manajemen pendidikan*. Bandung: Alfabeta.

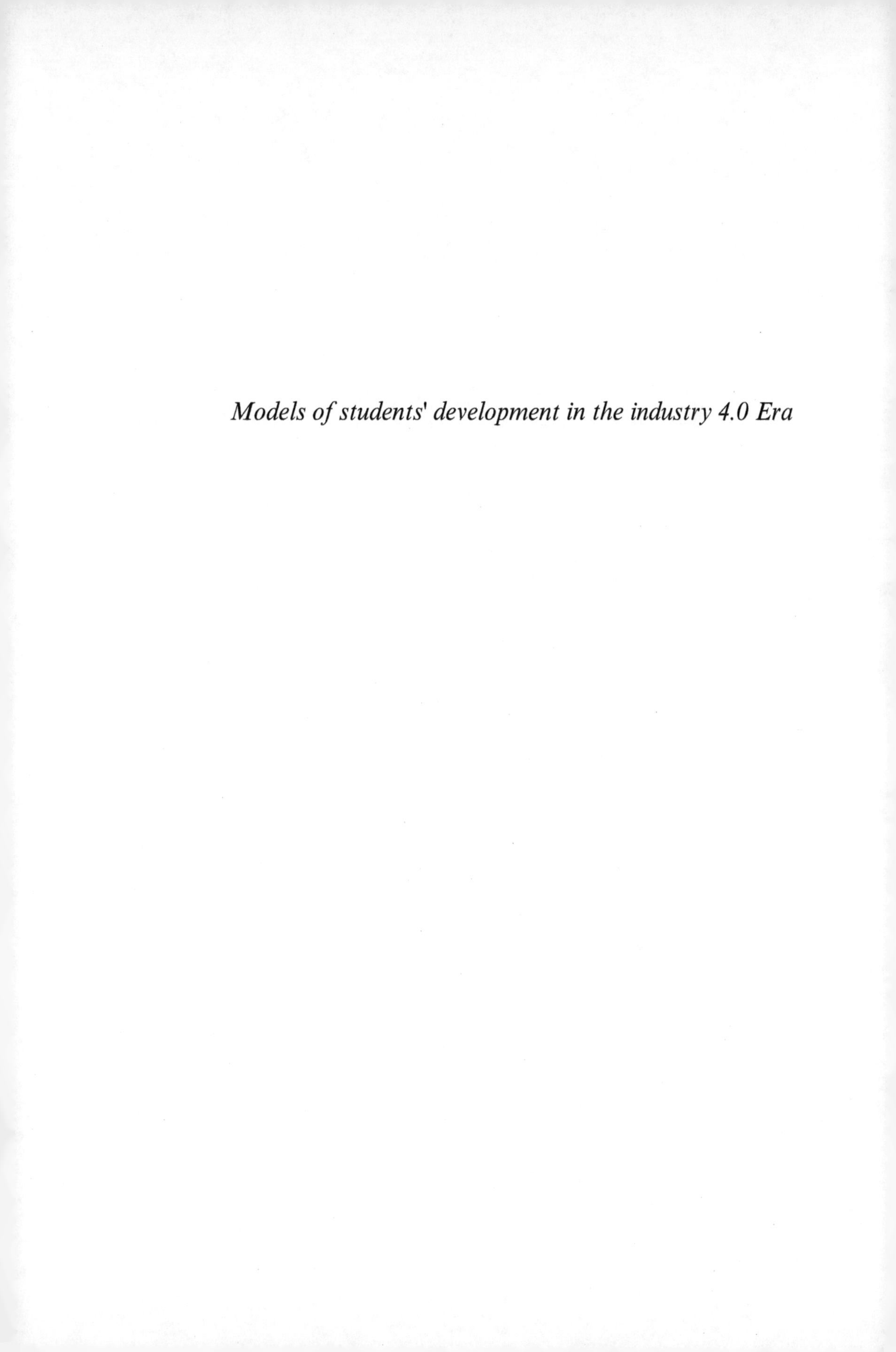

Models of students' development in the industry 4.0 Era

Borderless Education as a Challenge in the 5.0 Society – Abdullah, Adriany & Abdullah (eds)
© 2021 Taylor & Francis Group, London, ISBN 978-0-367-61960-2

Developmentally Appropriate Digital Practice (DADP): Integration of ICT with game-based self-care learning in early childhood education

T. Lestari, N.I. Herawati, E. Permatasari & P.R. Ariningrum
Universitas Pendidikan Indonesia, Bandung, Indonesia

ABSTRACT: This research tends toward a digital-generation, constructive-education movement based on digital practices in accordance with the psychological development of children or referred to as DADP. Now, parents and millennial teachers can rely on the ability of digital-generation alpha literacy in directing children character development. One method is the completion of the task of developing self-care, which consists of daily life activities. This research looks at ICT as supporting contextual learning media in ECE, conducted in three kindergartens in the Greater Bandung area, with purposeful sampling considering the effect of information technology development between the city center and the rural areas. A qualitative approach with non-participant observation, focus group interviews, and audiovisual materials is used to understand student attitudes, and teacher and parent perceptions about the effectiveness of DADP. The results show that the use of digital technology in early childhood in downtown, urban, and rural areas is influenced by parents as the center of determining the pattern of gadget use in children. Parents' backgrounds and parenting patterns at home affect various digital literacy abilities that affect various aspects of child development, including understanding development, fine motor skills, social emotional, social cognitive, language development, social culture, creativity, etc. In the psychopedagogics aspect, teachers understand ICT-based learning to be a media collaboration between teachers and children who can realize healthy digital literacy-based learning. ICT-based learning processes are seen as effective in increasing children's attention and learning interest if accompanied by creativity and learning interactions that pay attention to the child's nature. Children are interested in learning something new and dynamic, directing themselves to the games they like, including the tendency to choose to play without technological media when learning is static. Learning in the digital age is expected to always uphold the nature and task of children's development in life, because ICT-based learning is not mere stimulation of cognition, but also motivational aspects in the formation of character behavior both at home and at school.

1 INTRODUCTION

Technology has a central role in human life in the world, with 49.8% (update June 2019) of internet users being from Asia (Internet World Stats 2019). Humans create inventions that facilitate life itself. Until now, technology and information developed in all areas of human life and the lives of children in Indonesia are no exception. The Ministry of Communication and Information of the Republic of Indonesia conducted a joint research by UNICEF (2014) that showed that the use of digital media is an integral part of the daily lives of Indonesian children. As many as 98% of the children and adolescents surveyed know about the Internet and that 79.5 percent of them are internet users.

The presence of ICT for children certainly affects the lives of children in various ways. As the results of researchers' interviews with some parents of kindergarten children aged 3–6 years, young children in the digital age are accustomed to technology and internet networks which ultimately affect their ability to regulate themselves in their daily tasks (activities of daily living). Curiosity develops children's cognitive abilities through sensory experiences and direct activities because children touch, touch, touch and participate in operating the gadget. On the other hand, the presence of digital media raises the protective attitude of parents and teachers, where digital media is considered to only have a negative side, so digital media users, especially teenagers and children, must be protected. In fact, digital media can be used in the context of empowerment, by using media as needed.

But, in general, the protectionist approach still dominates people's perceptions so a healthy digital literacy movement is needed to change the community's approach to dealing with massive technological change.

This research intends to move towards a healthy digital literacy movement in the context of early childhood education based on digital practices appropriate to the child's development. Digital media can be an educational inspiration, Although early childhood education often emphasizes the application of learning practices according to development (DAP), in today's digital era, it is important for teachers and students to think about how to expand technology-based DAP. As a result of Rosen & Jaruszewicz's (2009) research entitled *Developmentally Appropriate Technology Use and Early Childhood Teacher Education*, the use of technology that matches the level of children's development becomes a unique challenge that can stimulate children's natural desire to actively build knowledge. This happens because technology provides a popular platform for children today (Edwards 2013). Therefore, this situation demands a way of thinking about the best way to bridge the gap between pedagogical understanding of play and the experience of alpha generation children with digital technology.

The focus of this research is on the importance of actively understanding the growth of digital media, using gadgets in a healthy manner, and encouraging the growth of creative-inspirational content for early childhood education. This contextual orientation to the problem of digital games is an opportunity for teachers to effectively engage children in a variety of critical thinking skills related to the "new learning" and "new literacy" movements.

Departing from the study of Millestone children's growth and development theory, this study will examine the digital practice of game-based self-care learning that is suitable for early childhood development.

Self-care learning involves daily tasks that are done to be ready to participate in life activities (including dressing, eating, cleaning teeth, etc.). For children, this activity is usually supported by adults. In this research, ICT is a supporter of contextual learning media for early childhood that can increase their interest in learning. Bruning et al. (1999), in his book titled *Cognitive Psychology and Instruction*, suggests cognitive themes for education. One of them is that knowledge gained and the strategies used must always be related to various things that develop in the environment, which are in accordance with the times. As mentioned by Ali bin Abi Talib ra, about the advice to educate children, "educate your children according to their time, because they live not in your day."

So this research intends to show the utilization of technology in creating effective learning for today's generation of children. Self-care learning becomes basic knowledge that must be mastered by every child. The independence of children in activities of daily life is determined by various factors, including chronological age, level of maturity, culture, and environmental factors. A study of normal children in Hong Kong found a variety of skills caused by certain cultural aspects, one of which is the presence of helpers at home.

With this digital self-care learning, children are expected to be able to easily understand the basic tasks or abilities that must be mastered in everyday life, and learning not only becomes mere cognition, but also motivational aspects and learning belief systems (Surya 2015).

2 RESEARCH METHODS

This research was conducted for 8 weeks in three kindergartens in the Greater Bandung area, including the City of Bandung, Kab. Bandung, and Kab. Sumedang, with the D&D method. This study intends to design and develop ICT-based learning that is appropriate for early childhood development by evaluating the development of digital literacy abilities of early childhood as well as evaluating ICT-based learning activities. At the beginning of the research, descriptive methods are used to collect data on digital literacy of young children and the development of appropriate ICT learning especially in early childhood self-development learning by using evaluative methods to evaluate the process of testing the development of learning designs. Subjects in this study were early childhood aged 3–6 years and class teachers in three different kindergartens in Greater Bandung namely Bandung City, Bandung Regency, and Sumedang Regency. This locational consideration is based on the relationship between rural–urban interaction zones with the dynamics of community behavior in the use of digital media. As mentioned by Eagle et al. (2009), there are differences in network topology and adaptation of mobile phone usage behavior in villages and cities. Sampling was done purposively, that is, based on consideration of the accuracy and adequacy of the information needed in accordance with the research objectives (Creswell & Poth 2016), as well as consideration of certain criteria, such as intense gadget use, and its effect on attitudes and behavior and typical development. Samples with these criteria are seen as able to provide relevant information and help researchers understand the phenomenon of gadgets in children, as well as the design of ICT-based learning that is appropriate for children's development. Data collection instruments in this study include observation, interviews, and documentation studies (Sukmadinata 2005). Researchers make passive observations, by acting as data collectors. Observation is done by directly observing the learning process at school. Researchers also conducted interviews to obtain the necessary data related to parents' observations of the child's independence in self-care at home as well as teacher's pedagogical understanding of learning with technology appropriate to the child's development; child interviews were also conducted related to children's interests and motivation during the learning process. So this research uses multi instruments.

Other tools used are mobile phones with game applications, social media, YouTube, and other applications, to review the development of early childhood digital literacy abilities. The technique used is triangulation by collecting data on the same subject from various sources, including participatory observation, documentation, and teacher and parent interviews. Researchers assisted participants during the use of gadgets. In the next stage, there is a Focus Group Discussion (FGD) of teachers and parents to explore information related to the development of children's independence with unstructured interviews. The independence of the child referred to in this study is the tendency of children to regulate themselves in learning digital media, which is also accompanied by the fulfillment of appropriate development tasks. Data analysis uses the interactive miles and Huberman model, with the following steps: (1) data reduction, summarizing and searching for themes; (2) presentation of data through brief descriptions, charts, and relationships between categories; (3) drawing conclusions.

3 RESULTS AND DISCUSSION

Based on the analysis of findings in the field, in general there are several aspects of development that were observed as follows:

- Fine motor skills: Hand gestures that are not stiff when using a cellphone, coordination of hands and eyes toward the same focus, play on a cellphone without help.
- Gross motor skills: Coordination of the body mimics the movements of the moving animation it sees.
- Non-linear cognitive abilities: Knowing and understanding navigation buttons such as menu buttons, back buttons, volume buttons, swipe, menus to open applications such as cameras (for children aged 3–4 have not been able to adjust the camera with photo objects), Playstore, WhatsApp, Snapchat, Shopee applications, and most of YouTube and games.
- Linear cognitive abilities with different levels of achievement in each child:
  - C1 includes knowledge, such as explaining the use of a favorite application such as Snapchat, Minecraft, or YouTube (which is done at home), reproducing the message of independence from the application/games that are educational, namely my activities (in children 3–4 developing with the help), understanding the game content as Play activities, exemplify making video calls, modeling opening games, modeling opening YouTube, mentioning symbols and images that are shown (in children 3–4 years developing with stimulus),

understanding digital domain problems such as loading, low bat, and main screen lock patterns on mobile phone etc. (in children 3–4 not yet developed), understanding the difference between losing and winning in the game (in children 3–4 not yet developed).
  - C2 includes the ability to explain and exemplify clothing activities, understand the aggressive content of legendary mobile games.
  - C3 includes the ability to demonstrate opening photo and games applications, opening Playstore, Shopee and YouTube applications, but in some children the search is done by looking at the pictures (symbols) of watching or games that are normal (the same as those used at home), so that searches are based on pictures/letters instead of spelling (in children 3–4 years choosing randomly), tells the ability (application) to read prayers before eating.
  - C4 includes analytical skills (puppet and human characterization).
  - C5 includes evaluation capabilities such as comparing (evaluating) self-development activities at home with the media of my activities in the media.
  - C6 includes the ability to connect the message of independence from games with daily activities at home (systematically).

- Language skills such as initiatives to build conversations with those around you such as talking about the video being watched, abilities and experiences such as explaining games played at home for example: Tom and Angela (in children 3–4 years building conversations with stimulus).
- Social abilities observed in three domains (1) how children build cognition through social interaction (social cognitive), (2) how children use gadgets in social norms/contexts (social culture), and (3) development of social emotions that show the ability to control yourself in adapting to other people around like grabbing cellphones and backing up the facilitator, orshowing impatient body gestures while waiting for loading by touching the screen repeatedly. As for the social behavior that arises, the behaviors that appear include the majority of girls playing cellphones without involving friends; involving friends in playing cellphones if reminded and indifferent to friends around; and tending to scramble when playing cellphones. Conversely, boys are observed to be willing to involve friends in voluntarily playing cellphones; sharing and playing cell phones in turn; waiting their turn when playing cellphones with friends; and exchanging ideas, opinions or information with friends on completing the game. In certain cases, the behavior of using gadgets is characterized by ignoring people around (even adults), not being interested in other activities in the surrounding environment, focusing on games, and ignoring the invitation of the facilitator.
- The development of attitude includes interest, attention, and understanding of the values of the use of gadgets. Most children show a high level of attention on the cell phone, choose the application games or features that they like, refuse to return the cell phone, and do not want to stop exploring and playing the cellphone: some even scream and focus on maintaining the cell phone, answering questions with focus on the cellphone/without looking at the facilitator, or not responding quickly when the facilitator repeatedly calls his name or even plays cellphone and masters it only by himself.
- Psychomotor development is shown by the ability to review the power/battery of the cellphone, the ability to quickly imitate the security pattern of the cell phone, the ability to name/label the game with its own name, and the ability to identify one by one Snapchat application filter feature.

In addition, there are also aspects of creativity development that appear only in certain children. Children try out various possibilities of app features that are of interest to them, until ideas emerge that are unknown to other children of the same age.

The results of this study indicate that children's digital literacy abilities are very diverse and largely determined by many things. Based on the results of interviews with three parents of study participants it was concluded that some of the factors causing excessive use of gadgets include: parenting/grandfather/grandmother patterns, providing gadgets as an alternative game early on, gadgets as an alternative to playing so children don't play a lot outside, gadgets as an effective solution when children are fussy, parental factors around work, giving gadgets to children out of pity, as well as unfavorable playing environment/cramped land, etc. In addition, excessive use of gadgets is also influenced by children's motives in using digital media. Most children use gadgets as a means of playing to experience firsthand new things that can fulfill their curiosity, In general, children are able to understand the messages conveyed in the media, however, the stimulation and role of adult guidance is seen to influence children's perceptions and behavior in using digital technology, which is largely incompatible with sociocultural rules (Gee 2010). This is as stated by teachers and parents:

"If he plays cellphone, he will not eat, if I don't feed, then just keep playing until the battery runs out, then he turns on the TV, or if I let me know I say the "quota" runs out, he asks me to fill the "quota" again until I'm right to fill it, if not he won't stop crying." (RS, parents of N)

From this statement, it can be concluded that the child does not understand the proper use of technology by itself. The child will continue to fulfill his curiosity, as, according to Piaget, children learn actively and continuously seek solutions independently (Santrock 2011). This means teacher and parent guidance is needed on how children can understand and regulate themselves in learning digital media while also fulfilling other development tasks. This learning must stimulate their thinking about how the gadget is used. According to Santrock (2011), children will learn from real-world conditions, so that they provide an overview of daily events related to the proper way of using digital media. However, healthy digital literacy education in early childhood education has not been looked at much. This is because, as the results of teacher interviews in several kindergartens in Bandung Raya state, healthy digital literacy in particular has not been taught. Several efforts have been made as a form of commitment to the digital age school including, in learning for certain themes, most teachers have used digital media/technology as learning media with a view to increasing children's motivation in learning, providing parents guidance about the dangers of gadgets for children, and providing guidance to children about the effects of excessive use of digital media on health.

3.1 *Psychopedagogical perceptions of teachers related to ICT in early childhood education*

The development-based digital literacy education movement plays an important role in developing ICT education in early childhood education. This research produces many findings related to the opportunities and challenges of how ICT education for early childhood is reviewed in psychological and pedagogical practice. Technology education in early childhood curriculums is still a new concept. There is an article comparing curriculum in Austria, Finland, France, Germany, and Scotland; it found that the guidelines for technology education must be explicit (Turja et al. 2009) and also need to be adapted to different learning environments and developmental needs among children (Li 2007). As in this study, the teacher has a different style in the implementation of ICT education for early childhood – either the implementation of DADP or the implementation of other ICT media in daily learning activities (See in Figure 1):

This difference certainly shows a tendency for children's attitudes and behavior during learning. Of the three learning approaches taken there is a tendency for children to build a focus of divided attention (Santrock 2011) by looking at the material in learning media and listening to the teacher simultaneously. This looks different in children in Kindergarten C who are not accustomed to learning with laptop and infocus media, so attention becomes selective, attention is greater in media learning, and they are distracted from the focus on the media compared to listening to the teacher, whereas during learning the teacher continues to provide guidance and learning stimuli in the form of questions. The introduction of technology is considered necessary for children, so that children can better understand the messages conveyed through the media compared to merely being curios about the new media.

During the learning process, most children obey the teacher's invitation to listen, build attention focus (visible from the eye contact of the child with the material on the screen) automatically and interactively, and show their curiosity about the material displayed on the screen. Children show their curiosity with gestures such as approaching a laptop, touching infocus, looking at the infocus screen at close range, sitting in the front row, or, for children who sit behind, moving forward/or even standing on their knees as a support. Learning satisfaction appears in children when learning

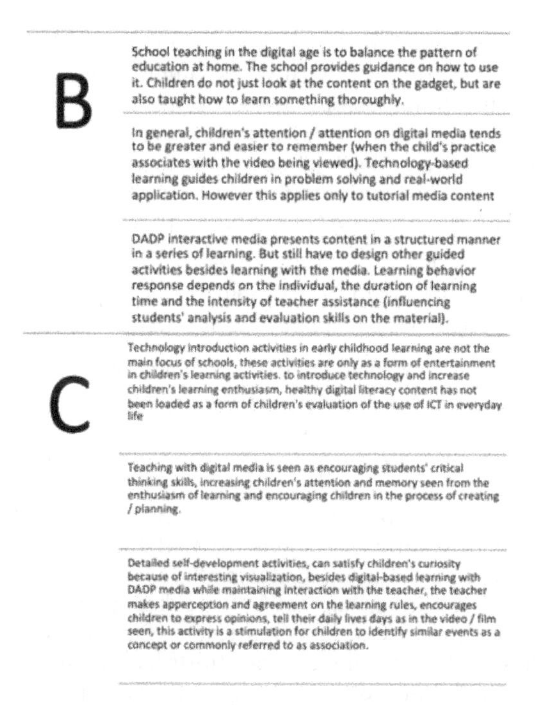

Figure 1. Comparison matrix of digital learning and DADP implementation in early childhood learning.

is over. Most of the children showed impatience to switch to playing outside, some who asked to play at the play center, and some even said that the learning was sufficient. In general, children say "happy" by watching, but do not want to watch a second time. This apparently shows that the child likes something new. This is as stated by the information process approach theory: cognitive development in early childhood is constructivist and children are regulators of their own cognitive development. In terms of learning, children's information processing speed can be seen from memory (Alexander 2006) and thinking competence (Bjorklund 2005), so it is not surprising that with digital learning, children easily remember material and display critical and evaluative thinking skills.

3.2 *Discussion*

The digital age has had a major impact on education, not only in the concept of children's education but also in the fostering and development of teacher competencies. Based on the law of the Republic of Indonesia No. 14 of 2005 concerning teachers and lecturers, the development of teachers should include four competencies. One of them is pedagogical competence, or the ability to manage student learning. In managing learning, the teacher workload includes the main activities of planning learning, implementing learning, assessing learning outcomes, guiding and training students, and carrying out additional tasks. Guidance is currently seen as one of the teacher development strategies that is in line with the demands of 21st-century professional and school teachers (Hargreaves & Fullan 2000). Guidance directing learning in the practice of applying knowledge in life, is also a teacher responsibility in efforts to improve the quality of education.

In practice, coaching is a form of assistance and collaboration. There is a process of sharing between mentors and mentees with a commitment to the development of effective learning implementation for students (Carr et al. 2005). Moreover, children's education in the 21st century is so complex and difficult that none of the teachers and parents can easily answer the problem or have the most correct answer. In other words, in mentoring, both parties learn from each other (Hargreavas & Fullan 2000), so that the benefits and results of mentoring are not only for individuals who are mentored, but also the mentors (Walkington 2005).

The concept of guidance or guided interaction was used more than 50 years ago to describe the ideal teacher–student relationship (Knowles 1950). In the context of technological learning mediation as used today, Plowman & Stephen (2007) state that guided interactions offer guidelines for the use of interactive media for students by teachers and parents.

Guided interaction illustrates the way children interact with technology while actively supported by people with better abilities, in this case teachers and parents. According to Plowman & Stephen (2007), parents and teachers provide supervision when they see the need for intervention, so children are rarely guided and assisted when using digital media. Lack of intervention by parents is caused by parents lacking confidence in understanding technology, children's desire to find out for themselves, rejection of approaches that are considered too instructional, and parental priorities in providing guidance. This activity often unwittingly becomes a free-play activity that causes children to become unproductive. According to Plowman & Stephen (2007) there are two forms of guided interaction that can support optimal development of students: distal and proximal. This concept is a form of scaffolding proposed by Vygotsky or ZPD that illustrates that a child's cognitive abilities are still in a developmental process that can be facilitated by people who have better abilities than the children (Santrock 2011).

Distal refers to the guided interaction that occurs at a certain learning distance and has an indirect influence on a child's learning process, while proximal refers to face-to-face interactions between adults and children and a direct influence on learning. The guided learning process allows teachers and parents to see how children exhibit cognitive abilities, form interactions, and increase awareness of the existence of technology for their lives, and not become isolated actions in completing their developmental tasks.

Based on Figure 2, in addition to paying attention to the management of digital learning, in this study the teacher pays attention to interactive guidance media DADP as psychological stimulation for children including cognitive abilities, learning motivation, completion of developmental tasks, and social interaction. This aspect develops optimally if teaching is in accordance with the child's development, such as not pressing, and not making it difficult, too easy, or even boring (Bredekamp & Copple 1997). The touch of technology in DAP is the focus of development and research related to DAP for today's generation. Although there are pros and cons to using technology at an early age, the analysis of researchers on DADP offers details of psychological skills that can be developed through guided guidance. Without adults, children will not know what is important to do (Beaty 2019). There are several digital literacy practices that are suitable for early childhood development (NAEYC 1998), one of which is the awareness of understanding messages from images. Children are expected to be visual literate by understanding pictorial subject matter, especially in the current digital era because understanding images requires specific explanations from mentoring parents and teachers to their students. In addition to the cognitive aspect, social interaction needs to be developed because this ability is not innate but rather a learning outcome. The social task at this age is to learn to deal emotionally with parents, siblings, and others (Hurlock

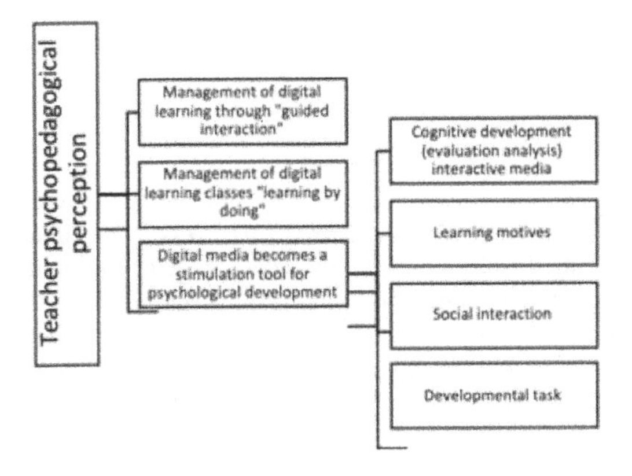

Figure 2. Psychopedagogic perceptions of digital learning in early childhood.

et al. 1990), through the process of socialization. According to Hurlock et al. (1990), the ability of social interaction is obtained by (1) learning socially acceptable behavior, (2) playing an acceptable social role, and (3) developing positive social attitudes.

Social attitude is an evaluative reaction (positive or negative) that is shown in social beliefs, feelings, or behavior. Sobur (2016) states that attitudes develop in the same way but differ in their formation. Every social group has standards for its members about acceptable behavior. Social values are first received through families. Currently, the phubbing phenomenon that has become the lifestyle of the millennial community has become a news trend, because this phenomenon is damaging to social interaction. Phubbing is a busy habit of communicating with other people via gadgets even though there are others in plain sight. Phubbing is quite alarming because it is done at a moment of togetherness, often even at home when having time together. People are more preoccupied with gadgets or smartphones than having to interact with the other person or build relationships with the environment. Sparks (2015) states that individuals who use their cellphones excessively will experience short attention spans or impaired concentration. At this level, children cannot completely comprehend the information that is delivered because technology such as smartphones causes distraction. This means that, in the context of digital learning practices, parents or teachers still have the obligation to provide guidance, so that the message delivered can be received in full by the child.

Motivation is an intrapsychic psychological aspect that involves the process of energizing, directing, and sustaining behavior (Santrock 2011). There are several perspectives on motivation, such as the behavioristic perspective that emphasizes the importance of extrinsic motivation in achievement; and humanistic and cognitive approaches that emphasize the importance of intrinsic motivation in achievement. Intrinsic/extrinsic motivational research involving parents and children said that children have higher intrinsic motivation when their parents engage in learning practices, that is, encouraging children's pleasure and involvement in learning rather than giving external rewards/consequences (Gottfried et al. 2009). This is the basis for the importance of assisting parents and teachers in the learning process. In addition to involved adults, energy is also obtained through contextual learning media (Bruning et al. 1999). Lin et al. (2017) conclude that digital learning presents a better positive effect on learning motivation than traditional teaching, and shows a better positive effect on learning outcomes than traditional teaching. Finally, the psychological aspects that can be developed in digital learning are developmental tasks. Success in doing so raises happiness and success in carrying out other tasks in the future, while failure leads to unhappiness and difficulties in carrying out other tasks in the future (Hurlock et al. 1990). The task of development becomes involved if a generation with a culture also changes. Millennial culture has its own impact on the pattern of forming simple concepts about everyday life skills. Despite the controversy, the pattern of planting life skills in digital concepts in millennial culture has several positive impacts, including learning to develop an autonomous personality, because in the digital era students can easily engage actively in independent learning by "preceding" learning activities (Keane 2012).

4 CONCLUSION

The development of digital literacy is influenced by various internal and external factors. The use of digital media does not merely have a negative impact on children: if parents and teachers monitor the environment, the child can learn to control their cognitive processes or metacognition abilities. The development of this ability is very highly dependent on classroom management and the competence and qualifications of the teacher to continue to guide and balance the characteristics of students. Sectoral training and mentoring is needed to ensure all teachers are truly able to design learning in the digital age.

Thus, the results of this study become a reference for the formulation of steps and strategies for digital literacy that are healthy for young children so that they develop digital literacy abilities which include when, where, how, and with whom to use gadgets for their benefit.

REFERENCES

Alexander, B. 2006. Web 2.0: A new wave of innovation for teaching and learning? *Educause Review* 41(2): 32.

Beaty, J. J. 2019. *Observasi perkembangan anak usia dini*. Jakarta: Kencana Prenada Media Group.

Bjorklund, D. 2005. *Children's Thinking: Cognitive Development and Individual Differences*. Belmont, CA: Thomson Wadsworth.

Bredekamp, S., & Copple, C. 1997. *Developmentally Appropriate Practice in Early Childhood Programs*. NW, Washington, DC: National Association for the Education of Young Children.

Bruning, R. H., Schraw, G. J., & Ronning, R. R. 1999. *Cognitive Psychology and Instruction*. Upper Saddle River, NJ: Prentice-Hall, Inc.

Carr, J. F., Herman, N., & Harris, D .F. 2005. *Creating Dynamic Schools through Mentoring, Coaching, and Collaboration*. Virginia: ASCD.

Creswell, J. W., & Poth, C.N. 2016. *Qualitative Inquiry and Research Design: Choosing among Five Approaches*. United Kingdom: Sage Publications Ltd.

Eagle, N., de Montjoye, Y. A., & Bettencourt, L. M. 2009. Community computing: Comparisons between rural and urban societies using mobile phone data. In *2009 International Conference on Computational Science and Engineering* 4: 144–150. Piscataway, NJ: IEEE.

Edwards, S. 2013. Digital play in the early years: A contextual response to the problem of integrating technologies and play-based pedagogies in the early childhood curriculum. *European Early Childhood Education Research Journal* 21(2): 199–212.

Gee, J. P. 2010. A situated-sociocultural approach to literacy and technology. In E.A. Baker (ed.), *The New Literacies: Multiple Perspectives on Research and Practice*: 165–193. New York, NY: The Guilford Press.

Gottfried, A. E., Marcoulides, G. A., Gottfried, A. W., & Oliver, P. H. 2009. A latent curve model of parental motivational practices and developmental decline in math and science academic intrinsic motivation. *Journal of Educational Psychology* 101(3): 729–739.

Hargreaves, A., & Fullan, M. 2000. Mentoring in the new millennium. *Theory into Practice* 39(1): 50–56.

Hurlock, E. B., Istiwidayanti, Sijabat, R. M., & Soedjarwo. 1990. *Psikologi perkembangan: Suatu pendekatan sepanjang rentang kehidupan*. Jakarta: Erlangga.

Internet World Stats. 2019. *Internet users in the world distributed by world regions—2012 Q2*. [Online]. http://www.internetworldstats.com [Accessed on June 2019].

Keane, T. 2012. Leading with technology: 21st century skills= 3Rs+ 4Cs. *Australian Educational Leader* 34(2): 44.

Knowles, M.S. 1950. *Informal Adult Education: A Guide for Administrators, Leaders, and Teachers*. New York: Associated Press.

Li, Q. 2007. New bottle but old wine: A research of cyberbullying in schools. *Computers in Human Behavior* 23(4): 1777–1791.

Lin, M. H., Chen, H. C., & Liu, K. S. 2017. A study of the effects of digital learning on learning motivation and learning outcome. *Eurasia Journal of Mathematics, Science and Technology Education* 13(7): 3553–3564.

National Association for the Education of Young Children (NAEYC). 1998. *Learning to Read and Write: Developmentally Appropriate Practices for Young Children*. Washington, DC: A joint position statement of the International Reading Association and the National Association for the Education of Young Children.

Plowman, L., & Stephen, C. 2007. Guided interaction in pre-school settings. *Journal of Computer Assisted Learning* 23(1): 14–26.

Rosen, D. B., & Jaruszewicz, C. 2009. Developmentally appropriate technology use and early childhood teacher education. *Journal of Early Childhood Teacher Education* 30(2) 162–171.

Santrock, J. W. 2011. *Perkembangan anak, Edisi 7 Jilid 2*. Jakarta: Erlangga.

Sobur, A. 2016. *Psikologi umum*. Bandung: CV. Pustaka Setia.

Sparks, G. G. 2015. *Media effects research: A basic overview*. Toronto, Canada: Nelson Education.

Sukmadinata, N. S. 2005. *Metode penelitian pendidikan*. Bandung: PT Remaja Rosdakarya.

Surya, M. 2015. *Strategi kognitif dalam proses pembelajaran*. Bandung: Alfabeta.

Turja, L., Endepohls-Ulpe, M., & Chatoney, M. 2009. A conceptual framework for developing the curriculum and delivery of technology education in early childhood. *International Journal of Technology and Design Education* 19(4): 353–365.

UNICEF. 2014. *Studi perilaku anak dan remaja dalam menggunakan internet*. [Online]. https://kominfo.go.id/content/detail/3834/siaran-pers-no-17pihkominfo22014-tentang-riset-kominfo-dan-unicef-mengenai-perilaku-anak-dan-remaja-dalam-menggunakan-internet/0/siaran_pers.

Walkington, J. 2005. The why and how of mentoring. *EQ Australia* (1): 12–13.

Measurement in education

Borderless Education as a Challenge in the 5.0 Society – Abdullah, Adriany & Abdullah (eds)
© 2021 Taylor & Francis Group, London, ISBN 978-0-367-61960-2

Development test measure of basic school cognitive ability

B. Susetyo
Universitas Pendidikan Indonesia, Bandung, Indonesia

ABSTRACT: Admission of new students at the level of basic education in Indonesia has not been based on abilities as a benchmark. In general the only parameter is age, namely a minimum age of 6 years. This is in accordance with government regulations and the Education Law that requires all children to take education at the Basic Education Level. Admission of new students is through a system that has been regulated by the government but it does not use a measuring device in the form of a school entrance test to determine the ability of pre-academic students. This research intends to arrange a measurement device that can measure students' basic cognitive abilities (pre-academic) in the field of arithmetic. Measuring devices for pre-academic abilities are few and not all are standardized. Dimensions to be measured in the domain of basic capabilities in the field of arithmetic are: classification, sequence or series, correspondence, and conservation. The study used an experimental research method prioritizing research development. Development research emphasizes continuous improvement in products, in this case, standardized test instruments. This study used a test to measure basic skills (pre-academic) totaling 61 items consisting of five dimensions. Respondents were students of 6 or 7 years of age in primary school grade 1 or kindergarten. The results of this study are in the form of a set of tests in the field of basic arithmetic skills that are in accordance with the provisions of good test preparation.

1 BACKGROUND

Basic education or elementary school is an educational institution intended for children between 6–7 and 12–13 years of age. Piaget's cognitive development scale puts primary school age at the basic operational stage, which consists of concrete operational stages (7–11 years) and formal operational stages (11–12 years and above). At the concrete operational stage, children are able think logically, rationally, and objectively, even if only on concrete subjects. At the formal operational stage (11–12 years and above), children can already think of something that might happen and something abstract (Bujuri 2018). Elementary school is the basis of all levels of education. The government is obliged to have primary schools in every village. The elementary school must accommodate all children who enter school age. Thus there is almost no screening of students who are going to school.

This causes problems in the implementation of learning, because the abilities of the students is generally not known. As a result, the learning process may not run according to the curriculum document. The application of a learning method will depend on the ability of the students. Students with high-level skills will find it easier to start learning, while those with lower abilities will be slow in accepting learning. Success in the learning process has many factors including the students themselves. Readiness in physical and psychological abilities is also a determining factor in students success. A healthy physical condition will be able to help fluency and success in learning; psychological and emotional maturity and motivation will contribute to learning success. Readiness in the field of basic abilities in arithmetic is one of the abilities necessary, as are verbal and spatial abilities. Verbal abilities are measured by using language. Spatial ability is measured by using pictures. These three components are used to reveal potential. On the other hand, an ability that appears in a person in the field of intelligence or intellect is classified as latent ability. This latent ability will only be known if measured by a measuring device or instrument, in this case, one called the academic potential test

(TPA). Instruments or test kits in the form of TPAs are widely made by various parties, but generally they are for teenagers or adults. Tests used to measure basic abilities for pre-school children or young children have not been widely circulated or made. Children between 2 and 7 years of age are generally at the stage of preoperational cognitive development, meaning their thinking is still symbolic, and not yet operational thinking. Therefore, through this research, a test kit was made that measures the basic abilities of young children before they enter school age.

Latent factors, such as abilities, skills, talents, interests, etc., can be discovered using a measuring instrument in the form of a test, though the testing process may not necessarily reveal all the hidden potential in a person. The test as a measuring device contains a series of questions or a series of tasks that must be answered, done, or carried out by respondents being tested. The results illustrate the ability of individuals, although not all abilities can be revealed. The test is a set of questions to correctly be answered verbally or in writing (Tinambunan 1988).

The Department of Special Education has a lot to do with the world of education, especially with children with special needs. Children with special needs are always given an assessment to find out the obstacles the children are experiencing. From the results of the assessment a program can be developed that, if carried out, will be effective and efficient. The assessment activities require the presence of test and non-test instruments in accordance with the aspects to be assessed: one of the test instruments needed is a test to measure basic ability in arithmetic. From the results of the assessment, a child at the age of 6 or 7 years with a high score can be concluded to have no disorder or be classified as a normal child, but conversely if the assessment results are low, then the child has a disorder or obstacle in basic abilities.

2 METHODS

This research used an experimental method with a form of research development. The study population was all children aged 6 or 7 years who were studying in elementary schools or kindergartens in the city of Bandung. The research sample used random sampling techniques. The research sample was 74 elementary school students. The test instrument was based on Piaget's theory of cognitive development, which consists of four periods (stages), namely, sensorimotor, preoperational, concrete operational, and formal operational (Suparno 2001). Developments in basic cognition in the field of arithmetic consist of four dimensions; classification, sequence or seriation, correspondence, and conservation (Mercer & Mercer 1989) developed into indicators and subsequently made into test items. According to Piaget, the basic abilities of the stages of cognitive development begin with the pre-operational stage, which runs from 2–7 years of age. Pre-operational thinking consists of the symbolic function stage and the intuitive thought stage, not yet operational thinking. The symbolic function stage occurs at the age of 2–4 years; at this stage, the child mentally begins to present objects that are not present. At the second stage – intuitive thinking (around the age of 4–7 years) – representation of an object is based on the perception of one's own experience, not on reasoning (Alhaddad 2012).

In the preoperational stage, concentration or focusing attention on one characteristic is accomplished by ignoring other characteristics. Conservation is the ability to understand that characteristics of the object remain the same even though the object changes its appearance. It is like understanding that the volume of water will remain the same even if it is put into a container with a different shape. Understanding operation is the ability of mental representation that can be reversed. A child of the age of seven understands $2 + 3 = 5$, and understands the opposite of $5 - 2 = 3$. At the preoperational stage there is also a phase of classification and sorting: classification means children can group or sort similar objects, and can sorting according to the length of the object, the weight of the object, and so on (Mu'min 2013).

Data processing was performed on test items with exploratory factor analysis. This technique is used to classify similar test items and eliminate items that are not the same. Processing techniques are carried out on each dimension separately, so items that are not of the same type or one group will be immediately removed or discarded (Susetyo 2015). Good instrument items required the fulfillment of requirements, namely the accuracy of the measured target (validity) and the determination of the test results (reliability) (Susetyo 2011). To determine

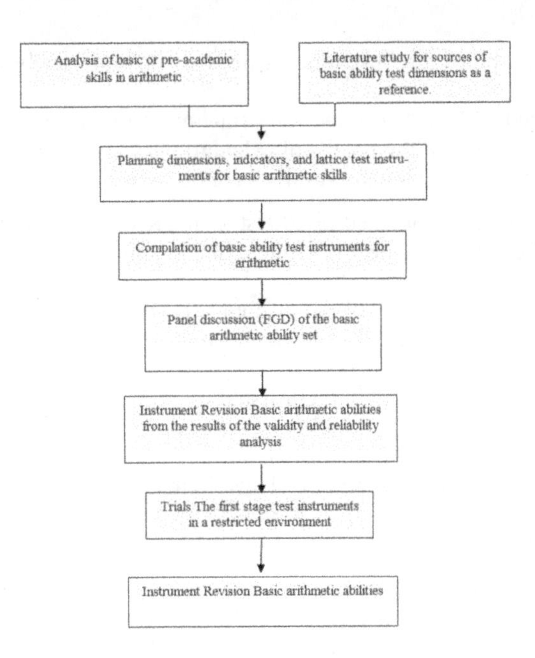

Figure 1. Brief stages of the research plan.

the requirements of a good test instrument, it is necessary to analyze the accuracy of each test item. Test items that are not on target according to most experts, are discarded. Analysis of the determination of the test results is carried out by testing the test items in a limited way, and from the trials it will be known the magnitude of the reliability coefficient. For more details, the stages of research can be seen in Figure 1.

3 RESULTS AND DISCUSSION

Testing the accuracy of the measurement target requirements (validity) is carried out with expert discussion by conducting analysis of each item on each dimension. The validity results obtained are as follows:

- The classification dimension consists of 20 test instruments. Based on the results of expert analysis rationally, all test items measure the classification dimensions. Thus, all 20 test items were declared valid and no test instruments were omitted.
- The sequence dimension consists of 15 items. Based on the results of expert analysis rationally, all test items measure sequence dimensions. Thus, all 15 test items were declared valid and no test instruments were omitted.
- The correspondence dimension consists of 10 items. Based on the results of expert analysis rationally, all test items measure the dimensions of correspondence. Thus, all 10 test items were declared valid and no test instruments were omitted.
- The sequence dimension consists of 21 test instruments. Based on the results of expert analysis rationally, all test items measure the conservation dimension. Thus, all 21 test items were declared valid and no test instruments were omitted.

Testing the requirements of the measurement objectives (reliability) was done by trying test participants or students who are different from the research respondents. The analysis technique used the Cronbach alpha formula (Susetyo 2015). From the results of the calculation, the reliability coefficient of 0.82 was obtained, so the test instruments are reliable.

Based on the results of testing the validity and reliability requirements, all 66 test instruments have met the requirements as good items.

All basic arithmetic ability test instruments need to be empirically tested using sufficient respondents.

This step was carried out to determine the attachment or grouping of test items to their respective dimensions. Based on the results of calculations with exploratory factor analysis for 74 respondents in grade 1, the following results were obtained:

- The classification dimension consisted of 20 test items, KMO value of 0.6, and no anti-image value below 0.5. Thus all test items constitute a single unit or one group in the classification dimension.
- The sequence dimension consisted of 15 test items, KMO value of 0.63, and one anti-image value below 0.5. Thus there are 14 test items that constitute a unit or a group in the sequence dimension.
- The dimensions of the correspondence consisted of 10 test items, KMO value of 0.59, and no anti-image value below 0.5. Thus all test items are one unit or one group in the dimensions of correspondence.
- The conservation dimension consisted of 21 test items, KMO value of 0.68, and four anti-image values below 0.5. Thus the 17 test items constitute a single unit or a group in the conservation dimension.

Based on the results of the exploratory factor analysis, we obtained items that match the dimensions of the basic abilities in the field of arithmetic taken from Piaget's theory there are 61 items that can be used as test kits to measure the basic females of 6-year-old students.

The test instrument used to measure basic skills in the field of arithmetic can be used as a primary school entrance selection tool. Basic skills are the foundation for learning at the elementary school level. By mastering basic skills, it is predicted students will not experience difficulties after entering school. Points of basic ability test instruments are arranged according to the development of children at school age. The instrument uses pictures and a few of them use sentence writing. This is a consideration to anticipate test participants who have not been able to read the writing, as with pictures test participants will find it easier to understand the questions. Thus this basic ability test instrument can be used, without worrying if there are students who cannot read and write.

4 CONCLUSIONS

The test instrument that has been compiled can be used to measure basic abilities in arithmetic consisting of four dimensions, namely classification, sequence, correspondence, and conservation, amounting to 61 test items. The preparation of test instruments that are ready for wide use needs to be rebuilt by selecting several items for each dimension that are adjusted to the time of the test and the conditions of the test taker. The test results can be used aat the time of admission of new students in elementary school and can also be used as a tool to detect whether a child is experiencing obstacles in the basic abilities of the cognitive field.

REFERENCES

Alhaddad, I. 2012. Penerapan teori perkembangan mental piaget pada konsep kekekalan panjang. *Infinity Journal* 1(1): 31–44.

Bujuri, D. A. 2018. Analisis Perkembangan Kognitif Anak Usia Dasar dan Implikasinya dalam Kegiatan Belajar Mengajar. *LITERASI (Jurnal Ilmu Pendidikan)* 9(1): 37–50.

Mercer, C. D., & Mercer, A. R. 1989. *Teaching students with learning problems*. USA: Merrill Publishing Co.

Mu'min, S. A. 2013. Teori Perkembangan Kognitif Jean Piaget. *Al-Ta'dib* 6(1): 89–99.

Suparno, P. 2001. *Teori perkembangan kognitif jean piaget*. Jakarta: Kanisius.

Susetyo, B. 2011. Menyusun Tes Hasil Belajar (dengan Teori Ujian Klasik dan Teori Responsi Butir). Bandung: CV. Chakra.

Susetyo, B. 2015. *Prosedur penyusunan dan analisis tes untuk penilaian hasil belajar bidang kognitif*. Bandung: PT. Refika Aditama.

Tinambunan, W. 1988. *Evaluation of Student Achievement*. Jakarta: Depdikbud.

Borderless Education as a Challenge in the 5.0 Society – Abdullah, Adriany & Abdullah (eds)
© 2021 Taylor & Francis Group, London, ISBN 978-0-367-61960-2

Development of evaluation tools to support improved critical thinking ability of student candidates

A. Hadiapurwa, L. Dewi & D. Suhardini
Universitas Pendidikan Indonesia, Bandung, Indonesia

ABSTRACT: This study aims to develop learning assessment tools that can measure the attitudes, knowledge, and performance of critically thinking students as ethical librarian candidates in CPM Professional Librarian Ethics courses and CPPS Library and Information Science Study Programs. This evaluation tool needs to be developed so that an educator has a standard reference on how to assess student attitudes, knowledge, and learning performance. Evaluation tools that have been developed will be trialed and measured in terms of validity and reliability, and include tests that measure critical thinking skills, a questionnaire to self-assess students attitude (self-assessment), as well as a guide to assessing student performance in developing texts and displaying role-playing about the attitude of librarian professionalism. Students' learning abilities that are critical, creative, and have a good attitude, need to be supported by a good learning process. For this reason, the assessment tools developed are adapted to the learning method used. The use of various kinds of learning methods can support the improvement of motivation to learn, increase positive attitudes, and improve students' knowledge and skills. The research used a design-and-development method, which is carried out systematically through the stages of study of the design, development, and evaluation process and the stated objectives. The results of this study are expected to formulate standardized evaluation tools that can assess attitudes, critical thinking abilities of students, and assessments of creativity on student performance in developing scripts and displaying role-playing as ethical librarian candidates.

1 INTRODUCTION

The learning component consists of objectives, materials, learning processes, and learning evaluation. The learning evaluation process consists of various activities, one of which is the development of assessment instruments. Developing assessment instruments must be done carefully because the assessment process can determine the effectiveness of the objectives set in the learning process.

Based on the National Higher Education Standards through Permenristekdikti No. 44 of 2015, each study program in tertiary institutions must develop CPPS as learning outcomes that students must achieve after attending lecture programs in higher education institutions. CPPS includes aspects of attitude, knowledge, general skills, and special skills.

Library and Information Science Study Program has determined the profile and CPPS of study programs following applicable regulations. The task of each lecturer is to realize the CPPS in the learning process, through (1) formulation of the CPM, sub-CMP, and indicators, (2) determination of the subject matter of lectures, (3) choosing an appropriate learning method, and (4) determining an appropriate assessment tool following set goals.

The Librarian Professional Ethics course studies attitudes and knowledge related to the ethics of the librarian profession. The learning process is carried out through the inculcation of attitudes towards the profession, providing knowledge about the ethics of the librarian profession through content analysis from various sources of reference, analyzing the problems faced by librarians, also through case studies relating to librarians in various types of libraries.

To design a planned librarian professional ethics learning that is oriented towards CPPS achievement, this course is developed through a learning design process using the ADDIE approach: Analysis, Design, Development, Implementation, and Evaluation (Branch 2009). The learning design process has been carried out using the ADDIE stages from the needs analysis process to the evaluation stage. The stages that have been realized at this ADDIE stage are analysis, design, and development.

The development of assessment instruments should include several stages of development. Mardapi (2008) mentions that nine stages must be carried out in developing learning tests, namely compiling test specifications, writing test questions, examining test questions, conducting test tests, analyzing items, refining tests, assembling tests, conducting tests, and interpreting test results.

This research will carry out the process of developing a comprehensive assessment instrument for the professional ethics course. Based on the results of the design, test techniques used include written assignments to assess the ability to think logically, critically, systematically, and creatively; attitude assessment conducted by lecturers and students themselves (self-assessment), observation and interview guides for conducting case studies in the field, as well as assessing student creativity in displaying performance in role playing. For this reason, the steps that will be carried out include (1) compiling the instrument lines, (2) making instruments (tests, questionnaires, interview and observation guides, role playing creativity assessment guides, and guidelines for writing field reports), (3) reviewing the instruments carried out by evaluation experts, (4) testing instruments, (5) analyzing instruments, and (6) improving instruments based on the results of trials. This study aims to develop an assessment instrument that can determine attitudes, and measure critical abilities and creativity of students in the Librarian Professional Ethics course concerning the CPPS that has been chosen and the CPM that has been set. This research will focus on developing assessment instruments to develop written tests to assess the ability to think logically, critically, systematically, and creatively. The student attitude questionnaire assesses student attitude (self-assessment) and lecturer assessment, observation and interview guides for conducting case studies to the field, as well as assessing student creativity in presenting performance in role-playing, and assessment rubrics for student reports that will be arranged in the form of articles.

In general, the problem that will be examined in this research is: "How to develop assessment instruments that can support the improvement of critical thinking abilities of ethical librarians?" Specifically, the research problem was formulated as:

- How to design assessment instruments that can support the enhancement of critical thinking skills of ethical librarian candidates?
- How does expert evaluation of the assessment instruments use instruments that support the improvement of critical thinking skills of ethical librarians?
- How are the results of the assessment conducted to students using instruments that support the enhancement of critical thinking skills of ethical librarian candidates?

This research is based on the need for quality assessment instruments developed by lecturers following the principles of developing assessment instruments. The assumptions of this study are as follows:

- Evaluation is one of the important components and stages for educators to determine the effectiveness of learning. The results obtained can be used as feedback for teachers in improving and perfecting programs and learning activities (Arifin 2012).
- CPPS in a study program is determined by the achievement of the learning process of each lecture that is appropriately planned starting from the formulation of objectives, determination of material, selection of methods, media, and selection of assessment instruments.
- The quality of a good learning assessment instrument will support the assessment of student learning outcomes following the targets set, which include assessments of attitudes, knowledge, and skills.

The main objectives in this study are as follows:

- Obtain analytical data on the design of assessment instruments that can support the enhancement of critical thinking abilities of ethical librarians.

- Obtain data on the results of expert evaluations of assessment instruments that use instruments that support the improvement of critical thinking skills of ethical librarian candidates.
- Obtain data on the results of student assessments by using instruments that support the enhancement of critical thinking skills of ethical librarian candidates.

This research is important because the results are expected to support CPPS study programs in improving their quality. It is important to have good quality assessment instruments, because through the assessment process comprehensive learning outcomes can be obtained, according to the needs and demands of future competencies and to assess attitudes and measure critical thinking skills and student creativity.

This kind of assessment had never been done, so it was researched in professional ethics courses. The assessment activities carried out also support 21st-century learning, namely collaboration by evaluating fellow friends.

1.1 *Learning outcomes*

Learning outcomes formulated by a study program are the objectives that must be achieved by all graduates of the study program. The formulation of learning outcomes must account for both the content and the results that will be demonstrated. The formulation of this learning achievement is following the mandate of the Higher Education Law No. 12 of 2012 and the National Higher Education Standard No. 44 of 2015 (Undang-Undang Nomor 12 2012, Kementerian Riset dan Pendidikan Tinggi 2015).

Learning outcomes must be compiled by each study program in tertiary institutions covering aspects of attitudes, knowledge, general skills, and special skills. The learning achievements of the aspects of general attitudes and skills have been prepared by the government listed in the Higher Education National Standard No. 44 of 2015 (Kementerian Riset dan Pendidikan Tinggi 2015). Learning outcomes of specific aspects of knowledge and skills must be developed by each study program.

If each study program has compiled a study program learning achievement (CPPS), then it is the duty of every lecturer who is in charge of a course to formulate course learning outcomes (CPMK) that refer to CPPS. After CPMK is composed, it is then reduced to sub-CPMK and indicators. Consumption of CPPS, CPMK, sub-CPMK, and indicators are the basis for lecturers in preparing semester lecture plans (RPS) for each course, and this RPS is the basis for lecturers in implementing the learning process.

1.2 *Authentic learning assessment*

1.2.1 *Learning assessment concepts*

Learning assessment is part of the learning component, serving as a controller toward achieving learning objectives that have been set in the learning process. Through the assessment process the strengths and weaknesses of a learning process can be discerned.

The term assessment clashes with two terms in English, namely assessment and evaluation. These terms are vey similar but have different scope. Assessment is a systematic and continuous process of gathering information about a process and results. Associated with learning assessment is a process of gathering continuous information to monitor the achievement of learning objectives. This information will be used to make decisions based on certain criteria and considerations (Asrul et al. 2014, Arifin 2012), including a person's ability related to intelligence, skills, speed, and accuracy in carrying out a particular task or job.

1.2.2 *Authentic assessment in the learning process*

Authentic assessment is an assessment of several actual and realistic abilities possessed by students. An authentic assessment can ideally reveal the abilities of students related to the attitudes, knowledge, and skills connected with the real life of students in the community (Asrul et al. 2014). An authentic assessment is a process of gathering information from a number of types of techniques carried out by an educator to assess student learning outcomes on aspects of attitude, motivation, knowledge, and skills. In connection with the learning process in tertiary institutions, it should be possible to carry out an overall assessment including the assessment of attitudes, knowledge, and skills.

1.2.3 *Attitude assessment*

Attitude assessment has two aspects, namely spiritual and social. The evaluation of spiritual attitudes is related to students who believe and are devoted, while the assessment of social attitudes is related to the formation of students who are noble, disciplined, responsible, honest, etc. (2013 Curriculum Assessment Guidelines). Attitude assessment is very closely related to the assessment of character education. Integrating character values in the learning process requires a variety of support such as the selection of methods, media, and types of learning evaluations that are compatible with character education (Darmansyah 2014). In this research, what will be developed is a guide to observing and evaluating oneself. Self-assessment is the process of assessing strengths and weaknesses related to attitudes. This instrument will be developed using a Likert scale questionnaire about attitudes related to learning.

1.2.4 *Assessment of critical thinking skills*

Critical thinking is reflective thinking that makes sense or is based on reasoning focused on determining what must be believed and done. In Ennis's (1993) opinion, "critical thinking is reasonable, reflective thinking that is focused on deciding what to believe or do." This definition of critical thinking emphasizes how a person makes decisions or considerations (Ennis 1993). Rational means having beliefs and views that are supported by standard, actual, sufficient, and relevant evidence.

1.2.5 *Assessing student creativity*

Creativity is widely known as a must-have skill in the 21st century. Creative thinking abilities are included as one of the skills to support learning and innovation (Plucker et al. 2015). Life in the 21st century is marked by uncertainty, namely by global, social, and economic changes. Guilford explains the importance of developing creative potential at school age (Beghetto 2010). Kaufman & Steinberg (2010) state that to be able to measure level of creativity tends to use the same instrument as measuring ability of divergent thinking. These instruments have been developed by several experts such as TTCT (Torrance test of Creative Thinking), CAMT (Creative Ability in Mathematic Thinking), and Guilford Alternative Uses Tasks (Fardah 2012). Although the content and systematicity of creativity testing varies, how test responses are categorized tends to be mostly the same, namely through the measurement of aspects of fluency, flexibility, originality, and elaboration.

1.3 *Librarian professional ethics code*

Librarian as a profession has an ethics code that must be understood and implemented by every librarian. There are still librarians who do not feel proud of their profession, and this also impacts on the lack of awareness of understanding the librarian code of ethics (Lasa 2009), compiled to be used as a guide for librarians in carrying out their duties.

The librarian code of ethics is issued by the Indonesian Librarian Association (IPI) which guides the behavior and performance of all IPI members in carrying out their duties in the field of librarianship. Librarian code of ethics consists of six major points, namely (1) the basic attitude of librarians, (2) relationships with users, (3) relations between librarians, (4) relationships with libraries (institutions), (5) relationships with professional organizations, and (6) relationships with community (Indonesian Librarian Association).

Based on the code of ethics, a librarian should have a good working attitude, professional competence, and good relations with various parties. For this reason, guidance to prospective librarians is needed so that when they are ready to work in the library world, they are familiar with the code of ethics and carry out their work following the established code of ethics.

2 RESEARCH METHODS

2.1 *Research time and place*

This research was conducted from May to October 2018. The research conducted was developing an assessment instrument to measure the achievement of CPM and CPPS students for Librarian Professional Ethics courses in the Library and Information Science Study Program Faculty of Education, Indonesian University of Education.

2.2 *Research approach method*

This research uses a design-and-development (D&D) research method because this study is by the characteristics of the type of research method. D&D research is research that conducts investigations carried out in the context of developing a product or program that can enhance "something" (Ellis & Levy 2010, Tracey 2007, Richey & Klein 2007). In this research, the product or program to be developed is an assessment instrument that supports the improvement of critical thinking skills.

2.3 *Research subjects*

Subjects in this study include:

- An evaluation expert needed to assess the validity of the appraisal instruments with a view to their construction.
- Content experts needed to assess the validity of the instrument in content.
- Learners needed to carry out the assessment process, after which the results are analyzed to obtain empirical validity data.

2.4 *Data collection instruments*

The instruments used in this study are:

- Assessment rubric for expert judgment.
- The developed test instruments include a written test, questionnaire, interview guide, student creativity assessment guide through role-playing, and article writing assessment guide that will be given to students.

2.5 *Research flowchart*

The research flowchart was developed by following each stage of D&D research (Figure 1).

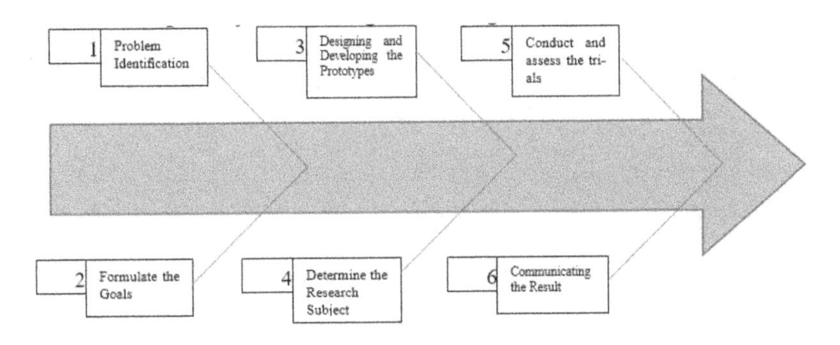

Figure 1. Research flowchart.

3 RESULTS AND DISCUSSION

The results of the study were obtained after conducting trials for students and then conducting an assessment. The assessment of learning outcomes was carried out by the Librarian Professional Ethics course lecturer and also by peer assessment by students. The role-playing learning assessment instrument consists of two parts, namely the implementation and closing of role-playing. In the first part of the learning assessment instrument, six aspects are to be assessed, namely: (1) The suitability of the material to be conveyed with the learning objectives to be achieved, (2) The suitability of the roles performed by the cast with the character that should be played, (3) How the techniques are carried out by actors to explore the core of the problem being played, (4) How is the way used by the actor to display the background/setting of librarians in the library so that role-playing activities can be felt in real terms, (5) Whether the problems conveyed can be solved by various ways of solving problems, and (6) Whether information can be interestingly displayed so that the audience can understand the role-playing role as a whole. In the second part of the learning assessment instrument, the aspects assessed are whether the problem can be resolved properly and how the cast can conclude the material at the end of the role.

The assessment instruments were made in the form of an online form and the peer assessment was based on an assessment interval of 1 to 4, with the assessment rubric as follows (see Figure 2).

After the assessment instrument in the form of an online form has been completed using the interval above, the data obtained from the assessment instrument questionnaire are as follows (see Figure 3).

Based on the results of the study, the material presented can be understood by students in the good category. This means that the material delivered is in line with the objectives formed so that the formation of learning achievements as mandated by the Higher Education Law No. 12 of 2012 and National Higher Education Standards No. 44 of 2015 is important. Playing a role requires a very deep appreciation. This is an effort to be able to portray a role to fit the expected character. Role-playing can also improve social skills and the application of character education can increase communication skills. The results showed that the suitability of the cast when playing a role was considered very good, proving the theory put forward earlier. In the third assessment aspect, the cast can explore the core of the problem being portrayed through a variety of

	Very Creative (4)	**Creative (3)**	**Ordinary/Routine (2)**	**Initative (1)**
Variety of Ideas and Contexts	Each actor explores the core of the problem who played through a variety of techniques/ways which are extraordinary by using various contexts and even various disciplines	Each actor explores the core of the problem who played through a variety of techniques/ways which are using various contexts and even various disciplines	Each actor explores the core of the problem who played through a variety of techniques/ways which are using the same contexts and disciplines	Each actor explores the core of the problem which they played
Variety of sources	Use a variety of sources to present the roles and the settings of the stage using a wide and unlimited variety of sources for example, students look in accordance with the characters but are "unique", the setting uses various sources, such as videos, writing, pictures that resemble real objects or present real objects.	Use a variety of sources to present the roles and the settings of the stage using a wide and unlimited variety of sources for example, students look in accordance with the characters, the setting uses various sources, such as videos, writing, pictures that resemble real objects or present real objects.	Use a limited source to present the roles and the settings of the stage using an available of sources for example, students look in accordance as the usually, the setting uses limited sources.	Only use the resources available in the class
Combining ideas	Problems presented in role-playing are solved in various ways with extraordinary and new original ideas	Problems presented in role-playing are solved in various ways with original ideas	Problems presented in role-playing are solved with other ideas	Problems presented in role-playing are solved in one ways with other ideas
Communicating something new	Information and ideas are displayed in accordance with the objectives, in an interesting, new, and original way so that the audience becomes understood as a whole, even things that were not previously understood become understood	Information and ideas are displayed in accordance with the objectives, in an interesting, new, and original way so that the audience becomes understood.	Information and ideas are displayed in accordance with the objectives, but normal, there is nothing new in solving problems	Information and ideas are displayed not in accordance with the objectives to be achieved

Figure 2. Role-playing assessment rubric.

No	Aspect	Score	Criteria
A1	The material presented by the objectives to be achieved	3,16	Good
A2	The cast acts according to the character	3,49	Very Good
A3	Each actor explores the core of the problem played through various techniques/ways (variety of ideas and contexts)	3,27	Good
A4	Use various sources to display the setting of librarians in the library. (Variety of sources)	3,06	Good
A5	Problems conveyed in role-playing are solved through various ways of solving problems (Combining ideas)	3,13	Good
A6	Information is displayed in an interesting, new and original way so that the audience understands the whole (Communicating ideas)	3,35	Good
B1	Resolve the problem well	3,17	Good
B2	Summing up the material at the end of the role	3,27	Good
	Average	**3,24**	**Good**

Figure 3. Data processing results.

techniques that are in either category indicating a smooth identification. The next aspect, namely, the use of existing resources to display the background of librarians, is in the good category. This shows that flexibility can be used to measure the ability to think creatively with the cast able to use existing resources to display the setting.

In the second assessment instrument, problem-solving at the end of the role-playing scene is also in the good category. This is consistent with the theory put forward by Edward Glaser, that the discovery of ways to deal with problems is a sign of critical thinking ability. The second aspect states that the conclusion material presented by the players is in a good category. This shows that there is a good elaboration between the material and the role played by students. So, from the two assessment instruments, it can be concluded that the role-playing activities in the Librarian Professional Ethics course are classified as good.

4 CONCLUSION

Based on the results of the research, it can be concluded that the development of this evaluation tool is important as an assessment tool in the implementation of the learning process especially in the professional ethics course of librarian studies. The reference used is the standard on how to assess aspects of student attitudes, knowledge, and performance in the learning process. The assessment process is carried out via role-playing activities in which students are required to develop scripts and play roles related to the attitude of librarian professionalism. In this activity, students are required to be critical and creative, supported by a good attitude displayed by students during the learning activities.

The instrument being tested was divided into two parts. The first part of the instrument is contained in the instrument for peer assessments, that is, assessments conducted by the students themselves. This assessment process is contained in the implementation and closing part of the role-playing activities. Meanwhile, the second part of the instrument is in the preparation part where the assessment activities are carried out by the lecturer. These instruments are made using Google forms so that the instruments can be accessed directly online.

After testing, the results of the study showed that the application of the instrument to the librarianship professional ethics roleplaying activity was in a good category. In general, students can convey the objectives of the roleplaying activities undertaken. Success in delivering this goal is supported by the success of students who are able to play in accordance with their assigned character so that the problems explored could be conveyed. The combination of ideas made to integrate various kinds of problem-solving is also done well. In addition, other supporting sources, such as the setting that is displayed, become more valuable and give the impression of information that is displayed in an interesting, new, and original light, so that the audience can understand the core of the story as a whole.

ACKNOWLEDGMENT

The compilation of articles cannot be separated from the support and assistance of various parties to facilitate the completion of this article. Therefore, the authors thank the lecturers of the Professional Ethics Library and other parties who have helped in the process of preparing the article and to the ICES conference committee in 2019.

REFERENCES

Arifin, Z. 2012. *Evaluasi Pembelajaran.* Jakarta: Direktorat Jenderal Pendidikan Islam Kementerian Agama RI.

Asrul, A., Ananda, R., & Rosnita, R. 2015. *Evaluasi Pembelajaran.* Bandung: Ciptapustaka Media.

Beghetto, R. A. 2010. Creativity in the classroom. *The Cambridge handbook of creativity* 447–463.

Branch R. M. 2009. *Instructional Design: The ADDIE Approach.* New York: Springer Science & Business Media.

Darmansyah, D. 2014. Teknik Penilaian Sikap Spritual dan Sosial dalam Pendidikan Karakter di Sekolah Dasar 08 Surau Gadang Nanggalo. *Al-Ta lim Journal* 21(1): 10–17.

Ellis, T. J., & Levy, Y. 2010. A guide for novice researchers: Design and development research methods. *In Proceedings of Informing Science & IT Education Conference (InSITE)* 10: 107–118.

Ennis, R. H. 1993. Critical thinking assessment. *Theory into practice* 32(3): 179–186.

Fardah, D. K. 2012. Analisis Proses dan Kemampuan Berpikir Kreatif Siswa dalam Matematika Melalui Tugas Open-Ended. *Kreano, Jurnal Matematika Kreatif-Inovatif* 3(2): 91–99.

Fisher, A. 2009. *Berpikir Kritis: Sebuah Pengantar.* Jakarta: Erlangga.

Kaufman, J. C., & Stenbergh, R. J. 2010. *The Cambridge Handbook of Creativity.* Cambridge: Cambridge University Press.

Kementerian Riset dan Pendidikan Tinggi, 2015. *Permendikti No. 44 Tahun 2015 tentang Standar Nasional Pendidikan Tinggi.* Jakarta: Dirjen.

Lasa, H. S. 2009. *Kode Etik Profesi Pustakawan dalam Perspektif Islam.* Yogyakarta: Makalah.

Mardapi, D. 2008. *Teknik Penyusunan Instrumen Tes dan Nontes.* Yogjakarta: Mitra Cendikia Press.

Plucker, J. A., Kaufman, J. C., & Beghetto, R. A. 2015. *What we know about creativity.* DC: Partnership for 21st Century Learning.

Richey, R. C., & Klein, J. D. 2007. *Design and Development Research.* New Jersey: Lawrence Erlbaum Association Inc. Publisher.

Tracey, M. W. 2009. Design and development research: A model validation case. *Educational Technology Research and Development* 57(4): 553–571.

Undang-Undang Nomor 12 Tahun 2012 *Tentang Pendidikan Tinggi.*

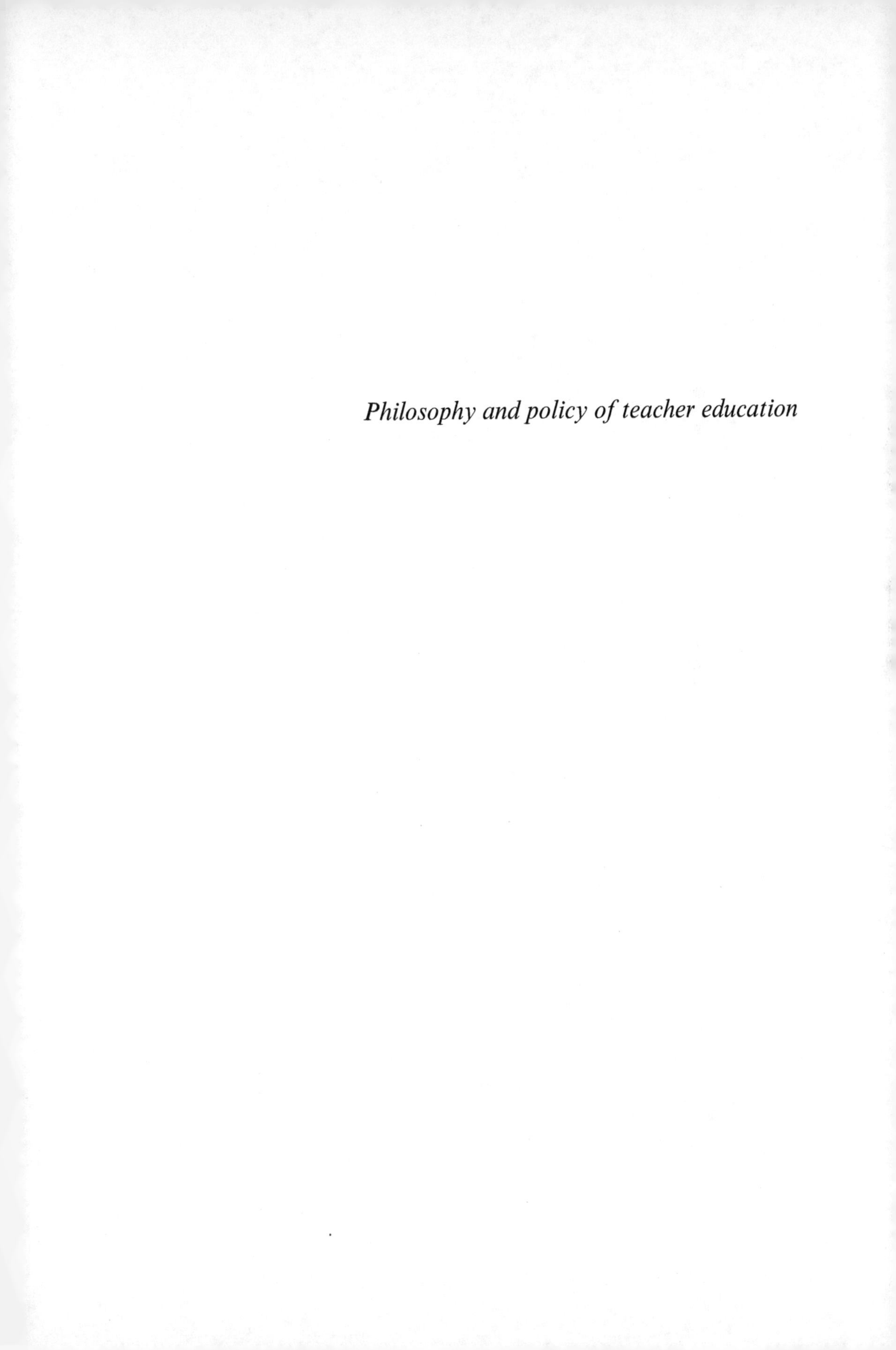

Philosophy and policy of teacher education

Borderless Education as a Challenge in the 5.0 Society – Abdullah, Adriany & Abdullah (eds)
 ISBN 978-0-367-61960-2

Consumer behavior based on lifestyle and economic literacy

S. Lestari, T. Yuniarsih, N. Fattah & E. Ahman
Universitas Pendidikan Indonesia, Bandung, Indonesia

ABSTRACT: Lifestyle and economic literacy are important factors in consumer behavior. This study aims to find out whether there is a correlation between lifestyle, economic literacy, and consumer behavior among university students in Banten Province. From a population of 3,877 students, a sample of 363 students was taken using random sampling. This research used the explanatory survey method. The results showed a positive and significant correlation between lifestyle and economic literacy compared to consumer behavior. Based on this study, policies aimed at boosting economic literacy would help to improve the rationality of consumer behavior, such as making shopping lists and prioritizing based on needs.

1 INTRODUCTION

In an increasingly globalized world, communication technology is making it easier for consumers to choose among various types of quality goods and services according to their wants and needs. This includes university students, who tend to be consumptive, as saving is not a priority. Students' consumption behavior is not based on priority needs but is influenced by trends, friends, and advertisements offered. The potential of students as consumers is very large: although most do not have their own income, many have a considerable expenditure allocation.

However, based on the results of direct interviews with respondents, many report the budget they received from their parents was not able to meet all their needs. The percentage of expenditure allocation for 45 respondents (based on the level of budget obtained) is shown in Table 1.

Table 1 shows respondents' behavior tends to be consumptive. Tertiery needs such as shopping or skin and body care are roughly equivalent for all categories, whether due to necessities, or because of invitations from friends, or being tempted by promotions or advertisements. Budgeting for pleasure categories is therefore very high.

Based on interviews and information from students in the questionnaire about internet data allocation, students buy regular credit and internet packages to access social networks such as Line, WhatsApp, Facebook, Instagram, and Path.

However, consumption behavior of students who do not pay attention to priorities and who tend to be irrational will lead to problems, especially if excessive spending is not

Table 1. Budget expenditure of student need (Lestari et al. 2017).

	Budget			
Need	Very high (%)	High (%)	Middle (%)	Low (%)
Food	35.2	31.9	37.1	26.5
Internet data	7.7	10	9.2	12.9
Books, photocopies, stationery	7.5	13.6	6.9	4.2
Shopping	13	16.5	11.4	16.8
Skin and body care	9.8	11.5	12.6	17.4
Savings	5.2	8.7	10.2	13.9
Etc.	21.6	7.8	12.6	8.3
Total	100.0	100.0	100.0	100.0

supported by adequate finance. Irrational behavior will make students become more consumptive, choosing not based on urgent basic needs but on secondary and tertiary needs.

In terms of lifestyle, humans tend to adapt to their environment to maintain standards of living of their peers and attempt to satisfy their desires. The research of Kanserina et al. (2015) showed positive significant influence between lifestyle and student consumption behavior at the Jurusan Pendidikan Ekonomi Undiksha.

The influence of globalization on the young generation is very strong, especially for clothing, shoes, accessories, and gadget products. Students are potential market segments who are easily influenced by friends and trends. One factor that influences consumption behavior is the reference group: for example, friends, shopping groups, virtual groups, or communities (Setiadi 2013).

As consumers, students are expected to always be rational in carrying out consumption actions so that consumer behavior problems can be avoided. The socioeconomic situation certainly has a role in the development of adolescent children, such that an adequately strong economy will provide wider opportunities to develop various skills (Gerungan 2002). University S1 students in Tangerang City who come from more affluent economic circles get more pocket money to spend. They therefore have higher purchasing power, which encourages irrational consumption.

Economic literacy is very important for students because they, like all consumers, are easily influenced by persuasive techniques and ideas of prestige, and so they tend to be wasteful in spending their money. Sometimes they consume not based on the use value of goods and services but only to show off consumption patterns to peers. If students are not good at managing money and have low economic literacy, it will result in irrational consumption actions. Thus students need to gain sufficient knowledge and understanding of the economy so they can meet their needs according to the priority scale: taking are of primary needs first rather than secondary needs or tertiary needs.

This research aims to describe lifestyle; to describe economic literacy; to describe student consumption behavior; and to identify the influence of lifestyle and economic literacy on students' consumptive behavior.

2 CONSUMPTION BEHAVIOR, LIFESTYLE, AND ECONOMIC LITERACY

Individuals request goods and services because they desire the satisfaction or utility that comes from consuming such goods and services.

2.1 *Consumers*

Consumers are users of marketed products (Assauri 2017). Up to a certain point, the more units consumed by individuals per unit of time the greater the total utility obtained (Salvatore 2006).

2.2 *Consumption behavior*

There are two factors that influence consumption behavior: internal factors and external factors. Internal factors include personality, IQ, emotions, ways of thinking, and perception (Erni & Basri 2013).

Consumption refers to how people use goods or services for their needs (Danil 2013). Indicators of consumer behavior include consumer preferences, budget constraints, and consumer choices (Pindyck & Rubinfeld 2012).

Consumer preferences refer to the reasons consumers prefer one item over another. Budget limitations refer to consumers considering prices in light of income that limits the amount of goods they can buy. Consumer choices means consumers buy combinations of items in order to maximize their satisfaction.

2.3 *Lifestyle*

Lifestyle is a pattern by which one lives one's life, as reflected in activities, interests, and opinions (Kotler & Keller 2002). Lifestyle reflects one's interactions with one's environment.

Consumer lifestyles indicate how consumers think, live, act, and behave (Kowel 2015). This is generally determined by an individual consumer's demographic background, experiences, current situations or actions, socioeconomic characteristics, and behavioral tendencies. A changing lifestyle makes shopping an important consideration for someone trying to make ends meet (Kosyu 2014). Consumers' buyer behaviour is influenced by four major factors: cultural, social, personal, and psychological (Rani 2014).

2.4 *Economic literacy*

Economic literacy is a useful tool for changing to smarter buying behavior (Sina 2012). Student economic literacy indicators are reflected in indicators of economic knowledge, rationality, and economic morality (Haryono 2009).

3 RESEARCH METHODS

The type of research is descriptive and quantitative in accordance with the purpose of research to describe the properties and correlation of lifestyle, economic literacy, and consumption behavior.

The study used an explanatory survey method with a Likert scale questionnaire as the research instrument (valid and reliable). The population was 3,877 and the sample was 363 students. The data analysis method used was path analysis.

4 RESULTS AND DISCUSSION

4.1 *Description of economic literacy, lifestyle, and consumption behavior*

Data analysis showed high levels of economic literacy, lifestyle, consumption behavior scores, with averages for economic literacy of 38.05, lifestyle 44.22, and consumption behavior 48.29.

Based on Table 2, data analysis showed high levels of economic literacy, lifestyle, and consumption behavior.

Figure 1 shows the frequency distribution of economic literacy, from low (17) to high (55), on a histogram, with a standard deviation of 7.782.

Figure 2 shows the frequency distribution of the lifestyle variable, from low (16) to high (63), on a histogram, with standard deviation of 7.969.

Figure 3 shows the frequency distribution of consumption behavior, from low (28) to high (66) on a histogram, with standard deviation of 7.685.

4.2 *Lifestyle, economic literacy, and consumption behavior correlations*

Table 3 shows that $p = 0.505$, sig. 0,000 for economic literacy means there was an influence of economic literacy on consumer behavior. The p value of lifestyle of 0.258 means there is an influence of lifestyle on consumer behavior.

Higher economic literacy therefore means increased levels of rationality in consumer behavior. Likewise, a more rational lifestyle encourages more rational consumer behavior. Economic literacy and lifestyle therefore have a significant relationship with consumer behavior.

Table 2. Description of economic literacy and consumption behavior.

Descriptive statistics					
	N	Minimum	Maximum	Mean	Std. deviation
Lifestyle	363	16	63	44.22	7.969
Literacy	363	17	55	3.05	7.782
Consumer behavior	363	28	66	48.29	7.685

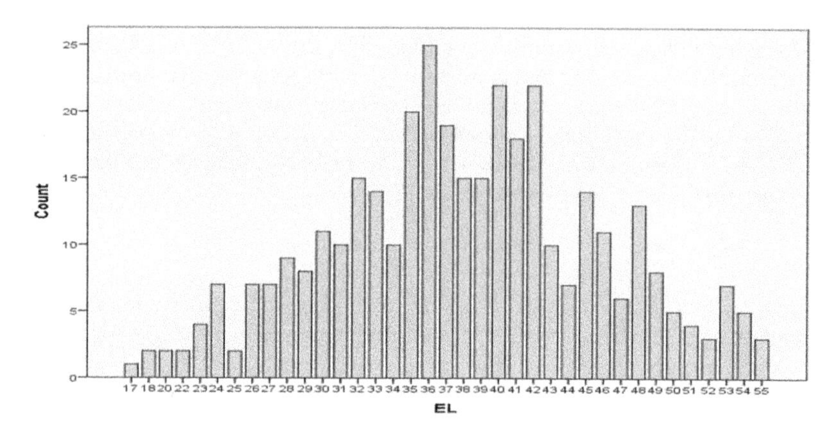

Figure 1. Economic literacy histogram.

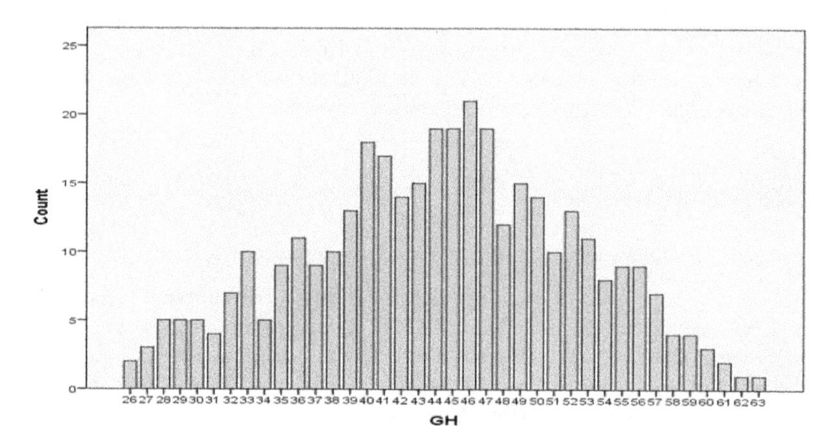

Figure 2. Lifestyle histogram.

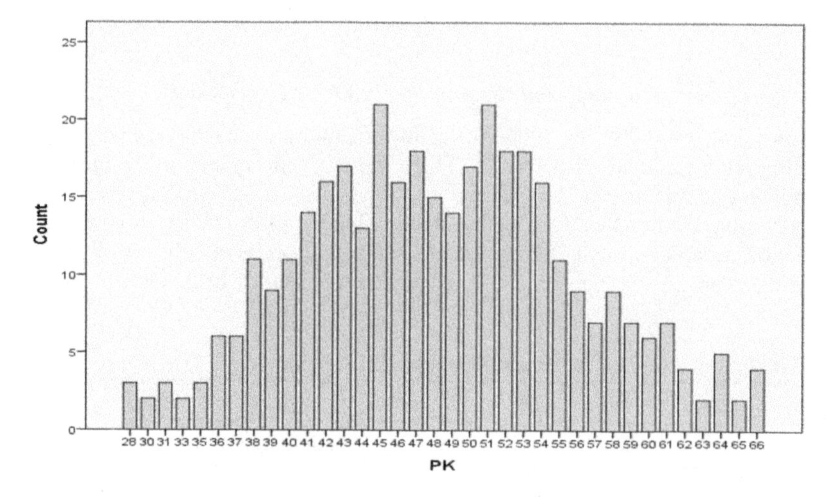

Figure 3. Consumer behavior histogram.

Table 3. Lifestyle, economic literacy, and consumption behavior correlations.

Coefficients[a]						
		Unstandardized coefficients		Standardized coefficients		
Model		B	Std. error	Beta	t	Sig.
1	(Constant)	15.872	2.089		7.596	0.000
	Economic social	0.090	0.041	0.091	2.220	0.027
	Economic literacy	0.496	0.042	0.503	11725	0.000
	Lifestyle	0.249	0.040	0.258	6.189	0.000

a Dependent variable: consumer behavior

The research supported that of Kanserina et al. (2015) showing positive significant influence of lifestyle on student consumption behavior.

5 CONCLUSION

Based on the findings, lifestyle, economic literacy and consumer behavior of students were at high levels. There was an influence of lifestyle and economic literacy on consumer behavior. To increase lifestyle and economic literacy therefore increases rationality of consumption behavior.

Leaders and teachers should therefore work to inculcate a mindset of economic literacy and rational consumption behavior. Researchers should also conduct research on other factors influencing student consumption behavior, such as gender.

REFERENCES

Assauri, S. 2017. *Manajemen Pemasaran*. Jakarta: RajaGrafindo Persada.

Danil, M. 2013. Pengaruh pendapatan terhadap tingkat konsumsi pada pegawai negeri sipil di kantor bupati kabupaten Bireuen. *Jurnal Ekonomika Universitas Almuslim Bireuen Aceh* 4(7): 33–41.

Erni, R., & Basri, M. 2013. Pengaruh Pembelajaran Ekonomi dan Status Sosial Ekonomi terhadap Perilaku Konsumsi. *Jurnal Pendidikan dan Pembelajaran Khatulistiwa* 2(7): 1–9.

Gerungan, W. A. 2002. *Psikologi Sosial*. Bandung: Refika Aditama.

Haryono, A. 2009. *Pengaruh Sistem Pembelajaran dan Status Sosial Ekonomi terhadap Economic Literacy Siswa SMA di Kota Malang*. Malang: Pascasarjana UM.

Kanserina, D., Haris, I. A., & Nuridja, I. M. 2015. Pengaruh Literasi Ekonomi dan Gaya Hidup Terhadap Perilaku Konsumtif Mahasiswa Jurusan Pendidikan Ekonomi Universitas Pendidikan Ganesha Tahun 2015. *Jurnal Pendidikan Ekonomi Undiksha* 5(1): 1–11.

Kosyu, D. A. 2014. Pengaruh Hedonic Shopping Motives Terhadap Shopping Lifestyle dan Impulse Buying (Survei pada Pelanggan Outlet Stradivarius di Galaxy Mall Surabaya). *Jurnal Administrasi Bisnis* 14(2): 1–7.

Kotler, P. 2000. *Consumer market and consumer behavior*. Upper Saddle River, NJ: Prentice-Hall.

Kowel, C. A. 2015. The influence of personality, lifestyle, money attitude on customer purchase decision (case study: manado grand palace convention hall). *Jurnal berkala ilmiah efisiensi* 15(5): 417–425.

Lestari, S., Yuniarsih, T., Fattah, N., & Ahman, E. 2019. Economic Literacy and Student Consumption Behavior. In *2nd International Conference on Educational Sciences (ICES 2018)*. Amsterdam: Atlantis Press.

Pindyck, R. S. & Rubinfeld, D. L. 2012. *Mikroekonomi edisi kedelapan*. Jakarta: Erlangga.

Rani, P. 2014. Factors influencing consumer behaviour. *International Journal of Current Research and Academic Review* 2(9): 52–61.

Salvatore, D. 2006. *Mikroekonomi*. Jakarta: Erlangga.

Setiadi, N. J. 2003. *Perilaku konsumen: Konsep dan implikasi untuk strategi dan penelitian pemasaran*. Jakarta: Prenada Media.

Sina, P. G. 2012. Analisis literasi ekonomi. *Jurnal Economia* 8(2): 135–143.

Borderless Education as a Challenge in the 5.0 Society – Abdullah, Adriany & Abdullah (eds)
© 2021 Taylor & Francis Group, London, ISBN 978-0-367-61960-2

The new institutional theory and governance reform in the educational system of Vietnam

N.Binh Ly
Faculty of Education, University of Social Sciences and Humanities (USSH), Vietnam National University, Ho Chi Minh City, Vietnam

ABSTRACT: This research focuses, on a macro level, on the transition to and development of a market-oriented economy in Vietnam, which has impacted the development of the country's education system ever since the Sixth National Congress of the Communist Party of Vietnam's adoption of a Renovation Program, called Doi Moi, in 1986. This study used the theoretical framework of the New Institutional Theory to explain problems in Vietnam's education system and to provide approaches toward rebuilding the Vietnamese school system, looking at educational philosophy, financial resources, and the democratization of the education system from 1986 to the present. This study mainly employs data collection methods of document review and secondary information analysis. The findings of this study indicate that Doi Moi introduced a new transition and development perspective for Vietnam. Changes in the new institution had a profound influence on Vietnam's education reform. The system of education has expanded following the establishment of new Vietnamese state institutions, and this new institutional change provides a way of viewing institutions outside of the traditional views of economics and education. Educational institutions in Vietnam have been established as part of processes of diversification, marketization, and privatization. The new institutionalism for education focuses on the rise of greater provider pluralism, marketization, privatization, technical accountability, and tighter coupling inside educational institutions. This finding could contribute to discussions among educators and governmental agencies involved in educational policy and administration.

1 INTRODUCTION

In December 1986, the Sixth National Congress of the Communist Party of Vietnam adopted a Renovation Program, called Doi Moi. Doi Moi was characterized by a shift from a "centrally planned" to a "socialist-oriented" market economy, allowing for private-sector economic development, international economic integration, and many legal reforms in various policy areas (Huan 2013). During this reform period, the education system in Vietnam experienced significant changes. These changes include the restructuring of the school system, education quality reform, and education financing reform. The goal of education renovation during this period was to change from a "planning-based to market-based model" and to prepare all young people to become more engaged in the state economy (Dang 1999, Huong 2004).

Since 1986, the Communist Party of Vietnam has expressed the aim of building an industry foundation based on strong scientific progress and improving the nation's development strategies, policies, and planning with a view to accelerating "economic growth linked closely with social progress and equity, educational and cultural development, and environmental protection" (Government of Viet Nam 1991). The national education system has undergone considerable improvements after the Renovation Program began, to meet the labor market demand in the rapidly changing socioeconomy of Vietnam (Hirosato & Kitamura 2009). Educational reform in Vietnam is closely linked to national goals. The existing socioeconomic, political, and cultural climate of the country has significantly affected the redesign of educational objectives, contents, and methods, to meet human-resources needs for the projected industrialization and modernization period (Huong 2004, Tran-Nam 2005). Following principles adopted in Doi Moi, the establishment of non-public schools was permitted in 1992, creting a new education structure, one that is more comprehensive and flexible in the context of a market economy (Brooks 2004).

This study mainly employs the data collection methods of document review and secondary data analysis. The major advantage of document review and secondary data analysis is that it eliminates the time and expense of gathering data and relies on high-quality, reliable data collected by experts in analyzing the impact of the new institutionalism on economic development and its effect on the development of educational reform since 1986 in Vietnam. Reviewed in this study are legal documents of the Vietnam Party, the government, and the Ministry of Education and Training (resolutions, laws, degrees, decision, and directives), as well as secondary information analysis focused on research papers, academic theses, public officials' statements, and leaders' policy statements, announcements, and interviews with press media, as well as other sources of information.

1.1 *The main components of the new institutionalism*

The new institutional economics take the transaction as the primary unit of analysis (Wells 1998). Central to the new institutional economics is the solution to the coordination problem of economic transaction between individuals by agreement under the assumption of transaction cost, property rights, and contractual relations (Furubotn & Richter 2005). Contracts to an exchange wish to economize on transaction costs in a world in which information is costly; some people behave opportunistically (Coase 1937).

Coase (1937), in *The Nature of the Firm*, and in his essay (1960) "The Problem of Social Cost," pointed out the role of organizations such as the state, the firm (public and private), and the market as the key institutions in transaction-cost environment (North 1981). The existence of firms implies that there are costs to market transactions. The new institutional economists no longer see firms as passive receivers of the economic signal (Hamilton 2006). Coase (1937) stated that the firms were considered as agents in actively setting prices, making markets for their products, and creating market organization. The development and cultivation in the interrelationships of peoples and firms in the economy begin to nurture a "self-reflexive" economic system (Mary & Victor 1998), and DiMaggio & Powell (1991) have called this "organizational isomorphism."

The new institutionalism in economics and political science has emphasized formal norms and the monitoring by third-party enforcers, the state, and the firms (Mary & Victor 1998). As exchanges among individuals grow more specialized and complex, contracts require third-party enforcement (Maurice & Paul 1998). States vary greatly in the ways to define property rights, and citizens may view political institutions as more or less legitimate, depending on their ideologies (Shepsle 1986). When ideological consensus is high, opportunistic behavior is curbed. When it is low, contracting costs are high, and more energy is expended on efforts at institutional change (North 1989).

The new institutionalism in organizational analysis takes as a starting point the striking homogeneity of practices and arrangement found in the labor market, in schools, states, and corporations (DiMaggio & Powel 1983, Meyer & Rowan 1977). The central importance of the new institutionalism for education is that it reestablishes and transforms new forms in the institutional reality of education in both the K–12 and higher education arena (Meyer & Rowan 2006). The new institutionalism brings insights about markets as institutions in which governance and market forces combine in institutional affairs to produce unique pressures on educational organizations, and it is now seen central to education reform in many other nations of the world (Rowan & Miskel 1999). The new institutionalism for education focuses on the rise of greater provider pluralism, marketization, privatization, and technical accountability and tighter coupling inside educational institutions (Meyer & Rowan 2006). According to Meyer et al. (1997), there is no longer a monopoly of accountability and control by states – education providers now come from the third sector and civil society and include private, market-oriented organizations. As a result, families, entrepreneurs, voluntary organizations, and corporate ventures take a stronger role in the governance of education, and the institutional landscape changes from a monistic to a pluralistic world (Levy 2004).

1.2 *Trends in the education system since 1986*

1.2.1 *Changes in the governing principles of the provision and payment for education*

The period of renovation known in Vietnam as Doi Moi dates from the Sixth Party Congress of 1986. Vietnam's leaders stated a commitment to developing a "socialist-oriented market economy" (Beresford & McFarlane 1995). The country has a regulated market economy that operates under market rules but with a socialist orientation that emphasizes social and educational equity and the leadership role of Vietnam's Communist Party (VCP) (Bang 1973). Figure 1 shows how Vietnam's formal education system incorporates both "market" and "state" elements.

The introduction of the "fee-paying" principle is active in the policy of education investment with the emergence of self-financing students and non-state education providers (including private and foreign providers). Educational institutions in Vietnam have been establishing the processes of diversification, marketization, and privatization (Jonathan 2004, Huong & Fry 2002). The policy of fee-charging was first introduced and applied to the higher education sector in 1987 and to students in grades four and five in primary schools and all students in secondary schools in 1989. In 1993, the government decided to waive the tuition fee for primary attendance, but instead it increased the charge for secondary school students. In addition to tuition fees, local governments also charge various compulsory payments, such as a registration fee, contributions to school maintenance and renovation funds, contributions to school purchasing, and so on (Jonathan 2004).

The distinguishing feature of education governance during the market transition is the state's withdrawal from its commitment, and Vietnam's state now provides a floor of basic education, principally by providing free access to primary education and by providing varying degrees of subsidies to secondary education. The adoption of state policy toward education under the market regime has shifted educational costs from the state onto consumers, and the state labeled these cost-shifting policies the "socialization" of education (*xã hội hóa*), which can be understood here as the broad-based social mobilization of resources to make contributions to national education (Jonathan 2004). Under the policy

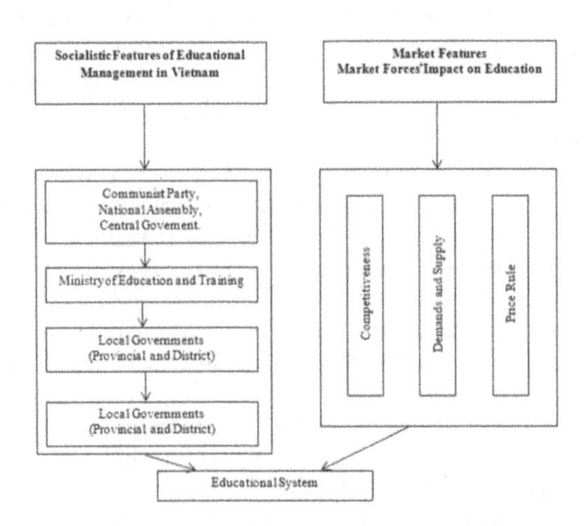

Figure 1. Socialist and market-oriented features of education in Vietnam (Tran 2011).

of "socialization" orientation, the state undertook and implemented various measures to initiate and expand formal "cost recovery" schemes in education (Jonathan 2004).

In 1989, the Council of Ministries issued Decree 44/HDBT to charge tuition fees to students in grade school. Students in first, second, and third grades were exempt because these grades were designed for the program to universalize primary education. In the public school system, decentralization in the administration and finance of public school led to differences in school quality across regions and in the level of economic development of the communities. All students are required to pay for tuition except at primary level. Aside from the tuition, families are responsible for all school-related fees, such as textbooks, transportation, parental association, school remodeling, and classroom equipment (Glewwe & Parinos 1998). New educational policies in this reform period caused an obstacle in schooling for children of low-income families. However, after the shocks of the new economic opportunities in the market economy, it seems that "families found out that education is the best strategy of human resource investment and the school enrollment rate started to recover from 1992 [levels]" (Lan 2005).

By 1993, expanding cost recovery policy through tuition fees and other charges was principally executed in lower and upper secondary and higher education. According to survey data on household education expenditure, by 1996–1997 school fees accounted for 46.1 percent and 61.7 percent of yearly education expenditures per lower and upper secondary students respectively, with other education expenditures including books, transport, and extra after-school study (General Statistics Office 1999). The characterization of socialization offered opportunities for private provision. Vietnam's state encouraged and permitted the development of private provisioning of education (Fforde 2019).

After more than 30 years of implementation from 1986 to present, socialization of education in Vietnam has been widely expanded throughout the country and gained certain achievements. The national education system has become increasingly developed, with diversified forms from public to non-public educational institutions. Among remote and isolated ethnic minorities – disadvantaged areas given attention at all levels, including the government, social organizations, businesses, and international organizations – enrolment has increased. According to Article 47 of the Vietnam Education Law of 2019, the objectives of socialization of education opened new disciplines, including not-for-profit schools; stakeholders of not-for-profit private schools committed not to withdraw capital or enjoy income, and the annual accumulated profits are continually reinvested, with stakeholders pouring all money back into the school. The increasing trend of private schools in the strategy of implementing various types of educational institutions in Vietnam in the school year 2016 to 2018 can be seen in Figure 2.

The new institutionalism in organizational analysis takes as a starting point criticism of the striking hegemony of practices and arrangements found in schools, states, corporations, and the labor market (DiMaggio & Powel 1983, Meyer & Rowan 1977). The new institutionalism for education focuses on the rise of greater provider pluralism, marketization, privatization, and technical accountability and tighter coupling inside educational institutions (Meyer & Rowan 2006). The expanding of private education's growth and function strengthens connections between education providers and private enterprises; it helps to equip students with skills needed for successful employment, thereby creating additional potential for development (World Bank 2016).

In the new institutionalism theory, the choice and standards movements are concentrated in the form of "consumer reports" that advertise school-by-school rankings according to standardized test scores and curriculum, giving to parents information needed to choose wisely from a menu of local schools (Meyer 1977, Meyer & Rowan 1977, Meyer & Rowan 1978, Meyer & Scott 1983). Schools collect funds directly from fee-paying parents and survive only by attracting clients. The free market, competitiveness, and entrepreneurialism improve educational performance and may force providers to make dramatic education changes to keep parents satisfied (Manhattan Institute 2005).

The rise of the education market in Vietnam since 1986 has seen funding power shift from central bodies to parents (Petersen 1993). Whereas public funding arrangements encourage

School	School Year 2016-2017			School Year 2017-2018		
	Total	Public	Private	Total	Public	Private
Elementary school	15,052	14,939	113	14,937	14,695	242
Elementary school classes	277,526	274,737	2,789	279,974	272,907	7,067
Elementary school pupils	7,801,560	7,733,318	68,242	8,041,842	7,882,145	159,697
Junior high school	10,155	10,124	31	10,091	10,068	23
Combined elementary-junior high school	773	749	24	848	819	29
Junior high school classes	151,669	149,622	2,047	153,582	151,380	2,202
Junior high school pupils	5,235,524	5,178,829	56,696	5,373,312	5,312,715	60,597
Senior high school	2,391	2,110	281	2,398	2,114	284
Combined junior-senior high school	420	266	154	436	279	157
Senior high school classes	65,094	59,963	5,131	65,806	60,455	5,351
Senior high school pupils	2,477,175	2,290,929	186,246	2,508,564	2,313,315	195,249
College (University)	235	170	65	235	170	65
Undergraduate students	418.991	348.832	70.159	437.156	352.982	84.174
Postgraduate students	44.469	42.707	1.762	48.106	41.908	6.198

Source: General Statistic Office of Vietnam Ministry of Education and Training

Figure 2. The increasing trend of private schools in the strategy of implementing various types of educational institutions in Vietnam in the school year 2016 to 2018.

schools to conform to legal conventions rather than providing effective service (Chubb & Moe 1990), market reforms release primary schools from the grip of central planning and create competitive pressures similar to for-profit firms. Many-choice arrangements in education markets weaken hierarchical regulations and force schools to survive by collecting funds directly from fee-paying clients (Hoxby 2003, Davies et al. 2006). Market conditions become widespread in education, and it is no longer so that schools are loosely coupled and governed by ceremonial conformity: the fusion of markets and test scores makes school environments more "technical" and less "institutional" (Rowan 1990).

1.2.2 *Market-oriented education and the role of private providers in education*

In 1992, the Constitution of Vietnam did not mention the state's monopoly on education. Article 41 of the 1980 Constitution was replaced by Article 36 in the new Constitution. This article confirms that the "state undertakes the overall management of the national system of education with regard to the objectives, intents, plans, the standards required of teachers, the regulations governing examination and competitions and the system of diplomas and certificates" (Dien 2004). Vietnam's increasing engagement with private-sector development has enabled greater flexibility for improving and strengthening its education system.

In September 1995, the government prepared and presented reports on the progress and future direction of education at a major donor meeting. Before 1990, Vietnam had relied on aid from the communist bloc led by the USSR (Union of Soviet Socialist Republics), while a limited number of donors such as UNICEF (United Nations Educational, Scientific and Cultural Organization) and the UNDP (United Nations Development Program) were active in aid delivery to Vietnam. After 1994, because of the normalization of diplomatic relationship between the USA and Vietnam, Western aid became to come in. The policy of the government was aided through the provision of several major loan projects provided by the World Bank and the Asian Development Bank. The World Bank had been supporting reform in primary schools within the World Bank Primary Education Project in 1995 and ongoing supporting in higher education (Hirosato & Kitamura 2009).

On education development, the Vietnam government used the 1990 World Declaration on Education for All (EFA) and Plan of Action at Jomtien, Thailand, as a framework for attracting education aid. The Plan of Action had six target areas: (1) early childhood care and education; (2) universalization of primary education; (3) basic education for youth and adults; (4) literacy, numeracy, and other life skills; (5) equitable access and achievement; and (6) quality and learning outcome (Hirosato & Kitamura 2009). From 2003 to 2015, Vietnam participated in the Forum of the New EFA National Plan of Action (EFA 2015). Vietnam began pursuing the twin education goals of expanding basic education and post-secondary education simultaneously. Both goals were pursued under the intertwined policy of "decentralization" and "socialization." Vietnam has a long-standing tradition of educational cost-sharing in the central government to pay for teacher salaries. Vietnam began to shift cost-sharing of school facilities and other costs to other stakeholders (parents, local resident, local corporations, etc.) –the so-called community contribution. The 1998 Law on Education stipulated "socialization of education activities" (Tan & Mingat 1992).

The EFA's 2015 major policy goals included shifting half-a-day schooling to full-day schooling in order to meet international standards. To do so, it was necessary to increase the number of classrooms and total spending on teacher salaries. However, due to the shortage of funds at the central government, there has been a strong trend toward promoting socialization (society supporting education) in the name of decentralization. In positive terms, this means diversifying funding sources, but in negative terms it imposes a financial burden on the "community" (Hirosato & Kitamura 2009, World Bank 2003).

In the introductory chapter to *Institutional Analysis and the Study of Education*, Meyer & Rowan (2006) outline three key changes in the field of education from the perspective of the new institutionalism, mainly emphasizing institutional changes in both the K–12 and the higher education arenas:

- Greater provider pluralism: Rapid growth in the private provision of educational services has dramatically altered basic schooling, and much of higher education, making it no longer a monopoly of government. Education providers now come from the third sector, including families, voluntary organization, and civil society, including private, market-oriented organizations.
- More tight coupling: The widespread need for more accountability has led to more tightly coupled and narrowly controlled practices in organizations that were previously exemplars of loose coupling.
- More central role for educational institutions in society: Schools and colleges take on a more central role in society's institutional web, and their performance has definite repercussions throughout society. Families, entrepreneurs, voluntary organizations, and corporate enterprises take a stronger role in the governance of education and the institutional landscape changes from a monistic to a pluralistic world.

According to the new institutionalism, there is no longer a monopoly of accountability and control by states, and education providers now come from the third sector: civil society, market-oriented organizations (Levy 2004). Institutional arrangements are the relationships and interactions among learners, parents, economic units, and social units. Schools and colleges take on a more central role in society's institutional fabric, and their performance has definite repercussions throughout society, families, entrepreneurs, and voluntary organizations, all of which take on a stronger role in the governance of education (Meyer & Rowan 2006).

In the new institutionalism theory, the choice and standard movements concentrate on "consumer reports" that advertise school-by-school ranking on standardized test scores and curriculum, giving to parents the information necessary to choose wisely from a menu of local schools. Schools collect funds directly from fee-paying parents and survive only by attracting clients. Free market, competitiveness, and entrepreneurialism improve educational performance and make dramatic education changes to keep parents satisfied.

2 CONCLUSION

The period of renovation known in Vietnam as Doi Moi dates from the Sixth Party Congress of 1986. Vietnam's leaders stated a commitment to develop a socialist-oriented market economy. The country has a regulated market economy that operates under market rules but with a socialist orientation that emphasized social and educational equity and the leadership role of Vietnam's Communist Party (VCP). The state began to allow the emergence of "non-state" provisions of social and educational services. In educational provision and reforms in the educational structure, Vietnam has made education a mixed economy of private and public consumption.

Educational institutions in Vietnam have been establishing processes of diversification, marketization, and privatization. The adoption of state policy toward education under the market regime has shifted educational costs from the state onto consumers, and the state labeled these cost-shifting policies the "socialization" of education. Under this policy, the state implemented various measures to initiate and expand formal "cost recovery" schemes in education. The characterization of socialization offered opportunities for private provision.

From the new institutionalism perspective, there is no longer a monopoly of accountability and control by the state: education providers now come from the third sector, including private, market-oriented organizations and civil society. Families, entrepreneurs, voluntary organizations, and corporate ventures play a stronger role in the governance of education, and the institutional landscape changes from a monistic to a pluralistic world. The change in educational organization introduces a new element of competition and forcies established institutions to become more market-oriented and entrepreneurial. The expanding of private education strengthens connections between education providers and private enterprises; it helps to equip students with skills needed for successful employment and creates an additional potential for development of human resource.

REFERENCES

Bang, V. 1973. *Aspects of Vietnamese History*. Honolulu: The University Press of Hawaii.

Beresford, M., & McFarlane, B. 1995. Regional inequality and regionalism in Vietnam and China. *Journal of Contemporary Asia* 25: 50–72.

Brooks, D. H. 2004. *Social sector issues in transitional economies of Asia*. Mandaluyong: Asian Development Bank.

Chubb, J. & Moe, T. 1990. *Politics, Markets and American Schools*. Washington, DC: Brookings Institute.

Coase, R. 1937. The nature of the firm. *Economica* 4: 386–405.

Dang Q. B. 1999. *Education and Its Development*. Hanoi: Information Press.

Davies, S., Quirke, L., & Aurini, J. 2006. The new institutionalism goes to the market: The challenge of rapid growth in private K–12 education. In H.-D. Meyer and B. Rowan (eds.), *New Institutionalism in Education*: 103–122. Albany, NY: State University of New York Press.

Dien, N. T. 2004. *Education in the Socialist Republic of Vietnam: 1976–2002*. London, ON: Faculty of Graduate Studies, University of Western Ontario.

DiMaggio, P. J., & Powell, W. W. 1983. The iron cage revisited: Isomorphism and collective rationality in organizational fields. *American Sociological Review* 48: 147–160.

Fforde, A. 2019. *From plan to market: The economic transition in Vietnam*. London: Routledge.

Furubotn, G. E., & Richter, R. 2005. *Institutions and Economic Theory: The Contribution of the New Institutional Economics*. Ann Arbor: University of Michigan Press.

General Statistics Office. 1999. *Vietnam Living Standards Survey, 1992–1993*. Hanoi: Statistical Publishing House.

Glewwe, P., & Parinos, H. A. 1998. *The Role of the Private Sector in Education in Vietnam: Evidence from the Vietnam Living Standards Survey*. Washington, DC: World Bank.

Hamilton, G. G. 2006. *Civilizations and the Organization of Economies*. London: Routledge.

Hirosato, Y., & Kitamura, Y. 2009. *The Political Economy of Educational Reforms and Capacity Development in Southeast Asia: Cases of Cambodia, Laos and Vietnam (Vol. 13)*. Berlin: Springer Science & Business Media.

Hoxby, C.M. 2003. *The Economics of School Choice*. Chicago, IL: University of Chicago Press.
Huan, V. D. 2013. *A New Approach to Explain Policy Reforms in Vietnam during Đổi Mới by Developing and Validating a Major Policy Change Model for Vietnam*. Portland, OR: Department of Public Affairs and Policy, Portland State University.
Huong, P. L. 2004. Education and economic, political, and social change in Vietnam. *Education Research for Policy and Practice* 3: 199–222.
Huong, P. L., & Fry, G. W. 2002. The emergence of private higher education in Vietnam: Challenges and opportunities. *Educational Research for Policy and Practice* 1(1–2): 127–141.
Jonathan, D. L. 2004. Rethinking Vietnam's mass education and health systems. In D. McCargo (ed.), *Rethinking Vietnam*: 127–142. London: Routledge Curzon.
Lan, P. N. 2005. *Educational stratification in Vietnam during the evolution and devolution of Socialism*. Manchester: University of Manchester.
Levy, C. N. 2004. The new institutionalism: Mismatches with private higher education's global growth. *Prophie Working Paper Series* 3: 1–34.
Manhattan Institute. 2005. *Manhattan Institute for Policy Research*. [Online]. http://www.manhattan-institute.org
Mary, C. B., & Victor, N. 1998. *The New Institutionalism in Sociology*. New York, NY: Russell Sage Foundation.
Maurice, M., & Paul, S. 1998. *Social Policy in a Changing Society*. New York: Routledge.
Meyer, D. H., & Rowan, B. 1977. Institutionalized organizations: Formal structure as myth and ceremony. *American Journal of Sociology* 83: 340–363.
Meyer, D. H., & Rowan, B. 1978. *The structure of educational organizations*. Lowa: University of Iowa.
Meyer, D. H., & Rowan, B. 2006. *The New Institutionalism in Education*. Albany, NY: State University of New York Press.
Meyer, J. W. 1977. The effects of education as an institution. *American Journal of Sociology* 83(1): 55–77.
Meyer, J. W., & Scott, R. W. 1983. *Organizational Environments: Ritual and Rationality*. Beverly Hills, CA: Sage.
North, D. C. 1981. *Structure and Change in Economic History*. New York, NY: Norton.
North, D. C. 1989. Institutions and economic growth: An historical introduction. *World development*, 17 (9): 1319–1332.
Petersen, T. 1993. Recent development in the economics of organization: The principle agent relationship. *Acta Sociologica* 36: 277–293.
Rowan, B., & Miskel, G. C. 1999. Institutional theory and the study of educational organizations. *Handbook of Research on Educational Administration* 2: 359–383.
Rowan, B. 1990. Commitment and control: Alternative strategies for the organizational design of schools. *Review of Research in Education* 16: 359–389.
Shepsle, K. A. 1986. Institutional equilibrium and equilibrium institutions. *Political science: The Science of Politics* 51: 51.
Tan, J. P., & Mingat, A. 1992. *Education in Asia: A Comparative Study of Cost and Financing*. Washington, DC: World Bank.
Tran, T. B. L. 2011. *Market-oriented education: Private (people-founded) upper-secondary schools in Hanoi*. Heng Mui Keng Terrace: ISEAS–Yusof Ishak Institute
Tran-Nam, B. 2005. Vietnam: Education law and economic growth. *Asian Analysis, July 2005*. [Online] http://www.aseanfocus.com/asiananalysis/article.cfm?articleID=863
Wells, A. S., & Oakes, J. 1998. Tracking, detracking, and the politics of educational reform: A sociological perspective. In Torres, C. A., & Mitchell, T. R. (Eds.). 1998. *Sociology of education: Emerging perspectives*. Albany: SUNY Press.
World Bank. 2003. Project Appraisal Document, Vietnam, Primary Education for Disadvantaged Children Project. Washington, DC: World Bank.
World Bank. 2016. Vietnam 2035: Toward Prosperity, Creativity, Equity, and Democracy. Washington, DC: World Bank.

Borderless Education as a Challenge in the 5.0 Society – Abdullah, Adriany & Abdullah (eds)
 ISBN 978-0-367-61960-2

Digital immigrants versus digital natives: A systematic literature review of the "ideal teacher" in a disruptive era

H. Helaluddin
Islamic State University of Sultan Maluna Hasanuddin Banten, Serang, Indonesia

H. Wijaya
Sekolah Tinggi Filsafat Jaffray Makassar, Makassar, Indonesia

M. Guntur
Institut Agama Islam Negeri (IAIN) Palopo, Palopo, Indonesia

Z. Zulfah
Sekolah Tinggi Agama Islam (STAI), Darud Dakwah Wal-Irsyad, Maros, Indonesia

S. Syawal
East Indonesian of University of Makassar, Makassar, Indonesia

ABSTRACT: Rapid technological development and the unique characteristics of digital native students can make teaching difficult, especially when teachers are often digital immigrants themselves. This study uses a systematic literature review (SLR) method to synthesize articles from several reputable international journals published from 2013 to 2019. The results of the analysis show that teachers of digital natives must possess several attributes/characteristics, including not keeping distance from students; being open to innovation and the development of the times; being able to make learning exciting; being able to assist each student; and being able to understand the needs of students.

1 INTRODUCTION

The presence of the industrial revolution (IR) 4.0 – marked by massive development of information technology – has influenced patterns of life worldwide. Digital technology has revolutionized all human activities, making social interaction highly dependent on it (Garba et al. 2015). These developments have also impacted the field of education and learning. In the past few decades, school learning has been based on "3R" teaching – reading, writing, and arithmetic, plus subjects such as social sciences and languages (Alismail & Mcguire 2015). In contrast, learning in the modern classroom is more dynamic, prioritizing a student-centered learning approach. Education in the era of IR 4.0 is no longer a process of transferring knowledge but is more directed at developing the experiences and talents of each student. In addition to transferring knowledge and developing skills, education should teach qualities such as how to be a good global citizen, as well as active, independent, self-confident, caring, and competent in cognitive, emotional, social, and technological aspects (Bautista & Ortega-ruiz 2015).

In recent years, however, there has been widespread concern that, in areas where this new model of education has not yet been adopted, the education system is failing to introduce skills needed by students for their future. Some experts argue that curriculum content, classroom practices, and learning environments continue to develop but that attention to teacher disposition is still minimal (Faulkner & Latham 2016).

Problematic in this regadr is the wide gap between teachers as digital immigrants and students as digital natives (Prensky 2001). The term "digital natives" refers to the idea of the "net generation" (Tapscott 1998) and to the group known as millennials (Howe & Strauss 2000). Several studies have looked at digital natives in the educational environment, especially in universities (Smith 2012). The difference between the younger an older generations (beween students and teachers, in other words) causes problems in the current education process. Digital immigrants and digital natives appear to have different characteristics, and this place them at odds with one another in an educational context (Uygarer et al. 2016). Generation Z and the alpha generation – residents of the digital natives group – learn differently from their digital-immigrant teachers; the mistake of some digital-immigrant teachers is to assume that students today are the same as those of previous generations and therefore that the same teaching and learning methods should remain active in effect (Prensky 2001). As a result, Poole argued that teachers must always be ready to embrace paradigm shifts in the world of education, especially in the use of technology and online activities, to keep up with students' educational needs (Maphosa & Mashau 2014).

1.1 *Digital immigrants vs digital natives*

The terms "digital immigrants" and "digital natives" have emerged in recent years in response to the rapid development of information and communication technology. Digital natives are those born and raised in an environment in which digital products and all-day online activities are standard (Bilgic et al. 2016). Digital natives have been given many nicknames, including students of a thousand years, the internet generation, cyber children, zapping humans, and grasshopper minds (Uygarer et al. 2016). They lived through such society-changing events as the birth of iPod devices (2001), the presence of the iTunes music store (2003), the launches of Facebook (2004) and YouTube (2005), cloud computing service (2006), and the development of smartphones (2007) and tablet devices such as the iPad (2010). In keeping with these experiences, digital natives share certain learning characteristics. They tend to enjoy collaborative activities or groupwork, are more interested in achievement-based tasks, learn in dynamic and interactive environments, and require rapid feedback (Coffman et al. 2007). On the other hand, they often do not spend time reading and reflecting on content, resulting in a weak ability to think critically.

Digital immigrants, in contrast, are members of the generation who encountered such technology when they were already adults, requiring them to adjust to its presence (the term "immigrant" refers to people adapting to a new environment). Although some digital immigrants have been able to master new technology, they tend to retain certain "digital immigrant accents" – behaviors such as printing/printing out emails; preferring to print documents to correct them rather than editing them directly on a screen; showing a website to students rather than sending them the URLs; and calling someone to confirm they have received a sent email (Helsper & Enyon 2009).

For the digital immigrant generation, learning new technologies is like learning to speak a new language they do not fully comprehend. In other words, immigrant digital teachers can only speak in outdated tongues when teaching students from digital natives, who do understand this new digital language (Prensky 2001). Teachers, as digital immigrants, must continue to learn better ways to remain effective in this disruptive era.

1.2 *Ideal teacher*

For the majority of teachers, many of whom are digital immigrants, many aspects and competencies must be improved in order to bridge the differences in characteristics and improve the learning process for the digital-native generation.

The term "ideal" refers to a predetermined standard and is related to perfection, beauty, and excellence. According to Schonwetter, the ideal teacher qualifications must be measured according to specific rules made locally (Sezer 2018). Asad & Hassan argue that, besides requiring academic qualifications, a teacher must also have other positive aspects to be suitable for the job (2013). Still, there are some general requirements that apply everywhere.

The concept of the ideal teacher has been examined in previous studies, with criteria ranging from teacher characteristics to teacher attitudes toward student achievement and motivation (Omar et al. 2014). Attributes of the ideal teacher include professional, personal, social, leadership, and interactive in-class attributes (Okoro & Chukwudi 2011). Other characteristics, not so different but stated in different terms, include: (1) professional roles and responsibilities, (2) professional values, (3) professional characteristics, (4) professional ethical principles, and (5) social responsibility (Tunca et al. 2015). Another study lists criteria such as substantive knowledge, pedagogical expertise, work-life skills, and developmental abilities (Määttä et al. 2015). Yet another study groups ideal teacher charactersitics into five aspects: (1) knowledge, intelligence, and ability in expertise; (2) professionalism; (3) ability to communicate; (4) openness; and (5) being nurturing and supportive (Zhang et al. 2015).

Across the literature, however, ideal teacher attributes include is interaction with students and management of learning processes specifically geared toward digital-native students . Journal articles were therefore analyzed in this literature review according to the following questions:

- What is the distribution of articles relating to ideal teachers in the millennial era, in terms of journal and method used?
- From the analysis in the findings section, what are the criteria or characteristics of an ideal teacher?
- In the discussion section, what aspects are listed for an ideal teacher seeking to better manage the learning process for digital natives?

2 RESEARCH METHODS

This study uses the Systematic Literature Review (SLR) method. This method is an approach adopted in recent studies which are widely published in high-quality scientific journals on various research topics (Danese et al. 2018). The purpose of this systematic review is to collect, report, summarize, and analyze studies that are relevant to the interests of researchers and that are not related to the specific characteristics of the research location or the background of the scientific discipline (Tan et al. 2016). SLRs are used to identify, evaluate, and interpret existing data in a specific time period and research area (Ramirez & Garcia-Penalvo 2018, Kasperiuniene & Zydziunaite 2019).

The articles analyzed in this study are international journal articles obtained through electronic databases such as Google Scholar, ScienceDirect, Researchgate, and Scopus by using the keyword "ideal teachers." Some of the criteria established for selecting articles in this study are: articles must be published in reputed international journals (indexed in Scopus or Copernicus); articles must be presented in English; journals must be open-access; articles must relate to ideal teacher attributes in managing learning for students; and articles must be published between 2013 and 2019.

This study uses several stages: identification; screening; eligibility; and inclusion (Gough et al. 2017). The initial phase, identification, was carried out by searching various articles from open-access electronic databases (2013–2019), which in this case yielded 42 articles. Next, the screening phase involved dividing those 42 into two groups: 20 journal articles and 22 non-journal articles. Third, in the eligibility phase, we analyzed abstracts and keywords for the 20 journal articles and from these chose 13 articles to carry to the next stage. Finally, the inclusion phase involved analyzing these 13 articles based on the indexation of the journal. This step resulted in six Scopus/Copernicus articles and seven articles from non-reputable journals. The six selected articles were then assessed in terms of the characteristics discussed that would be ideal for teachers of digital-native students.

3 RESULTS AND DISCUSSION

3.1 *Classification of articles based on year of publication, type of journal, and method used*

After screening articles obtained through several electronic database sources, the next step is to select articles using several predetermined criteria. Following screening, six articles met requirements for further analysis; these can be seen in Table 1.

Based on the research questions, the journal articles were then grouped based on the indexation of each journal. The indexation was used as a marker that articles have been published in reputed international journals, as shown in Table 2.

The articles were analyzed to determine the research methods used: three used the phenomenology method in qualitative studies (A1, A2, and A3); A4 used narrative approaches; A5 used descriptive studies; and A6 used a case studies approach.

3.2 *Attributes/characteristics of the ideal teacher*

In answering the second research question, the six articles were analyzed to determine suggested attributes or characteristics of the ideal teacher; the findings are listed in Table 3.

Table 1. Journal articles analyzed.

No	Article title	Author(s) name	Journal name
A1.	Qualities of ideal teacher educators	Nihal Tunca, Senar Alkin Sahin, Aytunga Oguz, and Halime Ozge Bahar Guner	*Turkish Online Journal of Qualitative Inquiry*
A2.	Identifying the qualities of an ideal teacher in line with the opinions of teacher candidates	Hatice Kandioglu Ates and Serkan Kandioglu	*European Journal of Educational Research*
A3.	Prospective teachers' cognitive construct concerning ideal teacher qualification: A phenome-nological analysis based on repertory grid technique	Ishak Kozikoglu	*International Journal of Instruction*
A4.	Ideal teacherhood in vocational education	Kaarina Maatta, Ane Koski-Heikkinen, and Satu Uusiautti	*British Journal of Education, Society, & Behavioural Science*
A5.	The ideal psychology teacher: Qualitative analysis of views from Brunei GCE A-level students and trainee psychology teachers	Nurul Azureen Omar, Sri Ridhwanah Matarsat, et al.	*Asian Social Science*
A6.	Evaluating views of teacher trainees on teacher training process in Turkey	Abdullah Oguz Kildan, Bilgin Unal Ibret, Murat Pektas, Duran Aydinozu, and Lutfi Incikabi	*Australian Journal of Teacher Education*

Table 2. Distribution of articles in the year of publication and research methods.

No	Article	Year published	Research method	Indexing by
1.	A1	2015	Phenomenology	Copernicus
2.	A2	2017	Phenomenology	Scopus
3.	A3	2017	Phenomenology	Scopus
4.	A4	2015	Narrative research	Copernicus
5.	A5	2014	Descriptive study	Scopus
6.	A6	2013	Case studies	Scopus

Table 3. Ideal teacher attributes in the disruptive era.

Article	Attribute of the ideal teacher
A1	(a) Professional roles, (b) professional values, (c) characteristics, (d) professional ethics principles, (e) social responsibility
A2	(a) Personal qualities, (b) professional competencies related to teaching and area of expertise
A3	(a) Communication skills, (b) student-centered, (c) sensitivity, (d) innovative, (e) sense of humanity and full of happiness, (f) teaching pedagogical skills, (g) leadership, (h) content knowledge professional, (i) personal value, (j) professional values
A4	(a) Experts in certain fields, (b) experts in all pedagogical aspects, (c) have extensive networks and are able to collaborate well, (d) foster equality and responsibility, (e) encourage and support students, (f) have positive basic values and the ability to act as a direction determinant
A5	Ideal teachers for students with low grades: (a) creating a living-learning environment, (b) apperception before learning, (c) providing in-depth knowledge and understanding, (d) increasing activities such as presentations and case studies, (e) giving many examples, (f) reviews the previous year's questions, (g) delivers content appropriately, (h) activities prioritize group work, (i) pleasant and friendly, (j) know students well, (k) carries out one-on-one sessions with students, (l) provides assistance to all students, (m) has respect for students' secrets Ideal teachers for students with moderate achievement: (a) teaching clearly, (b) no bias and discrimination, (c) recognizing student needs and learning styles, (d) being punctual and assertive at the time of the exam, (e) consistent, (f) easily approachable and pleasant, (g) providing real examples, (h) fluent in English, (i) following the progress of student performance, (j) being a good motivator, (k) using a fixed learning approach Ideal teachers for high-achieving students: (a) fluent in English, (b) providing real and contextual examples, (c) clear description, (d) appropriate teaching style, (e) strict class management, (f) interesting learning, (g) not biased and not discriminatory, (h) giving many examples, (i) establishing positive teacher and student relations, and (j) becoming a motivator
A6	Characteristics of an ideal twenty-first-century teacher: (a) patient, tolerant, open to constructive criticism, (b) has sufficiently mastered technology, (c) high ability to relate to others and communication skills, (d) possesses sufficient knowledge and skills, (e) curiosity and openness to innovation, (f) creative, (g) idealist, (h) high culture, (i) democratic and humanistic, (j) has exemplary character and is a good role model

Analysis of the six articles shows many lists of characteristics or attributes of ideal teachers. However, these attributes share similarities across articles: the difference comes predominantly in the the use of terms and the grouping of attributes.

3.3 *Ideal teacher attributes related to student interaction (digital natives)*

In all of the six articles, the ideal teacher attributes identified relate to the ability to interact with and serve students. In A1, one of the ideal characteristics that relates to student interaction is that of personal character. This characteristic can be divided into several sub-characteristics: neat, social, enthusiastic, humorous, well-disciplined, smiling, sincere, not selfish, self-evaluationing, unique, and courageous (Tunca et al. 2015). Several studies report that digital natives respond well to teachers who act as friends, do not maintain distance, and have good senses of humor.

A2 divides ideal teacher attributes into two main parts: personal qualities; and professional competencies related to teaching and area of expertise. Digital natives favor teachers who use modern methods and avoid traditional learning methods (Ates & Kadioglu 2017). This sub-attribute is found in professional group competencies related to teaching and area of expertise. Digital natives prefer attractive learning models, lessons presented through audio/video, teamwork, and so on.

A3 identifies ideal teacher attributes as those related to the ability to manage classes and interact with students: the attributes of student-centeredness and innovation. Student-centeredness refers to ability to attract student attention, handle student problems, and provide individual assistance (Kozikoglu 2017). Innovation refers to openness to technological developments and ability to improve with the times.

A4 describes two ideal teacher attributes: ability to master certain fields; and ability to act as a reformer (Määttä et al. 2015). An ideal teacher must continue to update his/her abilities on an ongoing basis, along with the progress of the times. As reformers, teachers must introduce innovations in learning so that students are motivated to develop their abilities.

A5 considers teacher attributes from the perspectives of students with low, moderate, and high levels of achievement. Broadly speaking, the ideal teacher attributes are identified from each perspective as those relating to presentations/case studies/teamwork, knowing the needs of students, providing assistance to each student, having an appropriate teaching style, and providing interesting learning (Omar et al. 2014). This indicates that an ideal teacher must adapt to a student's digital-native learning style and avoid using conventional methods.

Finally, A6 suggests ideal teachers must be able to master technology (Kildan et al. 2013). Teachers, most of whom are digital immigrants, must improve their skills in using technological devices, as this will help students who prefer learning with the use of technological tools. Other necessary attributes are curiosity and openness to innovation.

4 CONCLUSION

Various attributes are necessary to be considered an ideal teacher in this disruptive digital era, a period in which technology and online devices have changed the patterns of human life, including learning habits and education. This study is a systematic literature review that discusses the ideal teacher for digital-native students. Of the six articles analyzed, several necessary characteristics emerged, relating to the ability of digital-immigrant teachers to interact with and manage the learning process for digital-native students. Some of the characteristics identified include: (1) not keeping distance from students; (2) being open to innovation and development; (3) designing interesting learning opportunities; (4) providing assistance to each student; and (5) understanding the needs of students.

REFERENCES

Alismail, H. A., & Mcguire, P. 2015. 21st-century standards and curriculum: Current research and practice. *Journal of Education and Practice* 6(6): 150–155.

Asad, M. M., & bin Hassan, R. 2013. The characteristics of an ideal technical teacher in this modern era. *International Journal of Social Science and Humanities Research* 1(1): 1–6.

Ates, H. K., & Kadioglu, S. 2017. Identifying the qualities of an ideal teacher in line with the opinions of teacher candidates. *European Journal of Educational Research* 7(1): 103–111.

Bautista, A., & Ortega-ruiz, R. 2015. Teacher professional development: International perspectives and approaches. *Psychology, Society, & Education* 7(3): 240–251.

Bilgic, H. G., Dogan, D., & Seferoglu, S. S. 2016. Digital natives in online learning environments: New bottle, old wine – The design of online learning environments for today's generation. In M. Pinheiro & D. Simoes (eds.), *Handbook of Research on Engaging Digital Natives in Higher Education Settings.* Pennsylvania: IGI Global.

Coffman, T., Campbell, A., Heller, E., Horney, E. M., & Slater, L. P. 2007. The new literacy crisis: Immigrants teaching natives in the digital age. *Virginia Society for Technology in Education Journal* 21(6): 1–10.

Danese, P., Manf, V., & Romano, P. 2018. A systematic literature review on recent lean research: State-of-the-art and future directions. *International Journal of Management Review* 20: 579–605.

Faulkner, J., & Latham, G. 2016. Adventurous lives: Teacher qualities for 21st-century learning. *Australian Journal of Teacher Education* 41(4): 137–150.

Garba, S. A., Byabazaire, Y., & Busthami, A. H. 2015. Toward the use of 21st-century teaching-learning Approaches: The trend of development in Malaysian schools within the context of Asia Pacific. *International Journal of Emerging Technologies in Learning* 10(4): 72–79.

Gough, D., Oliver, S., & Thomas, J. (2017). *An Introduction to Systematic Reviews.* London: Sage.

Helsper, E., & Enyon, R. 2009. Digital natives: Where is the evidence? *British Educational Research Journal* 1–18.
Howe, N., & Strauss, W. 2000. *Millennials Rising: The Next Great Generation*. New York: Vintage.
Kasperiuniene, J., & Zydziunaite, V. 2019. A systematic literature review on professional identity construction in social media. *Sage Open* (January–March): 1–11.
Kildan, A. O., Ibret, B. U., Pektas, M., Aydinozu, D., & Incikabi, L. 2013. Evaluating views of teacher trainees on teacher training process in Turkey. *Australian Journal of Teacher Education* 38(2): 51–68.
Kozikoglu, I. 2017. Prospective teachers' cognitive constructs concerning ideal teacher qualifications: A phenomenological analysis based on repertory grid technique. *International Journal of Instruction* 10(3): 63–78.
Määttä, K., Koski-heikkinen, A., & Uusiautti, S. 2015. Ideal teacherhood in vocational education. *British Journal of Education, Society & Behavioural Science* 5(3): 276–288.
Maphosa, C., & Mashau, S. T. 2014. Examining the ideal 21st-century teacher-education curriculum. *International Journal of Educational Science* 7(2): 319–327.
Okoro, C. O., & Chukwudi, E. K. 2011. The ideal teacher and the motivated student in a changing environment. *Journal of Educational and Social Research* 1(3): 107–112.
Omar, N. A., Matarsat, S. R., Azmin, N. H., Chung, V., & Wei, A. 2014. The ideal psychology teacher: Qualitative analysis of views from Brunei GCE A-level students and trainee psychology teachers. *Asian Social Science* 10(12): 184–194.
Prensky, B. M. 2001. Digital natives, digital immigrants. *On the Horizon* 9(5): 1–6.
Ramirez, M. S., & Garcia-Penalvo, F. J. 2018. Co-creation and open innovation: Systematic literature review. *Comunicar* XXVI(54): 9–18.
Sezer, S. 2018. Prospective teachers' cognitive constructs related to ideal lecturer ualifications: A case study based on repertory grid technique. *Akdeniz Egitim Arastirmalari Dergisi* 12(25): 255–273.
Smith, E. E. 2012. The digital natives debate in higher education: A comparative analysis of recent literature. *Canadian Journal of Learning and Technology* 38(3): 1–18.
Tan, A. H. T., Muskat, B., & Zehrer, A. 2016. A systematic review of quality of student experience in higher education. *International Journal of Quality and Service Sciences* 8(2): 209–228.
Tapscott, D. 1998. *Grown up Digital: How the Next Generation Is Changing Your World*. New York: McGraw-Hill.
Tunca, N., Sahin, S. A., Oguz, A., & Guner, H. O. B. 2015. Qualities of ideal teacher educators. *Turkish Online Journal of Qualitative Inquiry* 6(2): 122–148.
Uygarer, R., Uzunboylu, H., & Ozdamli, F. 2016. A piece of qualitative study about digital natives. *Anthropologist* 24(2): 623–629.
Zhang, S., Fike, D., & Dejesus, G. 2015. Qualities university students seek in a teacher. *Journal Of Economic Education Research* 16(1): 42–54.

Borderless Education as a Challenge in the 5.0 Society – Abdullah, Adriany & Abdullah (eds)
 ISBN 978-0-367-61960-2

Character education development through ethnomathematics-based mathematics learning

T. Purniati & T. Turmudi
Universitas Pendidikan Indonesia, Bandung, Indonesia

M. Evayanti
SMP Laboratorium Percontohan UPI, Bandung, Indonesia

D. Suhaedi
Universitas Islam Bandung, Bandung, Indonesia

ABSTRACT: Character education needs to be developed in all subjects, including mathematics. Ethnomathematics is a program that presents mathematical concepts related to the cultural background of students. This study aims to assess attitudes and characteristics of students after lessons based on ethnomathematics. Data were obtained through a questionnaire given to 25 Grade 8 junior high school students (12 female, 13 male). Results showed that, in general, students behaved positively toward ethnomathematics and that, by learning mathematics based on ethnomathematics, they improved their love of Indonesian culture as well. By learning mathematics based on ethnomathematics, character education can therefore be developed.

1 INTRODUCTION

The development of character is one of the priorities of Indonesian national development. The National Long-Term Development Plan 2005–2025 states that "the realization of a strong, competitive, noble, and moral character is based on the Pancasila, which details the range of Indonesian people's character and behavior, under God Almighty: virtuous, tolerant, with mutual cooperation, patriotic, dynamic, and science- and technology-oriented" (Dianti 2014).

Education has two goals: to help people to be smarter; and to help them become better human beings. Making people smarter is easier than helping them become better people (Sudrajat 2011). Culture occupies a very important place in people's lives, influencing economic, social, religious, and educational activities. Education, in general, is influenced by cultural values (Fouze & Amit 2017). Therefore, education and culture are as one, and they cannot be considered separately.

Character education in schools involves all components of education: curriculum content, learning and assessment processes, handling or managing subjects, school management, implementing activities or cocurricular activities, empowering infrastructure, financing, and the work ethic of all school residents (Suyitno 2012).

Character education used to be limited to two classes: religious education and citizenship education. The two classes tend to provide knowledge about values in their subject matter, but the learning activities in general do not adequately encourage the internalization of values by each student to help them behave with strong characters. It is therefore not enough to leave character formation up to these two subject areas only (Khusniati 2012).

Student character education therefore needs to involve all subject areas, including mathematics. This can be done by incorporating the values of character education into the learning of mathematics, through both material and the learning process. By this it is hoped that the values of good character can be embedded in students.

1.1 *Character education*

Character is innate, encompassing such areas as heart, soul, personality, behavior, personality, nature, and temperament (Suyitno 2012). Good character relates to knowing what is good, loving what is good, and doing what is good, and the three are closely related (Sudrajat 2011). Thus, character can be described as having good personal qualities, knowing better, wanting to do good, and manifesting good behavior: it is the result of aspects related to the heart, the mind, sports and kinesthetic activities, and tastes and intentions (Hartoyo 2015).

Character education is a deliberate effort to teach virtue. Virtues are good for individuals (they help a person live a good life) and they are good for the whole human community (they allow us to live together harmoniously and productively). Justice, honesty, and patience are some examples of virtues. Virtue is an objective moral standard that transcends individual time, culture, and choice (Lickona 1997).

Character education in Indonesia refers to the teaching of noble values derived from the culture of the Indonesian people themselves (Julaiha 2014) and is imbued with the five principles of Pancasila (Khusniati 2012). Mutual cooperation is especially prized among Indonesian people (Bowen 1995, Davidson & Henley 2007, Irawanto 2015): the Javanese term *ngotong* refers to a group of people who lift or carry objects together, showing the importance of cooperative and reciprocal relations in Java (Beard 2005). Mutual cooperation helps unite the Indonesian people despite their different religions and cultures (Nisvilyah 2013). For the development of character-based education and national culture, input is needed concerning, among other things, models of national character and culture development as an inseparable part of the national education system (Suyitno 2012).

There are three reasons why schools must be involved in character education: (1) we need good character values to be fully human, such as honesty, empathy, caring, perseverance, self-discipline, and moral courage; (2) school is conducive to teaching the values of good character; (3) it is important to build a moral society, because it is very clear that people throughout the world suffer social and moral problems, such as the destruction of the family, violence, materialism, dishonesty, rudeness in everyday life, and so on (Lickona 1996).

The formation of a strong national character is important so that every student can face the challenges of life. The character education movement must be a proactive and deliberate effort by parents, schools, and governments to instill in students ethics and values such as caring, honesty, perseverance, creativity, innovation, fairness, fortitude, responsibility, and respect of one's self and others. Character education teaches students how to be the best version of themselves and how to do their best work possible (Hartoyo 2015).

Through education, students should be directed to excel in multiple competencies so that they can grow and develop into community members who are able to solve their life problems. In this case, the key is learning activities in schools (Suyitno 2012). Learning characterized by character education means activities, inside and outside the classroom, aimed at helping students not only master material but also recognize, realize, and internalize values and improve behavior (Julaiha 2014).

To develop education based on national character and culture, educational development strategies need to conceive of students as whole beings, emphasizing the importance of moral aspects. Values must be taught from an early age and continue through life. Educational programs and curriculum must be developed in an integrated manner in accordance with the sociocultural background of students, placing moral values into their spirits (Suyitno 2012).

1.2 *Ethnomathematics*

People who master mathematics will have better opportunities for their future (NCTM 2020, NRC 1989). Mathematics tends to be presented as a set of universally applicable rules and is generally considered to be free of culture, so there is a widespread view that there is no need to pay attention to the cultural diversity of students who learn it (Bishop 1994, D'Ambrosio 1985, Rosa & Orey 2011). However, the use of mathematics outside of school is often different from its use in schools (Naresh 2015). Therefore, it is necessary to consider what mathematics is and how it relates to cultural values (Rosa et al. 2016).

Mathematics is a language that describes patterns, both those in nature and those found by the human mind (De Lange 2006). The result of cultural development (Ernest 2012, Orey & Rosa 2007, Rosa & Gavarrete 2017, Waller & Flood 2016), mathematics is a part of students' daily lives (Eglash et al. 2006). In teaching mathematics, teachers should begin by exploring the mathematical knowledge students have obtained in their home environments. Concrete matters related to the daily experiences of students can be used as a source of learning. One aspect that can be developed in such learning is culture (Marsigit et al. 2018).

Ethnomathematics is a term coined by D'Ambrosio in 1985 (El-Kafafi 2011); it presents mathematical concepts associated with student culture so that connections and understanding can be increased (Rosa et al. 2016). The word itself derives from three Greek words: *ethno* (a group in a natural, social, and cultural environment), *mathema* (understanding, explaining, learning), and *techne* (way, art, technique) (D'Ambrosio 2016). Thus, ethnomathematics is a way, art, and technique for the understanding, explaining, and learning of natural, social, and cultural problems (Rosa & Orey 2013). (The suffix *tics* is used as a simplification of *techne*.)

Teaching mathematics through culture can help students to learn about reality, society, the environment, and themselves, and in this way they can master mathematics better. Ethnomathematics-based learning is a pedagogical means to achieve these goals (Rosa et al. 2016). Bishop proposes that the design of school mathematics curricula should be linked to the practice of ethnomathematics (Shirley & Palhares 2016); in this way, incorporating cultural aspects in the curriculum will have long-term benefits for students' mathematical achievements (Rosa et al. 2016).

2 RESEARCH METHODS

In this study, learning was carried out at eight meetings. The lesson topics were number patterns and Cartesian coordinates. Ethnomathematics was used, with lessons involving batik cloth, a traditional game, and traditional houses. The research subjects were 25 Grade 8 junior high school students (12 female, 13 male), with different cultural backgrounds. At the last meeting, students were given a questionnaire containing questions or statements assessing attitudes toward mathematics learning based on ethnomathematics; loving Indonesian culture; and respecting cultural differences.

3 RESULTS AND DISCUSSION

Indonesia consists of various tribes, each with different cultures. Students in the class also come from various tribes. D'Ambrosio (1985) states that everyone develops unique and different mathematical knowledge that is incorporated into his or her culture, as can be seen in ways of arranging, measuring, using numbers, combining geometric shapes, and more. Thus, mathematical thinking is influenced by cultural diversity (Rosa & Gavarrete 2017).

Mathematics, to be socially just, must be applicable to students from all different cultural backgrounds (Rosa & Gavarrete 2017, Dornoo 2015, Rosa et al. 2016). Educators (both school management and teachers) must learn about and respect the diverse cultural traditions of their students, in order to apply principles of cultural suitability (Rosa et al. 2016).

At the last meeting, students were given a questionnaire assessing their attitudes toward learning mathematics based on ethnomathematics, love of Indonesian culture, and respect for different cultures (see Table 1).

Table 1. Questionnaire results.

No	Aspect	Percentage	Explanation
1	Attitudes toward learning mathematics based on ethnomathematics	96	Positive
2	Love of Indonesian culture	96	Positive
3	Respect for different cultures	96	Positive

As shown in Table 1, the attitude of students toward learning mathematics based on ethnomathematics is positive. Students also said that they loved Indonesian culture and appreciated that learning mathematics was associated with different cultures.

This positive attitude of students toward learning mathematics based on ethnomathematics – the use of cultural concepts in mathematics – will make mathematics more relevant, meaningful, and enjoyable, increase learning motivation, and help students learn mathematics (Dornoo 2015, Furuto 2014, Fyhn et al. 2016, Mogari 2017). Ethnomathematics research is also important for teachers, in that culture-based mathematics helps them develop learning that can influence the development of students' mathematical knowledge (Rosa & Gavarrete 2017).

In addition, learning mathematics based on ethnomathematics, the students reported a love of Indonesian culture and a respect for different cultures. Ethnomathematics has the potential to help students feel accepted, become more accepting of others, and even fight against racism (Brandt & Chernoff 2015). Ethnomathematics practices in schools also encourages respect, solidarity, and cooperation among students (Rosa & Gavarrete 2017, D'Ambrosio 1998). The main purpose of ethnomathematics is to build a civilization free from cruelty, pride, intolerance, discrimination, injustice, bigotry, and hatred (D'Ambrosio 1985).

Ethnomathematics can help students learn about the cultures represented in their classrooms. Students from underrepresented groups can show contributions from their groups. This will show students that their own culture contributes to mathematical thinking and exposes students from majority cultures to others, building respect (Shirley & Palhares 2016).

It is time for ethnomathematics to be integrated into every mathematics class, in accordance with the constructivist theory, in which students build understanding and knowledge based on what they have learned before. Mathematics is practiced by every culture, and this fact needs to be reflected in the school curriculum (Brandt & Chernoff 2015). Thus, ethnomathematics can aid in character education.

4 CONCLUSION

Character education helps students face life challenges, and it therefore must be integrated into all subjects of instruction, including mathematics. This needs to happen in an integrated manner in accordance with students' cultural backgrounds – the best way to do this is through ethnomathematics, which presents mathematical concepts in ways tailored to students' cultural backgrounds. The study showed students have a generally positive attitude to mathematical learning based on ethnomathematics; it also confirmed that love of Indonesian culture and respect for different cultures can be developed as part of classroom mathematics lessons.

ACKNOWLEDGMENT

We would like to express our gratitude to Universitas Pendidikan Indonesia for granting us a research findings-based public service fund, through which we were able to write this article.

REFERENCES

Beard, V. A. 2005. Individual determinants of participation in community development in Indonesia. *Environment and Planning C: Government and Policy* 23(1): 21–39.

Bishop, A. J. 1994. Cultural conflicts in mathematics education: Developing a research agenda. *For the learning of mathematics* 14(2): 15–18.

Bowen, J. R. 1995. The forms culture takes: A state-of-the-field essay on the anthropology of Southeast Asia. *Journal of Asian Studies* 54(4): 1047–1078.

Brandt, A., & Chernoff, E. J. 2015. The importance of ethnomathematics in the math class. *Ohio Journal of School Mathematics* 71(71): 31–36.

D'Ambrosio, U. 1985. Ethnomathematics and its place in the history and pedagogy of mathematics. *For the Learning of Mathematics* 5(1): 44–48.

D'Ambrosio, U. 1998. In focus ... mathematics, history, ethnomathematics and education: A comprehensive program. *Mathematics Educator* 9(2): 34–36.

D'Ambrosio, U. 2016. An overview of the history of ethnomathematics. In M. Rosa et al. (eds.), *Current and Future Perspectives of Ethnomathematics as a Program*: 5–10. Cham, Switzerland: Springer.

D'Ambrosio, U. 2018. The program ethnomathematics: Cognitive, anthropological, historic and socio-cultural bases. *PNA* 12(4): 229–247.

Davidson, J., & Henley, D. 2007. *The Revival of Tradition in Indonesian Politics: The Deployment of Adat from Colonialism to Indigenism*. New York, NY: Routledge.

De Lange, J. 2006. Mathematical literacy for living from OECD-PISA perspective. *Tsukuba Journal of Educational Study in Mathematics* 25: 13–35.

Dianti, P. 2014. Integrasi Pendidikan Karakter dalam pembelajaran Pendidikan Kewarganegaraan untuk mengembangkan karakter siswa. *Jurnal Pendidikan Ilmu Sosial* 23(1): 58–68

Dornoo, M. 2015. Teaching mathematics education with cultural competency. *Multicultural Perspectives* 17(2): 81–86.

Eglash, R., Bennett, A., O'Donnell, C., Jennings, S., & Cintorino, M. 2006. Culturally situated design tools: Ethnocomputing from field site to classroom. *American Anthropologist* 108(2): 347–362.

El-Kafafi, S. 2011. Why is it hard to engage students? Investigating epistemological theories underlying teaching and learning mathematics. *World Journal of Science, Technology and Sustainable Development* 8(1): 41–53.

Ernest, P. 2012. What is our first philosophy in mathematics education? *For the Learning of Mathematics* 32(3): 8–14.

Fouze, A. Q., & Amit, M. 2017. On the importance of an ethnomathematical curriculum in mathematics education. *EURASIA Journal of Mathematics, Science and Technology Education* 14(2): 561–567.

Furuto, L. H. 2014. Pacific ethnomathematics: Pedagogy and practices in mathematics education. *Teaching Mathematics and its Applications: An International Journal of the IMA* 33(2): 110–121.

Fyhn, A. B., Nutti, Y. J., Nystad, K., Eira, E. J. S., & Hætta, O. E. 2016. "We had not dared to do that earlier, but now we see that it works": Creating a culturally responsive mathematics exam. *AlterNative: An International Journal of Indigenous Peoples* 12(4): 411–424.

Hartoyo, A. 2015. Pembinaan karakter dalam pembelajaran matematika. *Math Didactic: Jurnal Pendidikan Matematika* 1(1): 8–22.

Irawanto, D. W. 2015. Employee participation in decision-making: Evidence from a state-owned enterprise in Indonesia. *Management – Journal of Contemporary Management Issues* 20(1): 159–172.

Julaiha, S. 2014. Implementasi pendidikan karakter dalam pembelajaran. *Dinamika ilmu* 14(2): 226–239.

Khusniati, M. 2012. Pendidikan Karakter Melalui Pembelajaran IPA. *Jurnal Pendidikan IPA Indonesia* 1 (2): 204–210.

Lickona, T. 1996. Eleven principles of effective character education. *Journal of Moral Education* 25(1): 93–100.

Lickona, T. 1997. The teacher's role in character education. *Journal of Education* 179(2): 63–80.

Marsigit, M., Setiana, D. S., & Hardiarti, S. 2018. Pengembangan Pembelajaran Matematika Berbasis Etnomatematika. *Prosiding Seminar Nasional Pendidikan Matematika Etnomatnesia* 20–38.

Mogari, D. 2017. Using culturally relevant teaching in a co-educational mathematics class of a patriarchal community. *Educational Studies in Mathematics* 94(3): 293–307.

Naresh, N. 2015. A stone or a sculpture? It is all in your perception. *International Journal of Science and Mathematics Education* 13(6): 1567–1588.

NCTM. 2020 *Principles and Standars for School Mathematics*. Reston, VA: NCTM.

Nisvilyah, L. 2013. Toleransi antarumat beragama dalam memperkokoh persatuan dan kesatuan bangsa (studi kasus umat Islam dan Kristen Dusun Segaran Kecamatan Dlanggu Kabupaten Mojokerto). *Kajian Moral dan Kewarganegaraan* 2(1): 382–396.

NRC. 1989. *Everybody counts: A Report to the Nation on the Future of Mathematics Education*. Washington, DC: National Academy Press.
Orey, D., & Rosa, M. 2007. Cultural assertions and challenges towards pedagogical action of an ethnomathematics program. *For the Learning of Mathematics* 27(1): 10–16.
Rosa, M., & Gavarrete, M. E. 2017. An ethnomathematics overview: An introduction. In M. Rosa et al., *Ethnomathematics and its Diverse Approaches for Mathematics Education*: 3–19. Cham, Switzerland: Springer.
Rosa, M., & Orey, D. C. 2011. Ethnomathematics: The cultural aspects of mathematics. *Revista Latinoamericana de Etnomatemática: Perspectivas Socioculturales de La Educación Matemática* 4(2): 32–54.
Rosa, M., & Orey, D. C. 2013. Ethnomodeling as a research theoretical framework on ethnomathematics and mathematical modeling. *Journal of Urban Mathematics Education* 6(2): 62–80.
Rosa, M., D'Ambrosio, U., Orey, D. C., Shirley, L., Alangui, W. V., Palhares, P., & Gavarrete, M. E. 2016. *Current and Future Perspectives of Ethnomathematics as a Program*. Cham, Switzerland: Springer.
Shirley, L., & Palhares, P. 2016. *Ethnomathematics and Its Diverse Pedagogical Approaches*. Cham, Switzerland: Springer.
Sudrajat, A. 2011. Mengapa Pendidikan Karakter? *Jurnal Pendidikan Karakter* 1(1): 47–58.
Suyitno, I. 2012. Pengembangan pendidikan karakter dan budaya bangsa berwawasan kearifan lokal. *Jurnal pendidikan karakter* 2(1): 1–13.
Waller, P. P., & Flood, C. T. 2016. Mathematics as a universal language: Transcending cultural lines. *Journal for Multicultural Education* 10(3): 294–306.

Borderless Education as a Challenge in the 5.0 Society – Abdullah, Adriany & Abdullah (eds)
© 2021 Taylor & Francis Group, London, ISBN 978-0-367-61960-2

Implementation of the tournament-type active method of learning in basic electrical installation subjects

S. Islami, O. Candra, U. Usmeldi & W. Sartiva
Universitas Negeri Padang, Sumatera Barat, Indonesia

Z. Imelda
SMA Negeri 12 Padang, Sumatera Barat, Indonesia

ABSTRACT: This article aims to improve student learning outcomes in Basic Electrical Installation subjects. The active learning method of tournament-type learning was applied, and a quasi-experiment research method with a pretest and posttest design was used. The subject of research was Grade X Electrical Engineering Installation at a vocational high school. Data collection in this study used pretest and posttest in the form of objective questions, and the data obtained were analyzed using the Gain score. The results of this study indicate an increase in student learning outcomes on the subject of research: the percentage of students graduating before carrying out the active method of learning was 3 percent, compared to 73 percent after. It can be concluded that applying active-learning tournament-type methods leads to an increase in student learning outcomes.

1 INTRODUCTION

In the current era of globalization, education needs to equip students with necessary skills. One educational institution aiming to produce students with independent skills and expertise is the vocational high school (Jama 2010).

Vocational schools are formal secondary education institutions that play an important role in the development of science and technology. SMKN 5 Padang is among the vocational schools aiming to prepare students to become skilled workers in their fields. One of the departments in the school is Electrical Power Installation Engineering (Electrical Engineering). This department teaches many classes, including Basic Electrical Installation, Analyzing Electrical Circuits, Understanding the Basics of Electronics, and so on.

Basic Electrical Installation aims to help students understand the planning and installation of electrical lighting in simple buildings. This Basic Electrical Installation course consists of two parts: theory and practice. Before conducting practice, students must understand the theories related to the practice that will be carried out: this lesson consists of several basic competencies, for instance drawing a lighting installation plan. This basic competency covers standardization of lighting installation drawings and planned power recapitulation in accordance to PLN's (Perusahaan Listrik Negara, Indonesia's power company) standard power capacity. Students are required to be able to apply standardization and normalization of images and to plan power requirements in accordance with PLN's power capacity.

Basic Electrical Installation requires students to be skilled in planning and installing electrical installations. Activeness and motivation are prioritized in this learning, because students must understand the principles of installation in accordance to the General Requirements for Electrical Installation. Students must be curious about this subject: by being active and motivated, learning objectives can be achieved properly.

Students must be active in the learning process, whether through asking questions, adding information, refuting incorrect answers, or providing solutions to problems that are not solved. Students can work collaboratively. Based on the observations of Grade X Electrical Engineering Installation, however, students showed a lack of motivation to learn. In general, students do not pay attention to the teacher during the learning process. They are busy with their respective activities, such as talking to friends next to them, being noisy, disturbing friends who are seriously studying, and being busy with electronic devices. Lack of student motivation affects the learning outcomes. This is shown in Table 1, which illustrates the many grades of Electrical Installation students under 75.

The Ministry of Education and Culture (in Trianto) explains that a class is said to have finished learning (classical completeness) if 85% of students have finished learning in that class (Trianto 2009). Based on the results of the daily tests, only 50% of students have completed their studies, while the remaining 50% have not been completed. Student learning outcomes are therefore still far from being mastered, and remedial efforts are necessary to improve student grades.

One of reasons students get low grades is that the learning process is still teacher-dominated: the teacher uses lecture and question-and-answer methods, so that learning process remains centered on the teacher, and students are required to sit down, take notes on the subject matter, and listen to what is conveyed by the teacher. Most teachers only have access to blackboards as teaching aids, meaning students are less actively involved in the learning process and most lack enthusiasm.

One aspect that plays an important role in learning is how the teachers teach in the class. A teacher must be able to choose the right learning method in delivering material so they can focus students' attention on the material being taught (Candra et al. 2019, Islami & Yondri 2016, Pernanda et al. 2018). The type of learning method used makes a difference when it comes to student learning outcomes. Teachers can apply improved learning methods that challenge students to learn actively, so that students will look for information themselves, with the teacher acting as communicator, facilitator, and motivator.

One learning method thought to improve motivation to learn is the active learning method of tournament-type learning. Winarsih (2014), Nurhidayat & Rahmawati (2012), and Nurmawati (2012) state that tournament-type learning leads students to be more active. This active method challenges students to share opinions and to obtain the right answers to questions that have been given. They are required to be more active in the learning process, digging up information and building their own insights, while the teacher acts as a facilitator. This leads to a sort of learning competition, with students formed into groups, usually consisting of two to six members of differing levels of ability (Winarsih 2014, Nurhidayat & Rahmawati 2012, Nurmawati 2014).

The teacher gives the subject matter, and students work in their respective groups to further understand the material. To ensure all group members have mastered the subject matter, they will be given challenges. In the first session of the tournament, students individually work on the questions given by the teacher without any help from the group members. Next, teacher and students discuss the questions to get the right answers. After that, each student records and shares the score obtained to calculate the average value of the group. In the tournament, each group tries to be the best. The group that can solve the most problems will be rewarded, while the loser might be given a penalty. This is seen to help increase student motivation.

Table 1. Percentage of completeness of student learning outcomes in the daily test of basic electrical installation subjects.

	Percentage					
Score	Group 1	%	Group 2	%	Group 3	%
≥ 75	19	65.5	11	37.9	11	47.8
< 75	10	34.5	18	62.1	12	52.2
Total	29	100	29	100	23	100

Basic Electrical Installation requires high levels of enthusiasm and willingness to learn, as it involves not just theory (odd semesters) but also practice (even semesters). To avoid making practical mistakes, students must understand the underlying theory properly first. To stimulate student enthusiasm and willingness to learn, the teacher must implement learning scenarios that challenge and respect each student's achievements. Tournament-type learning is expected to improve student learning outcomes: everyone wants to be a winner, so they will try harder as a result. This article aims to assess the impact on student learning outcomes of the active method of tournament-type learning among Basic Electrical Installation students.

2 RESEARCH METHODS

Experimental research is used in this study, as Sukardi finds it to be the most productive research method: if the research is conducted well, it can answer hypotheses that are primarily related to causal relations (Sukardi 2008). Experimental research is research conducted to determine the causal relationship by giving treatment to one or more groups. This study aims to identify an increase in student learning outcomes among those learning with the tournament-type method.

This quasi-experimental research used the One Group pretest and posttest design. In this design, a group of subjects was used. Measurements are conducted, then the subjects are treated for a certain period of time, then the measurement is made for second time (Suryabatra 2010).

A group of subjects were treated for a certain period of time using an active method of tournament-type learning. Measurements were made twice, the first before the treatment (the pretest) and the second after (posttest). To find out whether student learning outcomes increased due to the active method of learning, the posttest results are compared with the results of the pretest (see Table 2, in which X is the active method of tournament-type learning).

Besides looking at the percentage of students completing their learning in the traditional format, to see the effectiveness of the media developed, it is necessary to calculate the improvement in student learning outcomes by using a gain score (Hake 1999), as can be seen in the formula

$$g = \frac{S_{post} - S_{pre}}{100 - S_{pre}} \% \tag{1}$$

where g is the gain score, S_{post} is the posttest score, and S_{pre} is the pretest score. The g factor category can be seen in Table 3.

Table 2. Research design.

Pretest	Treatment	Posttest
T_1	X	T_2

Table 3. Gain score category.

No.	Gain score	Category
1	$g > 0{,}70$	High
2	$0.30 \leq g \leq 0.69$	Medium
3	$g < 0.29$	Low

3 RESULTS AND DISCUSSION

In this experimental study, a pretest was conducted to determine the students' initial abilities prior to being given treatment. From the results of the pretest, the average of experimental class was obtained (55.125). From this it can be concluded that student learning outcomes did not meet the minimum completion criteria in SMKN 5 Padang, which is 75. For the next assessment activity, the experimental class was given treatment in the form of tournament-type active learning. After 8 weeks of the experimental class, the final test was given. The questions given to the experimental class at the final test were first tested on Grade X TITL 3 of SMKN 5 Padang.

After the experimental class was given treatment, the final test data (posttest) was administered. The results of the posttest mean score for the experimental class was 73.75. The next step is to determine whether there was an increase in scores (gain score) to see whether learning outcomes improved.

Before testing the improvement of learning outcomes using gain score, a normality test was conducted on the pretest and posttest results, to see whether or not the data was normally distributed, which they were. After the class, scores were tested normally with a maximum score = 100, so the average increase in learning outcomes was obtained using the formula gain score = 0.41, including the medium category.

Figure 1 shows that student learning outcomes increased following active learning methods, in which students interacted during the learning process, the teacher conducted a pre-test, then the teacher gave a summary of the material. Students discussed the challenge questions in groups. After finishing the discussion, students worked on the tournament questions individually. The score obtained was used as a group score. The group that got the highest score was given a reward and the loser was given a penalty. At the end of the lesson, the teacher conducted a posttest as an evaluation of the final learning outcomes.

The percentage of students meeting the learning outcomes before the active learning method was 3%, whereas the percentage of students who scored above 75 afterward was 73%. This result is strengthened by research suggests active learning strategies of the tournament type guide students to become more active in learning (Winarsih 2014, Nurhidayat & Rahmawati 2012, Nurmawati 2014). In addition, this active method can provide opportunities for students to exchange opinions to get the right answer in accordance with the questions that have been given, helping improve student learning outcomes (Yulastri et al. 2017, Islami & Yondri 2016).

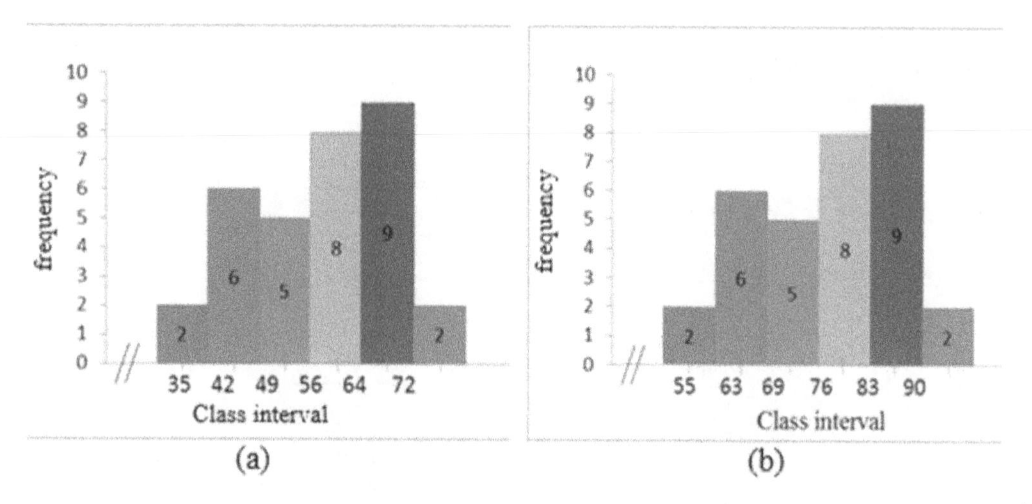

Figure 1. (a) Pretest score; (b) Posttest score.

Obstacles found during research included students' lack of discipline during learning; students coming late; students who were absent; and students who fussed, especially when the teacher was dividing students into discussion groups. To overcome the latter commotion, at the next meeting students were asked to be ready in their respective groups when learning began.

After several meetings had been held, students began to show interest in this strategy. The tournaments included rewards (pens, books, chocolates, candy) for the winners, which was seen a great motivator for students to work more actively in their groups to better understand the material and win the tournament.

4 CONCLUSION

Based on student learning outcomes data before being given treatment with active learning of the tournament-type method, the pretest average was 55.13; after being given treatment, the average was 73.75. After the gain score test, student learning outcomes increased by 0.41 in the medium category. The percentage of students' learning outcomes before implementing the active method of learning tournament type was 3%; after implementing the active method of tournament-type learning, it became 73%. It can be concluded that an increase in student learning outcomes resulted from using these active learning methods.

REFERENCES

Candra, O., Eliza, F., Islami, S., & Alisman, Y. 2019. Pengembangan Multimedia Interaktif Mata Diklat Memperbaiki Motor Listrik Guna Peningkatan Hasil Belajar. *Perspektif* X(2): 7–15.

Hake, R. R. 1999. *Analyzing Change/Gain Scores*. Woodland Hills, CA: Indiana University.

Islami, S. & Yondri, S., 2016. Perbedaan Hasil Belajar Siswa dalam Pembelajaran Kooperatif Tipe Jigsaw dengan Konvensional. In: *National Conference of Applied Engineering, Business and Information Technology, Politeknik Negeri Padang. Padang*, Indonesia: Politeknik Negeri Padang.

Jama, J. 2010. *Transformasi Teknologi pada Pendidikan Kejuruan*. Padang, Indonesia: Aptekindo.

Nurhidayat & Rahmawati, Y. 2012. *Upaya Peningkatan Hasil Belajar Siswa Melalui Pembelajaran Aktif Tipe Learning Turnament pada Mata Pelajaran Matematika*. Kampar: UIN Suska.

Nurmawati, P. 2014. *Penerapan Strategi pembelajaran Aktif Tipe Learning Tournament untuk meningkatkan Motivasi dan hasil belajar Matematika*. Sukoharjo: UMS.

Pernanda, D., Zaus, M.A., Wulansari, R.E., & Islami, S. 2018. Effectiveness of instructional media based on interactive cd learning on basic network at vocational high school: Improving student cognitive ability. In: *International Conferences on Education, Social Sciences and Technology*, 440–444. Padang, Indonesia: Universitas Negeri Padang.

Sukardi, 2008. *Metodologi Penelitian Pendidikan: Kompetensi dan Praktiknya*. Jakarta: PT Raja Grafindo Persada.

Suryabatra, S. 2010. *Metodologi Penelitian*. Jakarta: PT Rajagrafindo Persada.

Trianto, 2009. *Mendesain Model Pembelajaran Inovatif-Progresif*. Surabaya: Kencana Prenada Media Group.

Winarsih, 2014. *Upaya Meningkatkan Hasil Belajar Fiqih pada Materi Zakat Melalui Strategi Learning Tournament pada Siswa Kelas IV MI Yaspi Kaponan Kec. Pakis Kab. Magelang Tahun Pelajaran 2013/2014*. STAIN Salatiga.

Yulastri, A., Hidayat, H., Ganefri, Islami, S., & Edya, F., 2017. Developing an entrepreneurship module by using a product-based learning approach in vocational education. *International Journal of Environmental & Science Education* 12(5): 1097–1109.

Borderless Education as a Challenge in the 5.0 Society – Abdullah, Adriany & Abdullah (eds)
 ISBN 978-0-367-61960-2

Management of the Community Learning Center (CLC) program in Kinabalu Sabah Malaysia

U. Wahyudin, A. Hufad & P. Purnomo
Universitas Pendidikan Indonesia, Bandung, Indonesia

ABSTRACT: Education is the right of every Indonesian citizen, including those in the border regions. Autonomy and community emphasis is a priority in implementation of education everywhere, which is why Kota Kinabalu, Sabah, established a community learning center (CLC) as a forum for helping all community members. This descriptive qualitative research method seeks to provide an overview of the management of the CLC program in Kota Kinabalu. The organization is a high-performing resource of choice for the government of Indonesia through the Ministry of Education and Culture. The results showed that management was sufficient, though more support is needed to boost infrastructure.

1 INTRODUCTION

Indonesia's National Education System Law of 2003, in Article 5, paragraph 1, states, "Every citizen has the same right to obtain quality education." Paragraph 5 further states, "Every citizen has the right to the opportunity to improve lifelong education" (Indonesia 2003). This includes citizens of the border regions: underdeveloped areas with limited social and economic facilities and infrastructure (Warsilah & Wardiat 2017).

Kota Kinabalu is the capital of Sabah, located in East Malaysia, across the border from Indonesia. The city is the seat of government for the West Coast Division of Sabah. The number of Indonesian citizens (WNI) in Sabah is officially recorded as 462,506 people. These include formal migrant workers (259,829), service sector workers (8,269), informal migrant workers (6,536), professionals (310), ABK 423, and troubled Indonesian citizens such as those subject to deportation, convicted of crime, undergoing prison sentences, and so on. It is not known how many live illegally in Sabah.

Mantra (1998) found that one reason for Indonesians to become migrant workers abroad, especially in Malaysia, is the prospect of receiving higher wages for the same type of work. Bank Indonesia (2008) notes that the majority of workers sent abroad have completed primary school (42%) and junior high school (34%), whereas only 24% are graduates of general secondary schools or higher. Border communities tend not to put a premium on education, however, as, even without going to school, they can earn a large income from work. The overall level of education is still relatively low, although some have graduated from high school (Irmawati 2017, Kementereian PPN 2011, Warsilah & Wardiat 2017). Indonesian citizens in Sabah generally work in the palm oil sector or on plantations, in animal husbandry, as housewives, as restaurant servants, as construction workers, or in other service sectors.

The number of WNI children in Kota Kinabalu in 2016 in primary and secondary education totaled 15,454, with 12,743 elementary students, 2,507 junior high students, and 204 high school students, accommodated in 174 CLCs (Community Learning Centers) or PKBM Centers under CLCs and Humana (92 CLC and 82 non-CLC). The total number of teachers is 461: 231 Ministries of Education and Culture teachers and 230 local teachers (166 Indonesian citizens and 64 non-Indonesian citizens).

The Kota Kinabalu CLC was formed by the Indonesian community in Kota Kinabalu to serve education for children who cannot study at the Kota Kinabalu Parent School (SIKK).

This CLC is coordinated by SIKK under the supervision and guidance of the Ministry of Education and Culture. The CLC's vision in Kinabalu is to provide education, protection, and empowerment services for children of Indonesian citizens in Malaysia.

To realize this vision, the CLC needs to analyse community needs and learning programs; develop strategy materials and diverse learning resources; implement materials, learning methods, and various learning resources in the learning process; and evaluate regularly. This research will focus on the management of the CLC in Kota Kinabalu, Sabah, Malaysia, which supports public education in the context of Education for Sustainable Development (ESD).

2 RESEARCH METHODS

This study uses a qualitative approach, using a descriptive study method to assess the CLC in Kota Kinabalu. As stated by Taylor, this research will provide a description of the object studied in full, being a research procedure that produces data in written or oral form by observing behavior (Moleong 1993).

The research was conducted at CLC Insan Inanam, Lorong Inanam Point No. 10 Blocks I Tk. 01–02, Kota Kinabalu. Sources of information are students, teachers, and managers at the CLC. Researchers conducted quantitative descriptive research using a questionnaire filled out by 45 respondents, who answered questions about the provision of learning facilities for families of Indonesian workers in Kota Kinabalu.

3 RESULTS OF THE STUDY

The CLC course on "The Progress of Inanam Insan (Inanam Individuals)" was studied in this discussion. The Malaysian government only allows CLCs in rural areas, but the "Progress" CLC program has been held in urban centers, as it has been informally permitted. Teaching is carried out by select outstanding educators chosen by the Indonesian government through the Ministry of Education. Due to the legal status of the program, for the sake of mutual security and comfort, this CLC does not cover Indonesian state symbols such as the white and red flag or the student uniforms that are a symbol of Indonesia.

3.1 *CLC program planning*

The CLC is managed by people from Indonesia, with the organizational structure as shown in Figure 1.

Facilities of the program are obtained from Education Assistance, as outlined in Table 1:

Using these facilities, the program serves 229 students, as detailed in Figure 2.

Interviews with students and managers show that planning of the program is satisfactory. It was carried out under the supervision of SIKK (Sekolah Indonesia Kota Kinabalu), for curriculum and learning tools, and supported by the Indonesian Embassy, which recruited educators. However, CLC activities benefit school-age students only, not adult learners, as local teaching is not yet of a level that can handle older learners. Figure 3 shows the results of the questionnaire survey, given to 45 respondents.

The interval with the greatest value is that most widely reported by students. The authors categorize the interval line as follows:

Minimum score = Minimum score x Number of respondents
= 1 x 45
= 45
Maximum score = Maximum score x Number of respondents
= 4 x 45
= 180
Difference = Maximum - Minimum score

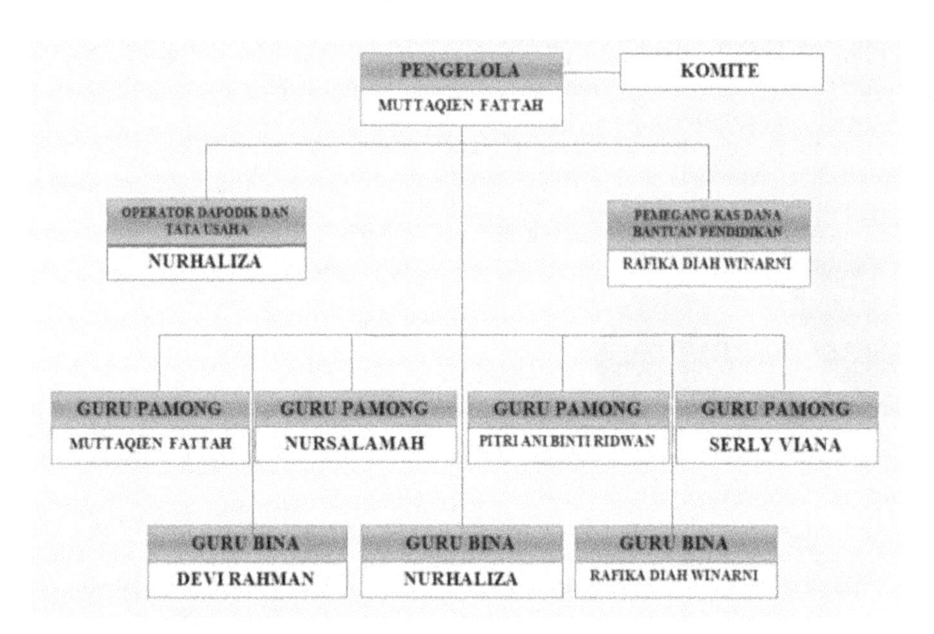

Figure 1. Organizational structure.

Table 1. Learning support facilities and infrastructure.

No	Type of facility	Total	Condition
1	Classroom	5	Good
2	Hall	1	Good
3	Bathroom	5	Good
4	Bookshelf	8	6 good, 2 not
5	Cupboard	4	Good
6	Teacher's desk	7	Good
7	Teache"s chair	7	Good
8	Tables and chairs	200	Good
9	Student desk	10	Good
10	Student chair	14	Good
11	Shoe rack	3	Not good
12	Speaker	2	Good
13	Microphone	3	2 good, 1 not
14	Laptop	2	Good
15	Computer	2	1 good, 1 not
16	Printer	5	3 good, 2 broken
17	Whiteboard	10	Good
18	AC	7	5 good, 1 not, 1 broken
19	Bulletin board	1	Good
20	Statistics board	1	Good
21	Clock	6	Good
22	Props	2	Good

No	Name Study group	Sex		Religion			Total
		M	F	Islam	Christian	Catholic	
1	Study group 1 A	13	13	7	6	13	26
2	Study group 1 B	16	8	6	9	9	24
3	Study group 2	19	15	6	4	24	34
4	Study group 3	14	21	5	11	19	35
5	Study group 4 A	13	15	8	9	11	28
6	Study group 4 B	8	18	7	9	10	26
7	Study group 5	15	18	11	8	14	33
8	Study group 6	8	15	3	8	12	23
Total		**106**	**123**	**53**	**64**	**112**	**229**

Figure 2. CLC students 2019–2020.

Feedback on Planning the CLC Program						
Question item	Value					Total score
	4	3	2	1	0	
1	43	2				178
2	13	24	8			140
3	28	12	3	2		156
4	17	25	3			149
5	18	25		1	1	148
6	23	14	2	3	3	141
7	20	9	14		2	135
8	11	24	9		1	134
9	33	9	3			165
10	7	28	8	1	1	129
11	17	24	2	1	1	145
Average score						147.3

Figure 3. Questionnaire responses.

= 180 - 45
= 135
Distance interval = Interval: Level 4
= 135: 4
= 34

With an interval of 34 for each category, the distribution of the total score of respondents can be shown in a continuum as in Figure 4.

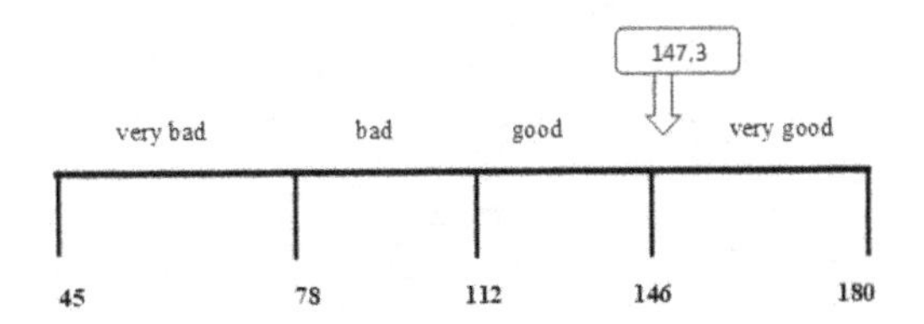

Figure 4. The distribution of the total score of respondents regarding CLC program planning.

The results of these responses are calculated in the maximum line categorization. The average score respondents gave to program planning was 147.3; in classifying the interval, the score fell into the "very good" category.

4 IMPLEMENTATION

Program teachers work 7–4 each weekday (see Figure 5) and are domiciled around the CLC in people's homes or even at the CLC itself. The educators are devoted to the CLC, so they want to improve not only the level of teaching but also the administration, to make the learning space better so that students feel comfortable.

As mentioned, for safety and comfort while learning in Malaysia, students do not use Indonesian state symbols and wear Malaysian school uniforms (see Figure 6).

Interviews with students and managers show that the current program implementation provides good results. Pride at being part of an Indonesian education program while in another country acts as a motivation: although infrastructure is limited, this limitation does not prove a barrier in desiring to learn. However, compared to the condition of schools in Indonesia, this CLC falls below the standard for program implementation, with inadequate facilities.

The students reported being happy with their teachers/tutors, finding them energetic, creative, and comfortable to speak to. As minorities, they felt the problem of learning facilities was not always a concern, as the most important thing was that they could learn at all. Survey results are shown in Figure 7.

Time	Program	
	Monday to Thursday	Friday
07.15-07.30	Morning parade 1. Line up 2. Make a student promise 3. Prayer 4. Morning motivation	Morning parade 1. Line up 2. Make a student promise 3. Prayer 4. Morning motivation
07.30-09.15	Classroom learning activities	Classroom learning activities
09.15-09.30	Rest, eat together	Rest, eat together
09.30-12.30	Classroom learning activities	Classroom learning activities
12.30-13.00	Rest, prayer in congregation	Go home
13.00-15.00	Classroom learning activities	
15.00	Go home	

Figure 5. Learning schedule.

Figure 6. CLC students wearing Malaysian school uniforms.

Question number	Feedback on program implementation					Total score
	4	3	2	1	0	
12	22	23				157
13	29	11	5			159
14	18	22	3	1	1	145
15	29	12	2	2		158
16	40	4			1	172
17	10	25	8		2	131
18	24	12	6	1	2	145
19	15	25	3		2	141
20	17	9	18		1	131
21	11	28	5		1	138
22	29	11	3		2	155
23	23	16	3	2	1	148
24	34	7	3		1	163
25	21	21	1	1	1	150
26	32	10	2		1	162
27	38	2	1	1	3	161
28	10	26	6	1	2	131
Average score						153.4

Figure 7. Response on CLC program implementation.

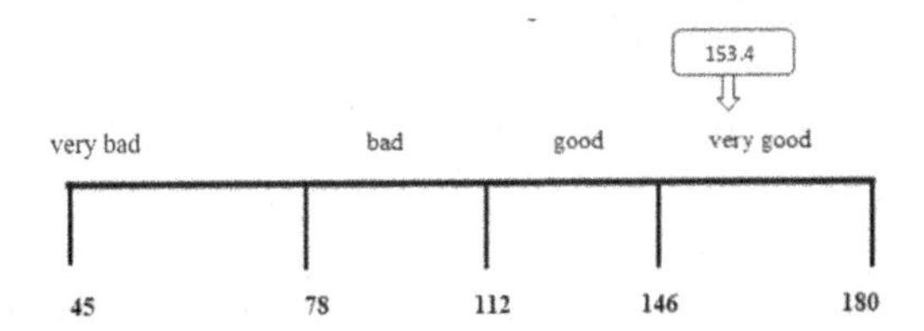

Figure 8. The distribution of the total score of respondents regarding CLC program implementation.

Following the categorization of interval lines detailed earlier, with an interval of 34 for each category, the distribution of scores can be shown in a continuum as in Figure 8.

As shown in Figure 8, the average score was 153.4, which can be classified in the "very good" interval.

5 EVALUATION

Interviews with students, teachers, and managers found that all involved felt program management was well handled. Learning outcomes have been implemented well by students and graduates, such that some graduates have even pursued further education and continued to study at Indonesian universities. Response on CLC program evaluation can be seen in Figure 9.

Feedback on CLC program						
Question item	Valuc					Total score
	4	3	2	1	0	
1	25	17	2		1	155
2	9	18	15	1	1	121
3	24	15	5		1	151
4	29	10	4	1	1	155
5	30	8	4	2	1	154
Average score						147.2

Figure 9. Response on CLC program evaluation.

Figure 10. The distribution of the total score of respondents regarding CLC program evaluation.

Using the same interval classification as before, the distribution in the form of a continuum is shown in Figure 10.

The average score of respondents evaluating the program was 147.2, classified as "very good."

6 DISCUSSION

Informal education provides educational services to communities who may not otherwise receive it. Lifelong access to education aims to:

- serve students so they are able to grow and develop as early as possible and throughout life to improve levels of dignity and quality of life;
- foster students in teaching knowledge, skills, and attitudes so they can work for a living or continue to a higher level of education; and
- meet learning needs within the community that cannot be met through formal education in schools.

The CLC studied here functions as a motivator, innovator, and facilitator of learning and community empowerment in its community; as an information center for the community/regional area; as a mediator between the community/region and CLC partners; as a distributor of contributions/CLC partner support to those communities that need it; as a marketing center for businesses students and community members; and so on. A CLC, especially in a developing region, shares certain characteristics and characteristics:

- It consists of three pillars: learning, entrepreneurship, and community mobilization.
- Each CLC is unique in terms of the learning programs organized, the PKBM management system, and so on.
- Each is flexible in terms of the learning models/methods used, the programs that are held, and so on.
- The institutions are established from, by, and for the local community.
- They involve various parties across different sectors (CLC partners).

For these reasons, informal education is an integral part of the national education strategy. It provides educational autonomy, making community-level planning a priority and making CLCs a forum for the improvement of every strata of community member. This particular CLC program, an example of Education for Sustainable Development (ESD), reflects the individual, local needs of the developing community (Kemdiknas 2010, Kamil 2009, AONTAS 2000).

A program will run properly if it is properly managed. In general, management encompasses all activities carried out by one or more persons in a group or institution to achieve a predetermined goal (Engkoswara & Komariah 2010, Usman 2009). Management is divided into three stages: planning, implementation, and evaluation. George R. Terry expresses the notion that planning is a process for selecting and linking facts into an assumption about the future; formulating activity steps helps to achieve the desired results (Hasibuan 2013). The initial stage of the planning process is to identify and define existing problems – unemployment, low education, etc. – as well as the sources of these problems.

Implementation of the program is the most important management stage. Organizing brings higher levels of efficiency and effectiveness to the implementation of activities and helps achieve determined objectives. Flexibility provides opportunities for change, such as development or modification; changes can come from inside and outside the organization.

Sompong & Rampai (2015) discuss CLCs as part of a participatory strategy from the community to emphasize perception, accessibility, and participation in the development of learning centers through communication with various parties. Community education works with social organizations (DuFour 2004, Hargreaves 2007). The government encourages the development of community education in collaboration with existing social organizations. It is important to note that cooperation with other groups must not change the position and goals of the organization, namely fighting for its members. Management must be the responsibility not only of the manager but also of the government and other partners, if objectives are to be achieved.

Finally, evaluation of community education programs help guide the organization's implementation. As evaluation assesses whether or not the program has reached its expected goals, it focuses on results achieved (output). The evaluation process performance measures against established objective standards, then the organizer or evaluation participant takes action based on the results (Djaali & Mudjono 2000). Thus, evaluation is closely related to valuation assessment – a systematic activity for collecting, processing, analysing, describing, and presenting information as inputs for decision-making (Warju 2016, Dunn 2000, Sawitri 2007).

7 CONCLUSIONS

Empirical data resulting from the study of the CLC in Kota Kinabalu shows that the quality of education management is still relatively "sufficient"; the score, however, remains hampered by limited infrastructure. That said, organization of learning is ranked highly, due to good work by the teachers chosen by the government selection process.

The CLC in Kota Kinabalu must continue to meet expectations as part of community-based education. However, this is not only up to the CLC itself: it also needs support from the Indonesian government in improving areas that are currently inadequate.

ACKNOWLEDGMENT

We want to recognize the LPPM UPI (Indonesian Research and Community Service Research Institute) for research assistance funds. We would also like to thank CLC Inanam Kinabalu Malaysia for partnering with us in this research from start to finish.

REFERENCES

AONTAS. 2000. *Community Education: AONTAS Policy Series*. Dublin: AONTAS.

Bank Indonesia. 2008. *Laporan Survey Nasional Tenaga Kerja Asing Tahun 2008*. Jakarta: Direktorat Statistik Ekonomi dan Moneter.

Djaali, H., & Mudjono, P. 2000. *Pengukuran dalam Pendidikan*. Jakarta: Program Pascasarjana.

DuFour, R. 2004. What is a professional learning community? *Educational Leadership* 61(8): 6–11.

Dunn, W. N. 2000. *Pengantar Analisa Kebijakan Publik*. Yogyakarta: Gadjah Mada Press.

Engkoswara, A. K., & Komariah, A. 2010. *Administrasi pendidikan*. Bandung: Alfabeta.

Hargreaves, A. 2007. Sustainable professional learning communities. In L. Stoll & K. S. Louis (eds.), *Professional Learning Communities: Divergence, Depth and Dilemmas*. London: McGraw–Hill Education.

Hasibuan, J. 2013. Manajemen penyelenggaraan pendidikan dan pelatihan Balai Diklat Keagamaan Medan. *Jurnal Tabularasa PPS Unimed* 9(2): 215–228.

Indonesia, P. R. 2003. *Undang-undang Republik Indonesia nomor 20 tahun 2003 tentang sistem pendidikan nasional*. Jakarta: Pemerintah Republik Indonesia.
Irmawati, A. 2017. Peran Pusat Kegiatan Belajar Masyarakat (Pkbm) dalam Mengurangi Buta Aksara di Kabupaten Karimun. *Jurnal Pendidikan dan Kebudayaan* 2(1): 81–98.
Kamil, M. 2009. *Pendidikan Nonformal Pengembangan Melalui PKBM di Indonesia (Sebuah Pembelajaran dari Kominkan di Jepang)*. Bandung: Alfabeta.
Kemdiknas, 2010. *Model Pendidikan untuk Pembangunan Berkelanjutan (Education for Sustainable Development/ESD) melalui kegiatan Intrakulikuler*. Jakarta: Pusat Penelitian Kebijakan, Balitbang Kemdiknas.
Kementereian, P. P. N. 2011. *Kajian Direktorat Kawasan Khusus dan Daerah Tertinggal*. Jakarta: Bappenas.
Mantra, I. B. 1998. *Indonesian Labor Mobility to Malaysia: A Case Study: East Flores, West Lombok, and the Island of Bawean. Artikel ini dipresentasikan pada National Workshop on International Migration, tanggal9–11 Maret 1998, penyelenggara The Population Studies Center, Un*. Jogjakarta.
Moleong, L. 1993. *Metodologi Penelitian Kualitatif*. Bandung: Remaja Rosda Karya.
Sawitri, S. 2007. *Evaluasi Program Pelatihan Ketrampilan Membuat Hiasan Busana dengan Teknik Pemasangan Payet Bagi Pemilik dan Karyawan Modiste di Kecamatan Gunungpati Semarang*. Semarang: PPS UNY.
Sompong, N., & Rampai, N. 2015. The development model of knowledge management to strengthen Thai ICT community learning centers. *Procedia – Social and Behavioral Sciences* 176: 139–147.
Usman, 2009. *Implementasi Manajamen Strategi Dalam Pemberdayaan Sekolah Menengah Kejuruan*. Bandung: Cita Pustaka Media Perintis.
Warju. 2016. Educational program evaluation using CIPP model. *Innovation of Vocational Technology Education*.12(1): 36–42.
Warsilah, H., & Wardiat, D. 2017. *Pembangunan Sosial di Wilayah Perbatasan Kapuas Hulu, Kalimantan Barat*. Jakarta: Yayasan Pustaka Obor Indonesia.

Borderless Education as a Challenge in the 5.0 Society – Abdullah, Adriany & Abdullah (eds)
© 2021 Taylor & Francis Group, London, ISBN 978-0-367-61960-2

The knowledge of village family planning cadres in implementing needs assessment

A. Hufad & P. Purnomo
Universitas Pendidikan Indonesia, Bandung, Indonesia

ABSTRACT: Village Family Planning is a program developed by the Government, in reducing birth rates, improving reproductive health and improving economic welfare. The implementation of the KB village program is closely related to community empowerment, so that the program's needs are based on the community's needs. This study reveals the knowledge of KB village cadres about need's assessment. This research uses a quantitative descriptive method presented in the form of a percentage by distributing questionnaires to 50 family planning cadres supported by interviews in revealing all data and information. The results show the fact that KB village cadres do not fully understand the need's assessment procedures, so consideration is to improve competence in properly planning empowerment programs.

1 INTRODUCTION

Community education programs are expected to further develop in the future, due to the process of economic globalization that demands resilient competition in productivity, efficiency, and quality. Those various demands will bring increasing consequences, widespread and diverse community needs for community education program services.

The process of identifying needs requires cautiousness in collecting and processing data. The Needs Assessment process applies both qualitative and quantitative approaches, which allow for the maximum and concrete collection of data, but this is difficult when it is done in large populations. The facilitator is coped with maximum analytical and data processing skills to be able to decide on program needs priorities (Barbazette 2006, Altschuld & Witkin 2000).

The process of identifying needs requires to apply appropriate methods so that all information can be obtained completely. Minzey and Le Tarte emphasize that "the ultimate goal of community empowerment in community education is to develop a process in which community members work together to identify problems and to find solutions to those problems" (Brookfield 1983). Thus, the process of identifying needs will assist the analyst in gathering information and mapping appropriate community needs.

One of the implementations of community empowerment programs in addressing population problems today is village family planning program implemented by the National Population and Family Planning Agency (BKKBN). Until 2018 in the Province of West Java, there were 1.595 village family planning. More specifically in Bandung Regency, there were 3 of the other regions that had reached the target of establishing KB villages according to the target. Bandung Regency has established 68 village family planning that achieved 100% of the target category.

In an effort to run a community empowerment program, village family planning program optimizes the data and information as a basis for organizing a program through an activity known as the "House of Data". The house of data aims to facilitate the project implementers in planning program activities, providing data and information, improving the synergy of BKKBN and work partners.

Problems found from the realization of village family planning program due to the asynchronous needs of the community and the implementation of the program, resulting in some programs not being fully perceived by the community. The Needs Assessment process carried out by stakeholders only relies on the key person to obtain the data through a discussion or interview process, while other techniques in data processing are not performed (Fatmawati 2017, Nintrafil 2018, Susanti 2016).

Referring to some problems that occur in the community, especially in the village family planning area, the researcher examines the village family planning cadres' understanding of needs assessments, especially in processing the data and information carried out by the cadres, so that community needs can be mapped and can easily obtain local potential data based on 8 functions of the family declared by village family planning.

2 RESEARCH METHODS

This study employed a descriptive quantitative approach, to uncover data based on the distribution of questionnaires to 50 cadres of village family planning relying on quota sampling so that it can reveal in detail through interviews with several cadres of village family planning, community leaders and government leaders.

3 RESULTS OF THE STUDY

Based on the results of the questionnaire distribution to 50 cadres of village family planning respondents in Citeureup Village, Dayeuhkolot District, Bandung Regency, several opinions were obtained which illustrated the initial knowledge of needs assessment. Four indicators were asked through this questionnaire i.e. 1) Self-preparation for need assessment; 2) Designing need assessments; 3) Establishing Priority Needs, and 4) Reporting and Utilizing the Identification Results.

The needs assessment process requires some preparation in carrying out community needs identification. The study revealed an overview of the cadre's knowledge in self-preparing for conducting needs assessment as it follows: at average answered correctly 24 people (48%) and 26 people (52%) answered incorrectly. Based on the above, cadre's knowledge regarding self-preparation is included in the moderate category (48%). The following Figure 1 distributes the scores.

No	Question	Correct answer		Incorrect answer	
1	The process of identifying community needs is closely related to ... and ...	30	60%	20	40%
2	The process of gathering information based on needs will provide a clear picture of ...	23	46%	27	54%
3	The process of gathering information based on needs will provide a clear picture of ...	21	42%	29	58%
4	Each cadre has the ability to identify community needs, so before the identification activity is carried out, it is necessary to prepare ...	22	44%	28	56%
5	There are various strategies that demonstrate the ability of the facilitator to conduct a needs assessment, including ...	23	46%	27	54%
	Total	119	48%	131	52%
	Average	24	48%	26	52%

Figure 1. Respondents' knowledge of the indicators in self-preparation for a needs assessment.

No	Question	Correct answer		Incorrect answer	
1	The way of how to work systematically to facilitate the implementation of an activity in achieving the specified goals, is the definition of	23	46%	27	54%
2	A very clear difference is that the method is used in a more .....context, while the technique is used in more...... contexts	21	42%	29	58%
3	In designing identification activities that lead to gathering information intended for program planning, according to the needs they specify, included to the objectives of ...	20	40%	30	60%
4	One of the specific coverage areas of community needs including, except	10	20%	40	80%
5	The design of activities to identify community needs is an activity that consists of stages by gathering various information, this definition can be classified as........	19	38%	31	62%
	Total	93	37%	157	63%
	Average	19	37%	31	63%

Figure 2. Respondents' knowledge of the indicators in designing needs assessment.

The needs assessment is a stage of program planning, it is necessary to be able to formulate the needs assessment design. Knowledge of needs assessment is the basis for preparing the design so that the realization of the program can really be used. Based on the questionnaire distribution to 50 respondents, it is known that on average there are 19 people (37%) answered correctly and 31 people (63%) answered incorrectly (see Figure 2). Based on the score above, the respondent's knowledge about designing needs assessment is included in the low category (37%).

The needs assessment phase is determining needs priority. This stage requires the ability to understand the specific needs of the community with various analysis considerations of environment, fund, time and so on. Based on the questionnaire distribution to 50 respondents, it is known that 22 people (44%) answered correctly and 28 people (56%) answered incorrectly (see Figure 3). From these scores, cadres' knowledge on setting priority needs is classified in a moderate category (44%). The following is the distribution of the questionnaire questions.

The last stage in the needs assessment activity is the stage of compiling the results of needs identification and reporting the results. This stage is the ability of cadres to provide an overview of the results of priority needs analysis and to conclude the results of community needs analysis. Based on the distribution of questionnaires to 50 respondents, it is known that 19 people (39%) answered correctly and 31 people (61%) answered incorrectly (see Figure 4). From these scores, then the cadre's knowledge of reporting and utilizing the results of identifying the need to turn to be a program is classified in low category (39%). Following is the distribution of the questionnaire questions.

From these four indicators, it can be concluded that the respondent's knowledge regarding needs assessment is still in the low category. Most cadres answered the question incorrectly. It is proven by the fact that an asynchronous program with the community needs is caused by a condition when village family planning cadres do not have enough knowledge to conduct a needs assessment.

No	Question	Correct answer		Incorrect answer	
1	The systematic process for gathering and analyzing information related to the educational needs of individuals, groups and organizations / institutions is the definition of a concept ...	21	42%	29	58%
2	The process of gathering, analyzing and determining the subject matter and finding solutions to these problems can be done by discussing with several community leaders, the method is also referred to as	20	40%	30	60%
3	The process of determining the needs of community empowerment programs, it is very important to know the priorities of actual needs in the community, the needs assessment, has a purpose for, except	29	58%	21	42%
4	The needs assessment is part of the evaluation process because its purpose is to determine and establish, except	22	44%	28	56%
5	The nature of needs assessment is directly related to identification, organization, and document information about	23	46%	27	54%
6	Determination of community needs assessment is directly related to a variety of opinions, differences in situations and ...	19	38%	31	62%
7	The process of determining priority needs is an analysis that is	20	40%	30	60%
8	There are four types of needs including normative needs, stated needs, anticipated needs, and ....needs	22	44%	28	56%
	Total	176	44%	224	56%
	Average	22	44%	28	56%

Figure 3. Respondents' knowledge of the indicators in setting priority needs.

No	Question	Correct answer		Incorrect answer	
1	Identifying a problem or deficiency is a general goal of ...	21	42%	29	58%
2	Community empowerment programs has to describe the needs of the community which can be done through	22	44%	28	56%
3	The results of the needs assessment of community empowerment programs can be utilized for	15	30%	35	70%
	Total	58	39%	92	61%
	Average	19	39%	31	61%

Figure 4. Respondents' knowledge of the indicators in reporting and utilizing the results of identification.

4 DISCUSSION

A needs assessment is the initial stage in planning community education programs. The process is a step to collect the data and information regarding program planning to be carried out. The identification of learning needs in community education programs is the most important part of community education program management. One of the activities in the process of identifying needs is to determine an appropriate method to ensure that all information can be obtained comprehensively.

The determination of needs assessment is directly related to various opinions, differences in situations and various types of needs. Therefore, the strategies and procedures used in the assessment activities will vary. There are various procedures and stages used in needs assessment activities, including:

4.1 *Self-preparing for conducting a needs assessment*

Like the inquiry method, it needs to prepare yourself with a variety of in-depth questions. Curiosity will lead the program makers to the actual needs of the community. Besides, according to Suarez in Khumalo (1999) "due to educational needs is based on the values of the institution or society, procedures must be integrated into the process to ensure they can represent these values.

4.2 *Designing a needs assessment*

Preparation of needs assessment design begins with an in-depth study through preliminary studies, clarity of the study/needs to be revealed. This stage includes the activities in describing the specific objectives of the program, areas with needs that must be met and the types of needs identified. According to Suaraez in Khumalo (1999) explains that the completeness of the design included in it is a procedure for analyzing the data and reporting the results.

4.3 *Setting priority needs*

Need assessment review through identification will produce various needs. To obtain maximum results, the results of needs identification should be sorted by their immediate nature/ urgency/importance. This priority setting process is a complex analysis. Lund and Mc Gechan in Khumalo (1999) suggest specific criteria for analyzing needs which include:

4.3.1 *How many people will be identified*

- What are the consequences if those needs are not met?
- Can this need be fulfilled by educational activities?
- What priority needs should be met before other educational needs are addressed?
- Are there sufficient resources (funds, staff) to meet these needs?
- Reporting and Utilizing the Identification Results

Minzey and Le Tarte emphasize that "the ultimate goal of community education is to develop a process by which community members learn to work together to identify problems and to find solutions to those problems" (Brookfield 1983).

Methods and techniques for identifying needs, as a basis for organizing community education programs. Several cases of community education projects/programs, both those carried out by the government and private sector/NGOs (Non-Governmental Organizations), have taken many approaches in determining community needs. Implementation of analysis and approaches in determining learning needs is carried out systematically, to obtain valid and reliable data.

The main purpose of the needs assessment in the community education process is the assessment of learning needs intended for the implementation of community education programs. In line with the above ideas Suarez in Khumalo (1999) suggests the following needs assessment goals:

Providing information for planning is the most common reason given for conducting needs assessment. The process of gathering information based on needs will provide a clear picture of program planning.

The diagnosis or identification of problems or weaknesses is another common purpose of needs assessment. Identifying a problem or deficiency is a general goal of need assessment.

A needs assessment is a component of several evaluation models. Need assessment is part of the evaluation process because the goal is to determine the needs/disadvantages/gaps before the program starts or determine the status of the performance at a certain level that will be achieved based on certain aspects of competence.

A needs assessment is also conducted to hold educational institutions accountable for their efforts. Need assessment is also carried out as a form of accountability of educational institutions for all efforts that they do.

The link between need assessment and community education, Delaney and Nuttall argues that needs assessment has an immediate goal and ultimate purpose (Khumalo 1999). The immediate goal is the identification of needs that has to be immediately fulfilled. While the final goal is to direct activities to gather information intended for program planning based on the needs they specify (Finnie 2012, Fuller et al. 2002, Altschuld & Witkin 2000, Barbazette 2006).

Need Assessment is directly related to identification, organization, and document information about the needs that have to be met. Therefore, it is necessary to have a program model that can improve the competence of village family planning cadres so that they can do needs assessments to carry out empowerment in their respective areas.

ACKNOWLEDGMENT

We would like to dispute the gratitude calculated by the LPPM UPI (Indonesian Institute of Education Research and Community Service) for research assistance funds. We also want to thank the cadres of the family planning village as partners who have helped the research from start to finish.

REFERENCES

Altschuld, J.W. & Witkin, B.R. 2000. *From needs assessment to action: Transformingneeds into solution strategies*. New York: Sage.

Barbazette, J. 2006. *Training needs assessment: Methods, tools, and techniques*. New York: J. Wiley & Sons.

Brookfield, S.D. 1983. *Understanding and Facilitating Adult Learning*. Buckingham: Open University Press.

Fatmawati, N.I.D. 2017. *Perbedaan fertilitas pada kampung kb dan bukan kampung kb di desa krucil kecamatan krucil kabupaten probolinggo tahun 2016*. Malang: jurusan geografi-fakultas ilmu sosial um.

Finnie, C. 2012. *Rural Workforce Development: Assessing Employer Needs and Improving Access to Training*. Calgary: Bow Valley College.

Fuller, T., Guy, D. & Pletsch, C. 2002. *Asset Mapping: A Handbook*. Ottawa: Canadian Rural Partnership.

Khumalo, F.T.E. 1999. *Methods of assessing learning needs for community education programmes*. Pretoria: Faculty of Education University of Pretoria.

Nintrafil, 1. 2018. *Implementasi program kampung keluarga berencana: studi pada rw 06 kampung mekarlaksana desa citaman kecamatan nagreg kabupaten bandung* Bandung: uin sunan gunung djati bandung.

Nurwahidah, L.S. 2017. Pembelajaran Literasi Berbasis Potensi Lokal Untuk Pengembangan Kearifan Lokal Dalam Upaya Pemberdayaan Perempuan. *Caraka: Jurnal Pendidikan Bahasa dan Sastra Indonesia serta Bahasa Daerah* 6(2): 1–10.

Susanti, P. 2016. Partisipasi masyarakat dalam upaya pengendalian penduduk melalui program kampung kb di kelurahan situsaeur kecamatan bojongloa kidul kota bandung. Bandung: universitas pendidikan indonesia.

Borderless Education as a Challenge in the 5.0 Society – Abdullah, Adriany & Abdullah (eds)
© 2021 Taylor & Francis Group, London, ISBN 978-0-367-61960-2

The readiness of special school in developing independence of students with special needs through vocational skills

I.D. Aprilia & T. Soendari
Universitas Pendidikan Indonesia, Bandung, Indonesia

ABSTRACT: Schools' lack of preparedness in managing vocational skills and providing skills needed by the community (hard skills and soft skills), as well as the lack of a system/pattern that regulates the implementation of vocational skills among related institutions, allegedly led to a lack of absorptive capacity and low competitiveness of special needs school (SLB) graduates in the business and industrial world (DUDI), until unemployment numbers increased. This study aims to formulate a pattern of SLB readiness in independent students with special needs through the management of vocational skills. This study used a descriptive qualitative method. Data collection was done through interviews, observations, and documentation studies. Data analysis techniques included data reduction, data presentation, conclusions, and verification. Through this research, the data obtained that the school plans, implements, and evaluates vocational skills programs with referring to the results of the analysis of students' talent and interest assessment, although not all types of skills that are suitable for persons with disabilities are identified, and the absence of certified skills teachers and the lack of infrastructure means that vocational skills programs do not run optimally. Schools also have not been able to sell students' products and lack coordination with business partners or the government. Thus, seriousness and consistency of schools is needed in implementing vocational skills by increasing the competence (understanding and skills) of teachers about vocational skills, and exploring opportunities for partnerships to work together in seeking fulfillment of the independence of students with special needs after graduating from school to become entrepreneurs or work in companies.

1 INTRODUCTION

There are 21 million people with special needs and disabilities in Indonesia, 11 million of whom are of the age to be in the workforce, and data shows that 96.31%of them have pursued careers in various employment sectors. The results of a survey from the Indonesian Association of Persons with Disabilities (PPCI) reported by the Tribunnews website, shows the ratio of people with disabilities employed in Indonesia is just under 0.5%. In Romania, the percentage of employed people with disabilities is still low at 12.7%, while the number of unemployed people with disabilities is double that of non-disabled people (Angela 2014). Competition between people with disabilities and non-disabled people in terms of work, makes the number of unemployed people with disabilities more than double that of non-disabled people (Blazquez & Malo 2005). The high unemployment rate for persons with disabilities is allegedly due to their low quality of life skills, which the World Health Organization is defined as the ability to behave adaptively and positively in handling the demands and challenges of daily life effectively, including in the workforce.

In this regard, limited learning opportunities; developmental and/or cognitive barriers that affect learning, communication or problem solving; and low commitment in decision-making (Aprilia 2010), demonstrated that the quality of life skills of people with disabilities is still low. A similar study shows that the quality of life of Americans with disabilities is lower than non-disabled people in general (Ihara et al. 2012). Persons with disabilities often lag behind their peers in developing life skills (Gall et al. 2006), and often lack opportunities or experience to learn about the world and themselves (King & Palmer 2010). The work productivity of people with disabilities is very low compared to non-disabled people (Angela 2014, Guenther et al. 2008), and even though they have completed their education at the secondary and upper levels, it is still difficult for them to seek jobs (Guenther et al. 2008).

Law number 8 of 2016 concerning Disabled Persons, states there are 22 rights that shall be addressed for persons with disabilities including education and employment. Minister of Education and Culture Regulation number 157 of 2014 concerning the special education curriculum explained that the curriculum for students with special needs meets general programs, special needs programs, and independence programs requirements. The independence program is closely related to the development of life skills, especially vocational skills. Life skills programs have effectively supported the skills development among students with special needs and helped them prepare for the transition to adult life. Vocational learning for students with special needs is certainly based on the necessity, talents, and potential of the students (Brueggemann & Burch 2006).

Some special needs schools have generally implemented vocational learning through extracurricular activities and are integrated in co-curricular or intracuricular activities, training, or internships in companies. Vocational learning practices in SLB still have several challenges, including (1) a mismatch between students' talents and interests and the availability of types of vocational skills in schools. (2) The unpreparedness of schools in equipping students with the skills needed in the community or company. Limited human resources (teachers) including system support in managing vocational skills programs from planning, implementation, to evaluation, and therefore, the role of schools as a place of training (shelter workshop) is not optimal. (3) schools still focus on providing hard skills while soft skills are not accommodated. (4) There is no system governing partnerships between special needs schools (SLB), corporations and related departments (Aprilia 2010).

The complexity of the problems certainly requires a systematic effort by considering the potential or opportunities oriented toward the readiness of schools in developing empirically proven vocational skills that can contribute to developing the independence of students with special needs. Availability of employment opportunities both to work in companies or to create their own jobs (entrepreneurship) as an effort towards independence in students with special needs is through complex processes. For obvious reason, systematic and synergistic efforts from various parties are exttremely necessary so that students with special needs can develop their potential.

2 LITERATURE REVIEW

2.1 *Independence*

The term independence or autonomy refers to Steinberg's concept, called independence to act, not dependent on others, not relying on the help or support of others, competent, and free to act. It is further explained that the term "independence" refers to a belief in one's ability to solve problems without special assistance from others, unwillingness to be controlled by others, able to carry out activities, and independently solve the problems. It is emphasized that in addition to the belief in self-ability, in independence, there is also an element of self-determination in the form of the need to master the tasks given. Independence also refers to psychosocial abilities such as freedom to act, not dependent on others, not affected by the environment, and free to regulate their own needs (Dacey & Kenny 1997). Also included are personal decisions based on complete knowledge about the consequences of various actions and the courage to accept the consequences of these actions, and the freedom to take initiative, overcome obstacles, perform something correctly, be persistent in business, and do everything without the help of others.

2.2 *Vocational education*

Vocational education is part of the education system that prepares a person to be better able to work in one filed or line of work than another. Vocational education serves to guide students to readiness in facing the workforce, by exploring their interests and potential and learning everything related to work. The purpose of skills-based education is to increase the relevance of education to real life values or prepare students to have the abilities and skills needed to survive and develop themselves.

Components of vocational skills include soft skills and hard skills. Soft skill is an ability, talent, or skill that exists in every human being (personal and interpersonal skills). Hard skills are mastery of technical skills from learning outcomes related to a particular field of science. Hard skills are very closely related to technical skills that are inherent or needed for certain professions. With regard to this issue, Levinson & Palmer (2016) state that special skills (hard skills and soft skills) are necessary to survive in the workplace and in the community and need to be taught explicitly as both are very important and must complement each other.

3 METHODS

Through a descriptive qualitative method, the current research process extracted data about the implementation of vocational programs including planning, implementation, evaluation, obstacles, and school efforts in establishing students' independence. Furthermore, several steps analyzed data and theory in order to seek the readiness of schools to make their students independent. Research data is extracted from school principals, skills teachers, and deaf students enrolled in vocational skills classes at SLB B Negeri Cicendo Bandung, through observation, interviews, documentation studies, and literature studies.

4 RESULTS AND DISCUSSION

4.1 *Results*

At this stage, the researchers sought to obtain data on abilities, strengths, and the school's ability in empowering the independence of students, including aspects of assessment and planning, implementation, evaluation, obstacles, and efforts.

4.2 *Assessment and planning of vocational skills*

The special needs school in the present study conducted a vocational assessment at the beginning of the VII grade school year of the Special Senior High School to find out the interests and potential of the students in vocational skills. From 20 types of skills offered, the school has finally determined four skill areas: catering, beauty, multi-media, and wood crafts. The determination of the four skills areas was based on the demands of the business world or industrial world (DUDI), geographical conditions, and local wisdom. The parties involved in the assessment process include skills teachers, class teachers, and parents. These skills are then correlated with assessments that have been carried out on students, thus skills can be identified that are in accordance with the potential and needs of students. The results of the vocational assessment form the basis for preparing the vocational learning plan design for students.

4.3 *The application of vocational skills*

The time allocation of vocational skills learning at the SLB Cicendo State B is for 18 lesson hours per week, and for SMALB level under grade X 24 lesson hours, and XI and XII 26 lesson hours per week. Students carry out an internship program in class XI for at least 3 months. The types of skills provided are academic, communication, social and interpersonal, and vocational skills. Academic skills include reading ingredients, computer operating skills, and so forth. Communication, social, and interpersonal skills are soft skill materials delivered along with hard skills. Even though the provision of soft skills materials only begins in this learning year so the results yet been determined, yet students are taught to be able to communicate with the work environment, how to be responsible for work, discipline for work time, and so forth.

4.4 *Evaluation*

Evaluation is mostly conducted to measure students' understanding of the lesson and mastery of the process by using performance tests. The results of the assessment will be used as follow-up or remedial material for students who have not met the standards or still have difficulties in certain skills, yet students who have met the standards or mastered certain skills will be provided enrichment activities. Most importantly, there has been collaboration between teachers and parents in following up the results of learning the skills delivered at school, thus the task of vocational learning is not only on the school, but parents may also actively participate in addressing it.

4.5 *Research problems*

Problems in aspects of assessment and planning can be addressed as follows: (1) parents who do not provide enough support for involvement in vocational learning planning for their children, (2) the absence of instruments to explore students' talents and interests, (3) the determination of types of skills is entirely not based on the potential and needs of students as it must accommodate the ability of teachers and the facilities they have.

Furthermore, in the aspect of vocational skills learned, the problem exists within the limited ability of teachers to master hard skills if they are not from such a background, and the integration of hard and soft skills.

Other obvious matters are the limited opportunities and chances for teachers' skills to develop academic abilities and practices outside the school, limited infrastructure, and the concern of parents that their children will get unfair treatment in the workplace (bullied or harassed). Parents are sometimes reluctant to continue providing learning skills at home, because they assume that it is enough with the learning provided at school.

With regard to marketing of students' creations for catering, schools have difficulty meeting the high demands of customers. Woodcrafts, for example, are rather limited due to high production costs. In partnership cooperation, the school has not yet cooperated with companies in the distribution of work, although apprenticeships have already been carried out. Many students cannot deal with the situation and demands of the company, thus the school no longer focuses on preparation for work, but rather on entrepreneurship.

4.6 *Efforts made by the school*

Schools provide opportunities for teachers to develop students' talents and interests by participating in various competitions, assessments and planning, and events as a form of socialization and development of experiences to compete. The school also cooperates with the business world in providing training to teachers of teaching skills. The SLB Cicendo State B has enrolled into an MoU with several vocational training centers (BLK) according to the type of vocational skills provided in schools. For multimedia, a collaboration with LKP Putra; for catering, Ny for woodcraft Liem, in collaboration with Brother Wood; and for beauty procedures, LKP Puspita. Regarding the marketing of products, specifically for catering, schools have produced superior cakes with the mark 'Deavy" and have got customers even though they are still limited to the closest environment. Collaborating with the business world in channeling student work, schools have enrolled into an MoU with KFC. The school also provides information on job formation.

4.7 *Discussion*

Implementation of the assessment, which is not entirely based on the potential and needs of the students, yet also on the ability of the teacher and the facilities they have, illustrates the lack of comprehensive assessment process. Unclear data about the condition of students might influence the decision to plan training and/or education addressing the fact of potential and barriers of persons with disabilities and shall ultimately affect the ability of students to make successful transitions from school to work and community life (Levinson & Palmer 2016) and helps set realistic work and independent life goals for students.

Inconsistency in implementing vocational skills makes the independence program not optimal. Whereas in the Ministry of Education and Culture Regulation of the Republic of Indonesia number 157 of 2014 concerning the Special Education Curriculum, the independence program in the special education curriculum has been developed as a reinforcement for students with disabilities to live independently, not dependent on others, and to prepare for work. Likewise, with curriculum content that must focus on the balance between hard skills and soft skills, positive results have not yet been interpreted. As well, Levinson & Palmer (2016) stated that special skills (hard skills and soft skills) are particularly necessary to survive in the workplace and in the community and need to be taught explicitly.

Furthermore, it is related to the provision of skills for students with disabilities to take part in internships in business or industry. This training is a necessity, considering prospective apprentices have different educational experiences, and hence, require training that directs them to work effectively and productively. In Malaysia, before enrolling in particular jobs, youth with disabilities acquire work skills through a technical vocational education system and training managed by the government and according to what is needed by the company.

The next independence program is an internship. The recruitment of internship students is mainly adapted to the analysis of workforce requirements (number) and job analysis (types of knowledge, competencies, and skills needed), such as what types of work can be done by students with disabilities. The assessment process has become effective, as was done in Brazil with the Workplace Adaptation program, which is conducting an assessment in advance of work in companies with disability programs, thus duties carried out by disabled people are adjusted to their abilities (Guimaraes et al. 2015).

Related to the effort of channeling students or graduates, schools find it necessary to emulate what SLB of Subang, West Java, did through implementing "Honey Ferocious Management." Schools are able to channel their students to internships and work in several companies and collaborate with several home industries in an effort to develop entrepreneurship. Some companies still encounter obstacles due to difficulties in finding information about people with disabilities who have the ability to work: so far, it is merely conducted through asking for information from the Manpower Office for the recruitment of employees.

The low absorptive capacity of people with disabilities in companies, as well as the complexity of the demands of highly competitive companies, mean involvement of companies in the utilization of students with disabilities is expected to encourage provision of programs in accordance with the abilities of students with disabilities. Among other programs the company can develop, is entrepreneurship programs for people with disabilities. Through corporate social responsibility (CSR) funds aimed at empowering people with disabilities in the entrepreneurship sector, the expectation of success will be greater because in its implementation it can involve CSR distribution companies as a companion or mentor. The participation of the private sector in empowering people with disabilities is expected to overcome various obstacles faced by the Government.

5 CONCLUSION

Based on the results and discussion, the readiness of schools in developing the independence of students with special needs, is as follows:

- In order to prepare for independence of students with special needs, an analysis and needs assessment are necessary as a starting point for planning education or learning addressing potential and obstacles of persons with disabilities.

- The implementation of vocational skills requires educators who have special skills – both in hard skills and soft skills – and in turn, this may affect the optimal provision of skills to students, optimization in independence training for students through the collaboration of parents and teachers, and collaboration with the business world in providing training to teaching teachers' skills.
- It is necessary for schools to collaborate with companies to bring trainers or instructors who will provide skills training to teachers who do not have certain skills. Schools may also make use of vocational schools that have complete infrastructure thus students get both practical experience and direct lessons.
- It is also necessary to provide optimal vocational skills learning for students from assessment and planning and hard and soft skills, and to continue the process of career maturity through internships to job vacancy training, carried out through synergistic partnerships between schools, parents, government, and DUDI.

6 SUGGESTIONS

Based on conclusions obtained in this line of research, a number of suggestions related to the above matters are described as follows:

6.1 *For special needs schools*

Lack of people with disabilities employed after graduating from school makes people with disabilities return to their parents, meaning they do not have independence. Therefore, the skills provided are not only focused on hard skills, but also soft skills as both become the main capital for students to survive working in the company. Schools must also collect data on their student graduates thus they can be an evaluation material for schools in the future.

6.2 *For companies*

Companies that have welcomed government rules regarding employment opportunities for people with disabilities have not been accompanied by optimal funding for employees with disabilities. The willingness of companies to accept employees with disabilities should be accompanied by an attitude of accepting all limitations possessed by persons with disabilities through the provision of compensation for everything that prevents people with disabilities from jobs. The company shall conduct a needs assessment of employees with disabilities thus the placement of workers is in accordance with their needs.

6.3 *For department of education*

The role of stakeholders such as the department of Education – which leads the school in implementing such curriculum – is very important in the work system of persons with disabilities. The limitations of each school make the implementation of the curriculum less than optimal. The Education Department should therefore encourage schools to create a flexible curriculum to give to their students, because, in fact, not everything in the curriculum can be implemented for students.

6.4 *For department of labor*

As a government agency that has functions in the distribution and placement of work, the Department of Labor should be able to carry out its functions optimally, especially those relating to persons with disabilities. The establishment of a task force to carry out these functions should be carried out right away; thus, various work-related problems for persons with disabilities might be immediately resolved.

REFERENCES

Angela, B. M. 2015. Employment of persons with disabilities. *Procedia-Social and Behavioral Sciences* 191: 979–983.

Aprilia, I. D. 2010. *Model Bimbingan dan Konseling untuk Mengembangkan Kemandirian Remaja Tunarungu di SLB-B*. Bandung: Sekolah Pascasarjana UPI.

Blázquez, M., & Malo, M. Á. 2005). Educational mismatch and labour mobility of people with disabilities: The Spanish case. *Revista de Economía Laboral* 2(1): 31–55.

Brueggemann, B. J., & Burch, S. 2006. *Women and Deafness; Double Visions*. Washington: Gallaudet University Press.

Dacey, J., & Kenny, M. 1997. *Adolescent Developmental (Second Edition)*. New York: McGraw-Hill, Inc.

Gall, C., Kingsnorth, S., & Healy, H. 2006. Growing up ready: a shared management approach. *Physical & Occupational Therapy in Pediatrics* 26(4): 47–62.

Guenther, J., Falk, I., & Arnott, A. 2008. *The Role of Vocational Education and Training in Welfare to Work*. National Centre for Vocational Education Research.

Guimaraes, B., Martins, L. B., & Junior, B. B. 2015. Workplace adaptation of people with disabilities in the construction industry. *Procedia Manufacturing* 3: 1832–1837.

Ihara, E. S., Wolf-Branigin, M., & White, P. 2012. Quality of life and life skill baseline measures of urban adolescents with disabilities. *Social Work in Public Health* 27(7): 658–670.

King, K., & Palmer, R. 2010. *Planning forTtechnical and Vocational Skills Development*. Paris: UNESCO, International Institute for Educational Planning.

Levinson, E. M., & Palmer, E. J. 2016. Preparing students with disabilities for school-to-work transition and post-school life. *Principal Leadership* 5(8): 1115.

Borderless Education as a Challenge in the 5.0 Society – Abdullah, Adriany & Abdullah (eds)
 ISBN 978-0-367-61960-2

A child-centered classroom program: An approach to promote child friendly schools

H. Djoehaeni, M. Agustin, A.D. Gustiana & N. Kamarubiani
Universitas Pendidikan Indonesia, Bandung, Indonesia

ABSTRACT: Child-Friendly Schools are schools that consciously strive to guarantee and fulfill children's rights in every aspect of life in a planned and responsible manner. The main principle is non-discrimination of the interests, rights of life and respect for children. Children have the right to be able to live, grow, develop, and participate appropriately according to human dignity and get protection from violence and discrimination. Schools must create a conducive atmosphere so that children feel comfortable and can express their potential. The teacher has an important role in organizing activities in the school. The conducive school environment is very closely related to learning atmosphere that exists between the teacher and child. A child-centered classroom programs is based on the belief that children will grow and learn well if they are naturally involved in the learning process. An environment designed using the concept of a child-centered class provides the broadest opportunities for children to explore, become pioneers and be creative. The teacher's role is to design goals and learning environments that are appropriate to the interests and needs of children, respecting the strengths and needs of each child. This article discusses the ability of teachers to facilitate children in learning processes, that can support children potential, through relevant literature. This study is expected to be a reference for teachers in creating child-friendly schools.

1 INTRODUCTION

The issue of hatred lately is so massive that it occurs in almost all parts of the world, including in Indonesia. Hatred is patterned in various forms and lately a lot has happened, hatred is contained in social media. Data sourced from the healthy internet team posted on 30-30-2016 stated that in 2016 the Ministry of Communication and Information had received 1,769 negative content reports on Twitter, Facebook, and YouTube until early December 2016. Furthermore, it was found that the number of sites and accounts social media that spread hatred and false news tends to increase and is quite worrying.

In terms of culture, hatred originates from the strong stereotyping and prejudice behavior. The forms of stereotypes and prejudices that tend to develop are generally based on differences in ethnicity, gender, religion, politics, aggression, and sex (Gordijn et al. 2001). Stereotypes and prejudices are both interconnected and mutually reinforcing. This means that stereotypes and prejudices that are used negatively will have the effect of conflict and harm, either individually or in groups, and in some cases make prolonged wars that cost many lives (Falanga et al. 2014, Amodio 2008).

In the context of learning in schools including early childhood education, forms of stereotyping and prejudice that tend to develop are often wrapped in tangible ethnic forms of cornering behavior, discriminating and insulting certain ethnic groups. Apart from being ethnic, stereotypes and prejudices in schools also appear in the form of gender such as harassing, insulting certain sexes, discriminating based on male and female groups and feeling stronger than certain groups. There are also forms of stereotypes and prejudices in the form of religion such as forgiving, assuming more noble religion, and defaming certain religious beliefs. Stereotypes and prejudices are also reflected socially such as humiliating certain professions, consider certain professions to be of a higher degree than other professions (Cottrell & Neuberg 2005, Amodio 2014).

The most obvious impact of stereotypes and prejudices is hatred. The most obvious fact in the world of education is the oppression and torture by seniors to juniors which led to deaths on several campuses, especially those labeled official, acts of fighting between students and also acts of violence in schools. According to the campus on June 14, 2016, there were 1880 acts of violence in schools, the most striking of which was the killing of an elementary school student with the initials A who was tortured by his friend, eight years old R, which caused bleeding in the brain and ended with death. Even though childhood education has the potential for stereotyping and prejudice to emerge clearly, especially in the form of violent behavior both verbally, emotionally and physically (Agustin et al. 2018).

So real is the negative impact of stereotypes and prejudices in children's learning activities at school, so a solution or an appropriate alternative is needed to handle it. One alternative that can be applied is to implement a child-friendly school-based learning model. This learning model moves from a number of principles including that children have the right to fulfill educational needs in order to develop all their potential optimally, and children also have the right to grow and develop normally without getting intimidation, discrimination and intimidation especially acts of violence.

Child-friendly school-based learning, namely teacher development, preparing teaching materials that are appropriate to children's needs, preparing school infrastructure that supports, strengthening life skills for children and involving parents.

The teacher has a very important role in creating a conducive learning environment or school environment. Related to its role as a facilitator, the teacher has a role to make it easy for children to interact with their learning environment. A conducive learning environment is an environment that provides a safe and comfortable feeling for children to move. There is no discrimination, intimidation and various treatments that discredit children. This paper will examine more about learning approach to promote child-friendly schools.

2 METHODS

The method used in this article is the library research, a series of studies relating to the method of data collection library, or research object of research explored through a variety of information literature (books, encyclopedias, journals, newspapers, magazines, and other documents). The research literature or review of the literature is a research that examines or critically review the knowledge, ideas, or findings contained in the body of academic-oriented literature. The data used in this research is secondary data. Secondary data is data obtained not from direct observation. However, the data obtained from the research that has been done by previous researchers. Secondary data sources are referred to in the form of books and scientific reports on the primary or original contained in the article or journal.

3 DISCUSSION

3.1 *Child friendly schools*

Conceptually, Child Friendly Schools are schools that consciously strive to guarantee and fulfill children's rights in every aspect of life in a planned and responsible manner. The main principle in creating Child Friendly Schools is non-discrimination of interests, rights to life and respect for children. Child Friendly Schools are schools that openly involve children to participate in all activities, social life, and encourage child growth and welfare. Furthermore, Child Friendly Schools are safe, clean, healthy, green, inclusive and comfortable schools/madrasas for physical, cognitive and psychosocial development of girls and boys including children who need special education and/or service education special.

Child Friendly Schools are schools that consciously strive to guarantee and fulfill children's rights in every aspect of life in a planned and responsible manner. The main principle is non-discrimination of interests, rights to life and respect for children. As stated in Article 4 of Law No.23 of 2002 concerning child protection, states that children have the right to be able to live to grow, develop, and participate fairly according to human dignity and dignity, and get protection from violence and discrimination. Mentioned above is one of them is participating which is described as the right to have an opinion and to hear his voice. Child Friendly Schools are schools that openly involve children to participate in all activities, social life, and encourage child growth and welfare.

The study by Kusdaryani et al. (2016) emphasized the need for friendly schools through the creation of a culture full of friendliness and respect. Furthermore, learning in child-friendly schools has the following indicators that are inclusive and proactive, learning creates a healthy, safe, and protective climate, involves active community participation, is effective and child-centered learning, and values gender equality. The application of learning also implements activities that invite children to be more active, innovative, creative and fun that are carried out jointly by all learning citizens.

In order to create a conducive atmosphere in learning at child friendly schools, there are several aspects that need to be considered, especially: (1) appropriate school programs; (2) a supportive school environment; and (3) aspects of adequate infrastructure. The teacher has a very important role in creating a supportive school environment. One aspect that plays an important role in creating a supportive school environment is implementation of learning strategy that suitable for young children.

3.2 *Early childhood curriculum*

Early childhood education is seen as a fundamental education for an individual. Many experts put forward opinions about the importance of early childhood education.

Hildebrand (1981) states that the developmental tasks serve as a basis for the ten major goals for nursery school and kindergarten aged children. Parents and children will strive to help children growing in independence. Learning to give and share as well as receive affection. Learning to get along with others, developing self-control, learning non sexist human roles, beginning to understand their bodies, learning and practicing large and small motor skills, beginning to understand and control the physical world, learning new world and understanding others, and developing a positive feeling about their relationship in the world.

What is the purpose of early childhood curriculum? The purpose of early childhood curriculum is to foster competence of young children, though not only competence in intellectual areas. Competence should have thought of a relating to all aspect of the self. Learning to live comfortable with others, learning to master and safely express one's feelings, and learning to love, live and welcome new experience. The purpose of education then is to foster competence in dealing with life.

According to Scott, a curriculum can be defined in the following way: A curriculum may refer to a system, as in a national curriculum; an institution, as in a school curriculum; or even to an individual school, as in the school geography curriculum (McLachlan et al. 2018).

In term of curriculum, educators must be able to answer the following questions:

- What educational purposes should the school seek to have?
- What educational experiences can be provided that are likely to achieve
- these purposes?
- How can these educational experiences be effectively organized?
- How can we determine whether or not these purposes are being achieved?

A curriculum can be organized specifically to include four dimensions: aims or objectives, contents or subject matter, methods or procedures, and evaluation and assessment. The first dimension refers to the reasons for including specific items in the curriculum and excluding others. The second dimension is content or subject matter and this refers to knowledge, skills or dispositions which are implicit in the choice of items, and the way that they are arranged.

Objectives may be understood as broad general justifications for including particular items and particular pedagogical processes in the curriculum; or as clearly defined and closely delineated outcomes or behavior; or as a set of appropriate procedures or experiences. The third dimension is methods or procedures and this Early Childhood Curriculum refers to pedagogy and is determined by choices made about the first of dimensions. The fourth dimension is assessment or evaluation and this refers to the means for determining whether the curriculum has been successfully implemented. So the four crucial elements which apply to curriculum in any teaching and learning setting from early childhood through to tertiary education are:

- Aims, goals, objectives or outcome statements it related to what do we want this curriculum to achieve, what would we expect to be the outcomes as a result of participating in the implementation of this curriculum?
- Content, domains, or subject matter - what will we include or exclude from our curriculum?
- Methods or procedures is related to what teaching methods or approaches will we use to achieve these goals or outcomes?
- Evaluation and assessment it is related to how will we know when we have achieved them?

In relation to the objectives to be achieved in the early childhood curriculum, Hildebrand (1981) states that there are Ten Major Goals, which is:

- Growing in Independence
- Learning to Give and share, as well as receive affection
- Learning to get along with others
- Developing self-control
- Learning no sexist human roles
- Beginning to understand their bodies
- Learning and practicing large and small motor skills
- Beginning to understand and control the physical world
- Learning new words and understanding others
- Developing a positive feeling about their relationship to the world.

Achievement for the purpose of education is depend on the planning made by the teacher. planning as an action plan that will be done by teachers in the learning process needs to be prepared in earnest. there are some important components that should be the concern of teachers in developing a plan.

In connection with the planning that will be made in implementing the curriculum in teacher learning in kindergarten, teachers need to direct on three main questions Katz (2009) namely:

- What should be learned?
- When it should be learned?
- How is it best learned?

Responses to the first question provide the goal of the program for which pedagogical practices are to be adopted. The second question is developmental. In that it draws upon what is known about the development of the learner. In other words, child development helps to address the question of program's design. The third question turns specially to matters of appropriate pedagogy itself; it includes consideration of all aspects of a program implementation by which the program's goals can be achieves, depending on what is to be learned, and when it is to be learn. In other words, responses to one of the three questions are inextricably linked to responses to the other two.

There are four types of learning goals Katz (2009) states that whatever specific learning goals and objectives are identified by clients and educators, they are all likely to fit into each of four types of learning goals:

- Knowledge/understanding
- Skills
- Disposition
- Feelings

Knowledge/understanding can be broadly defining as ideas, concepts, constructions, schemas, fact, information, stories, customs, myths, song and other such contents of mind that come under the heading of what is to be learned. Skills are easily observable forms of behavior such as cutting, drawing, counting, classifying, make friends and solve problems. Unlike an item of knowledge or a skill, a disposition is not an end-state to be mastered once and for all. It is trend or consistent pattern of behavior and its possesion is established only if its manifestation is observed repeatedly. Thus a person's disposition to be reader, for example, can only be ascertained if he or she is observed to read spontaneously, frequently and without external coercion (Katz 2009).

3.3 *Developmentally appropriate practices*

Today there is a tendency to increase teaching that is formal, structured and systematic in early childhood education institutions especially in kindergartens. Learning practices are characterized by teaching that is more teacher-oriented, academic-oriented abilities such as teaching reading, writing and calculating formally, as well as providing worksheets and homework assignments. Many factors influence the emergence of such learning practices. One of them is the demands of parents who expect their children to have academic abilities. Such conditions are certainly not relevant to educational practices that are appropriate to the characteristics of children

According to Elkind in Masitoh & Djoehaeni (2005) this tendency is related to the increasing understanding of society that early age is a very important period, and early intervention can provide greater benefits for children. In addition, this also arises because of the desire to be able to compete with other nations and because of the belief that providing education early is better. As a result, young children find ways to learn by just sitting on a bench, listening, taking notes, memorizing and taking tests. Child education experts see this as "the erosion of childhood and the education of child community members". Both of these will be very dangerous for the future of children. Doctors in America report that from such learning practices many children experience stress or mental distress.

The National Association of Education for Young Children (NAEYC) defines the concept of Developmentally Appropriate Practices (DAP). This paper illustrates examples of appropriate practices and practices that are not appropriate for educational programs that serve children from birth to age 8 years (Bredekamp 2014). The purpose of DAP is to develop curriculum for early childhood education programs from curriculum oriented to academic skills, exercises and practical approaches to teaching, to curricula that are oriented to child development.

Knowing how children learn and develop is essential for teachers of young children. The more they know about and sensitive to the way children think and learn, the more effective their teaching a the more satisfying their work. Bredekamp & Copple (1997) state that to successfully engage in developmentally appropriate practice teacher need to:

- Meet children where they are, as individual and as a group.
- Use variety of intentional strategies to help each chodran attain challenging and achievable goals that contribute toi his or her ongoing development and learning.

In Developmentally Appropriate Practices, teachers must give encouragement to children to be able to go through each stage of their development in a meaningful, optimal, and learning manner that is fun, attractive, and relevant to their experiences. Developmentally Appropriate Practices gives more opportunities for children to learn in appropriate ways, for example through direct experience, exploring and other meaningful activities

Referring to the opinion of Bredekamp (2014) there is a set of Developmentally Appropriate Practices for children aged 3-5 years in terms of several dimensions as follows:

- Creating a positive climate for learning.
- Assist group closeness and meet individual needs.
- Environment and schedule.
- Learning Experience.
- Language and Communication.

- Teaching Strategy.
- Motivation and Guidance.
- Curriculum.
- Evaluation.

What learning experiences allow children to develop all aspects of their development? According to Pestalozzi, kindergarten education should provide experiences that are fun, meaningful, and warm as provided by parents in the home environment (Masitoh & Djoehaeni 2005). In line with this, Solehudin (1997) revealed that:

"In general, preschool education is intended to facilitate the overall growth and development of children in accordance with the norms and values of life. Preschool education should not be academically oriented, but should be able to provide learning experiences for children. Besides that, preschool education programs must be adapted to the needs, interests and development of children. "So the efforts to optimally facilitate children's growth and development are actually implemented through the provision of learning experiences that are not academic oriented, in the sense of not emphasizing the mastery of certain abilities, but rather emphasizing learning experiences that are appropriate to the interests and needs of children. So what kind of approach is appropriate to the interests and needs of the child?

3.4 *A child-centered classroom program*

According to Coughlin (2000), a child-centered classroom programs are development-oriented learning approaches that seek to develop all aspects of children's development optimally. Child-centered classroom programs, as one of the child-centered approaches, place great emphasis on the individualization of the child's learning experience, providing opportunities for children to make decisions or choose activities that are in line with their interests in activity centers, and family participation through activities which is prepared.

A child-centered classroom programs is based on the belief that children will grow and learn well if they are naturally involved in the learning process. An environment designed using the concept of a child-centered class provides the broadest opportunities for children to explore, become pioneers and be creative. The teacher's role is to design goals and learning environments that are appropriate to the interests and needs of children, respecting the strengths and needs of each child.

A child-centered classroom program is an approach that is in harmony with constructivist theory, because this approach provides the broadest opportunity for children to construct their knowledge through learning experiences designed by the teacher. Children's freedom to choose activities that are in accordance with their needs and interests as well as the existence of activity centers or areas in the class that children are free to explore are one embodiment of this theory.

This methodology is based on knowledge of child development. All children develop through common stages, however at the same time children are unique individuals. For this reason, teachers are expected to be able to know the growth and development in children so that they can facilitate and serve the needs of different children. Katz (2009) states that: "In a developmental approach to curriculum design...(decisions) about what should be learned and how it would be best learned depend on what we know of the learner's developmental status and our understanding of the relationship between early experience and subsequent development".

Bredekamp (2014) also revealed that there are 2 dimensions of the term "in accordance with development, namely suitability of the age and suitability of the individual. Human development research shows that children undergo a series of growth and universal changes that are expected during the first nine years. These predictable changes occur in all areas of physical, emotional, social, cognitive and linguistic development. The knowledge provided by this program regarding typical developments in the age range provides a framework that teachers can use to prepare learning environments and plan experiences accordingly (Bredekamp 2014).

To implement activities that are in accordance with development, the teaching team must be aware of the normal range of development. The teacher must realize that even though there is a series of developments that can be predicted, the developments are not the same situation every time and every individual. Related to individual suitability Bredekemp (2000) revealed that each child has a unique pattern and time of development, such as personality, type of learning and family background. Both the methodology and the interaction of adults with children must be consistent with children's individual differences. The learning process occurs because of the interaction between children's thoughts and experiences with teaching materials, ideas and people. These experiences must match the abilities of the developing child and also challenge the interests and understanding of the child.

As an implication of the above understanding, Coughlin (2000) revealed that class activities must be individually appropriate. For that the instructors must observe each child carefully and determine their abilities, needs, interests, temperament and ways of learning. To be able to adjust activities for each child, strong knowledge is needed about the activities that are potential for each learning center. The child-centered classroom program is a developmental approach. Activities are fully designed with reference to the characteristics of child development. The existence of centers of activities in the classroom, in essence is one of the efforts to be able to facilitate all aspects of child development while paying attention to individual differences. The child-centered classroom program emphasizes the aspect of individualization of the child's learning experience, providing opportunities for children to make decisions or choose activities that are in accordance with their interests in activity centers, and family participation through activities prepared.

Basically every individual has different characteristics from each other. This is what underlies the concept of individual differences in students. These individual differences should be considered by educators in developing learning. The belief that each child is a different individual is expressed by Bredekamp (2014), that: It is variations that make the world of early childhood attractive, and that is precisely what makes teachers love teaching activities, because every child, every group of children is different. In line with that, Solehuddin (1997) states that: children will learn well if:

- Children feel psychologically safe, and their physical needs are met.
- Children construct knowledge.
- Children learn through social interaction with adults and other children.
- Children learn through play
- Children's interests and needs to know are fulfilled.
- Elements of individual variation are considered.

This opinion implies that learning will be more meaningful if children can do something according to their interests, needs and capacities. The element of individual differences will directly impact the approach chosen by the teacher. A variety of approaches can facilitate different children's characters.

A child-centered classroom program lay a strong foundation for children to become adults who are insightful, active, successful and care for others. The teacher is very concerned with matters relating to childhood. They value the process of individualization by respecting the different stages of development of each child. In a child-centered approach there is a center of activity as a vehicle for children to explore various kinds of teaching materials and games. The variety of activity centers and available materials, provides a great opportunity for children to choose activities that suit their interests. The role of the teacher is to help provide a variety of interesting activities in the center of activity and look for material that is always interesting to explore.

According to Coughlin (2000), activity centers vary greatly from one class to another, however there are major centers of activity that should be shared by each class, namely mathematics/arithmetic, science, reading and writing recognition, art, role plays, music, cooking, playing as well as sand and water. In the 2013 Curriculum Implementation Guidelines for Early Childhood Education (2015) it is stated that the center or area learning model facilitates children's activities both individually and in groups to develop all aspects of development. This learning model provides opportunities for children to develop in accordance with their talents and interests.

4 CONCLUSION

Child-friendly schools are a learning environment that is the dream of all academics. Child-friendly schools are actually learning environments that provide a sense of comfort and security free from discrimination and treat students according to their circumstances. Including a conducive environment so that students especially early childhood will be stimulated all aspects of their development perfectly.

Learning will be more meaningful if children can do something according to their interests, needs and capacities. The element of individual differences will directly impact the approach chosen by the teacher. A variety of approaches can facilitate different children's characters.

A child-centered classroom programs is based on the belief that children will grow and learn well if they are naturally involved in the learning process. An environment designed using the concept of a child-centered class provides the broadest opportunities for children to explore, become pioneers and be creative. The teacher's role is to design goals and learning environments that are appropriate to the interests and needs of children, respecting the strengths and needs of each child.

A child-centered classroom programs lay a strong foundation for children to become adults who are insightful, active, successful and care for others. The teacher is very concerned with matters relating to childhood. In a child-centered approach there is a center of activity as a vehicle for children to explore various kinds of teaching materials and games. The variety of activity centers and available materials, provides a great opportunity for children to choose activities that suit their interests. The role of the teacher is to help provide a variety of interesting activities in the center of activity and look for material that is always interesting to explore.

A child-centered approach through the provision of activities in the center of interest or area provides the opportunity for children to carry out activities in accordance with the interests of learning while providing opportunities for children to interact with their social environment, fostering a sense of empathy for the different needs of others.

REFERENCES

Agustin, M., Saripah, I. & Gustiana, A.D. 2018. Analisis Tipikal Kekerasan pada Anak dan Faktor yang Melatarbelakanginya. *Jurnal Ilmiah Visi* 13(1): 1–10.

Amodio, D.M. 2008. The social neuroscience of intergroup relations. *European review of social psychology* 19(1): 1–54.

Amodio, D.M. 2014. The neuroscience of prejudice and stereotyping. *Nature Reviews Neuroscience* 15(10): 670–682.

Bredekamp, S. 2014. *Effective practices in early childhood education: Building a foundation*. Upper Saddle River, NJ: Pearson.

Bredekamp, S. & Copple, C. 1997. *Developmentally Appropriate Practice in Early Childhood Programs (Revised Edition)*. Washington, DC: National Association for the Education of Young Children.

Cottrell, C.A. & Neuberg, S.L. 2005. Different emotional reactions to different groups: a sociofunctional threat-based approach to" prejudice". *Journal of personality and social psychology* 88(5): 770.

Coughlin, P. 2000. *Menciptakan kelas yang berpusat pada anak*. Wahington D.C.: Children Resources International, Inc.

Falanga, R., De Caroli, M. & Sagone, E. 2014. The relationship between stereotypes and prejudice toward the Africans in Italian university students. *Procedia-Social and Behavioral Sciences* 159: 759–764.

Gordijn, E.H., Koomen, W. & Stapel, D.A. 2001. Level of prejudice in relation to knowledge of cultural stereotypes. *Journal of Experimental Social Psychology* 37(2): 150–157.

Hildebrand, V. 1981. *Introduction to early childhood education*. New York: Macmillan.

Katz, L.G. 2009. A developmental approach to the curriculum in the early years. In *Key issues in early years education*. London: Routledge Taylor and Francis Group.

Kusdaryani, W., Purnamasari, I. & Damayani, A.T. 2016. Penguatan Kultur Sekolah untuk Mewujudkan Pendidikan Ramah Anak. *Cakrawala Pendidikan* (1): 85532.

Masitoh, O.S. & Djoehaeni, H. 2005. Pendekatan belajar aktif di taman kanak-kanak. *Jakarta: Departemen Pendidikan Nasional.*

McLachlan, C., Fleer, M. & Edwards, S. 2018. *Early childhood curriculum: Planning, assessment and implementation*. New York: Cambridge University Press.

Solehudin, M. 1997. *Konsep Dasar Pendidikan Prasekolah*. Jakarta: Departemen Pendidikan dan Kebudayaan.

Teacher education qualification framework

Borderless Education as a Challenge in the 5.0 Society – Abdullah, Adriany & Abdullah (eds)
 ISBN 978-0-367-61960-2

Coaching of individual learning model for primary school teachers providing inclusive education

E. Rusyani, T. Hernawati & R. Akhlan
Universitas Pendidikan Indonesia, Bandung, Indonesia

ABSTRACT: The problem faced by teachers when implementing inclusive education is closely related to the competences and skills required to optimize the potential of students with special needs. The aim of this research is to develop effective coaching to promote primary school teachers' competences and skills when providing inclusive education. Pre- and posttest procedures are used to clearly evaluate the outcomes before and after an intervention. There are four variables' competencies and skills to develop in the coaching: inclusive education, children with disabilities, accommodation curriculum, and an individualized education program/plan. The results show that average points obtained by teachers increase from 20 pt (pre) to 62 pt (post). Furthermore, coaching impacts teachers' professional satisfaction, teaching practices, and self-efficacy. Teachers learned through the process to improve their practice and beliefs about teaching. Group approaches seem to be promising strategies to solve each school's problems with discussions to get other perspectives.

1 INTRODUCTION

The implementation of inclusive education has not met expectations in elementary school. Sunanto (2009) shows that elementary piloting providing inclusive education for students with special needs generally still relies on the assistance of experts or special needs education teachers. The inclusion index, which is an illustration of the extent to which the learning process in the class shows degree of inclusiveness, only reached 38.58%. This shows that the implementation of inclusive education is still far from the principles of inclusive education and learning is still not in accordance with the basic elements of inclusive education in terms of planning, process, and evaluation.

In general, learning in most elementary schools is still classical and largely consists of homogeneous classes. As a result, students' potential may not develop optimally. Though the development of learning methods has advanced, learning innovations continue to be developed. The method used in learning is highly dependent on the teacher factor. Teachers become one of an important element in evolving inclusive practices. The problem is related to how they accomplish individual learning in diverse classes. Teacher skill and competence need to get developed (Setiawati 20015, Yasa & Julianto 2018). Eventually, teachers have to be competent in developing teaching practices.

Research and literature can be found that coaching is an approach to developing general education and special education teachers (Biancarosa et al. 2010, Neuman & Wright 2010, Shanklin 2006). Cornett & Knight (2009), in their review of the coaching literature, suggested that coaching impacts teachers' professional satisfaction, teaching practices, and self-efficacy. The literature also indicates that, over time, teachers learned through coaching to improve their practice and beliefs about teaching.

Coaching programs, offer a promising approach for increasing teachers' capacities to integrate best practices within their teaching contexts. Coaching refers to a form of development in which an experienced person, called a coach, supports a learner or client in achieving a specific personal or professional goal by providing training and guidance (Passmore 2016). Coaching is about enhancing knowledge and skills.

1.1 *Individualized educational program/plans*

An individualized educational program/plan (IEP) is developed by teachers (special education and general education) and parents to specify a student's academic goals and the tools and actions needed to help achieve those goals (Watson 2019). IDEA requires that the individualized education program (IEP) of students with disabilities "must be aligned with the State's academic content standards for which the child is enrolled" (Ali 2010). Effective learning environment and instructional decisions are made at the individual student level, rather than making blanket decisions based upon a disability category (Bratshaw 1997).

The basic characteristics of the IEP included in the Code of Inclusive Practice are careful focusing on the specific child's learning difficulties: take into account what the child has achieved by following the curriculum; set clear goals that a child needs to master over a period of time; ensure the participation of parents and the child (if the child's condition allows it) in the development of IEP; and include specialists in the preparation of IEPs (Warnock 2008).

2 METHODS

2.1 *Design*

Pretest–posttest procedures are commonly used to evaluate learner outcomes of educational programs. This procedure clearly indicates results before and after an intervention. Likewise, this method can be helpful in determining growth toward achievement of specific standards or a blend of standards within a course. In this case, one-group, pretest–posttest design is used to determine the effect of a coaching on a given sample.

This research design is characterized by two features (Cranmer 2017). The first feature is the use of a single group of participants all part of a single condition, meaning all participants are given the same treatments and assessments. The second feature is a linear ordering that requires the assessment of a dependent variable before and after treatment is implemented (i.e., a pretest–posttest design). Within pretest–posttest research designs, the effect of a treatment is determined by calculating the difference between the first assessment (pretest) and the second assessment (posttest).

2.2 *Participants*

The coaching was implemented in a district in West Java, Indonesia. Before the sample was collected, the researchers had to determine the population. The population was 326 teachers from elementary schools. Purposive sampling technique was used, so the expected criteria for the sample was completely in accordance with the research to be conducted. The sample was chosen based on purposive sampling, dependent on the criteria. The criteria used as consideration in selecting a sample were based on experience, in this case, facing children with special needs in their class and a teacher who will support evolving inclusive practices in their schools. The sample for this study was 30 teachers.

2.3 *Data collection*

The participants filled out the four-item questionnaire before and after coaching with the same set of questions concerning inclusive education, children with disabilities, accommodation curriculum, and individualized educational program/plan.

3 RESULTS AND DISCUSSION

Table 1 shows the pretest and posttest scores on four variables encountered by teachers. Scores increased significantly on all variables posttest. For inclusive education, the variable increased 47 points, the disability children variable increased 35 points, the accommodation curriculum variable increased 39 points, and the individualized educational program increased 47 points. The average score increased about 42 points from 20 points (pretest) to 62 points (posttest) in four variables. Change score is about 210%.

Table 1. Result of pretest and posttest in four variables.

Variables	Pretest score (Average point)	Posttest score (Average point)	Difference	Change (%)
Inclusive Education (N=30)	23	70	47	204%
Disability Children (N=30)	25	60	35	140%
Accommodation Curriculum (N=30)	15	54	39	260%
Individualized Educational Program/Plan (N=30)	17	64	47	300%
Total Average	20	62	42	210%

The data shows that an individualized educational program/plan has a higher change score than others. Likewise, on accommodation curriculum variable, teachers have to analyze curriculum when devising an individualized educational program/plan (IEP). This could be due to the addition of practical material to the teaching during the coaching. Therefore, the addition of practical material affects the result.

The result shows that the coaching is increasing knowledge and skills of participants, especially when participants can practice their knowledge directly because they can explore their knowledge into the practical. In IEP, for example, teachers' practice directly impacted IEP based on their student's condition.

Within this study, the question analyzed was how coaching improves teacher's knowledge and skills. Our analyses show significant associations between procedures and success in teachers gaining knowledge and skill. Coaching procedure in this study was teachers learning through teamwork, propounding problems and discussing with others. Teachers can promote their critical thinking and creative thinking. Learning in groups can promote individual learning on problem-solving tasks (Laughlin 2008).

REFERENCES

Ali, R. 2010. *Dear Colleague Letter: Harassment and Bullying*. Washington, DC: US Department of Education.

Biancarosa, G., Bryk, A. S., & Dexter, E. R. 2010. Assessing the value-added effects of literacy collaborative professional development on student learning. *Elementary School Journal* 111(1): 7–34.

Bratshaw, M. L., Pelligrino, L., & Roizen, N .J. 1997. *Children with Disabilities* (6th edition). Baltimore, MD: Paul H. Brooks Publishing Company.

Cornett, J., & Knight, J. 2009. Research in Coaching. In J. Knight (ed.), *Coaching: Approaches and Perspectives* (pp. 192–216). Thousand Oaks, CA: Corwin Press.

Cranmer, G. A. 2017. One-group pretest-posttest design. In M. Allen (ed.), *The SAGE Encyclopedia of Communication Research Methods*, 1125–1126. Thousand Oaks, CA: SAGE Publications.

Laughlin, P. R., Carey, H. R., & Kerr, N. L. 2008. Group-to-individual problem-solving transfer. *Group Processes & Intergroup Relations* 11(3): 319–330.

Neuman, S. B., & Wright, T. S. 2010. Promoting language and literacy development for early childhood educators: A mixed-methods study of coursework and coaching. *Elementary School Journal* 111(1): 63–86.

Passmore, J. 2016. *Excellence in Coaching: The Industry Guide*. London; Philadelphia.

Setiawati, E. 2015. *Profil sekolah penyelenggara pendidikan inklusif di sekolah dasar negeri tamansari 1 yogyakarta*. Yogyakarta: eprints UNY.

Shanklin, N. L. 2006. *What are the characteristics of effective literacy coaching?* Urbana, IL: Literacy Coaching Clearinghouse.

Sunanto, J. 2009. *Profil Implementasi Pendidikan Inklusif Sekolah Dasar Di Kota Bandung*. Bandung: Universitas Pendidikan Indonesia.

Warnock, M. 2008. Special educational needs: A new look. In *Challenges for Inclusion* (pp. 43–65). Brill Sense.

Watson, S. 2019. *IEP-Individual Education Program*. [Online] https://www.thoughtco.com/iep-individual-education-program–3111299.

Yasa, R. B., & Julianto, J. 2018. Evaluasi Penerapan Pendidikan Inklusi di Sekolah Dasar di Kotamadya Banda Aceh dan Kabupaten Pidie. *Gender Equality: International Journal of Child and Gender Studies* 3(2): 120–135.

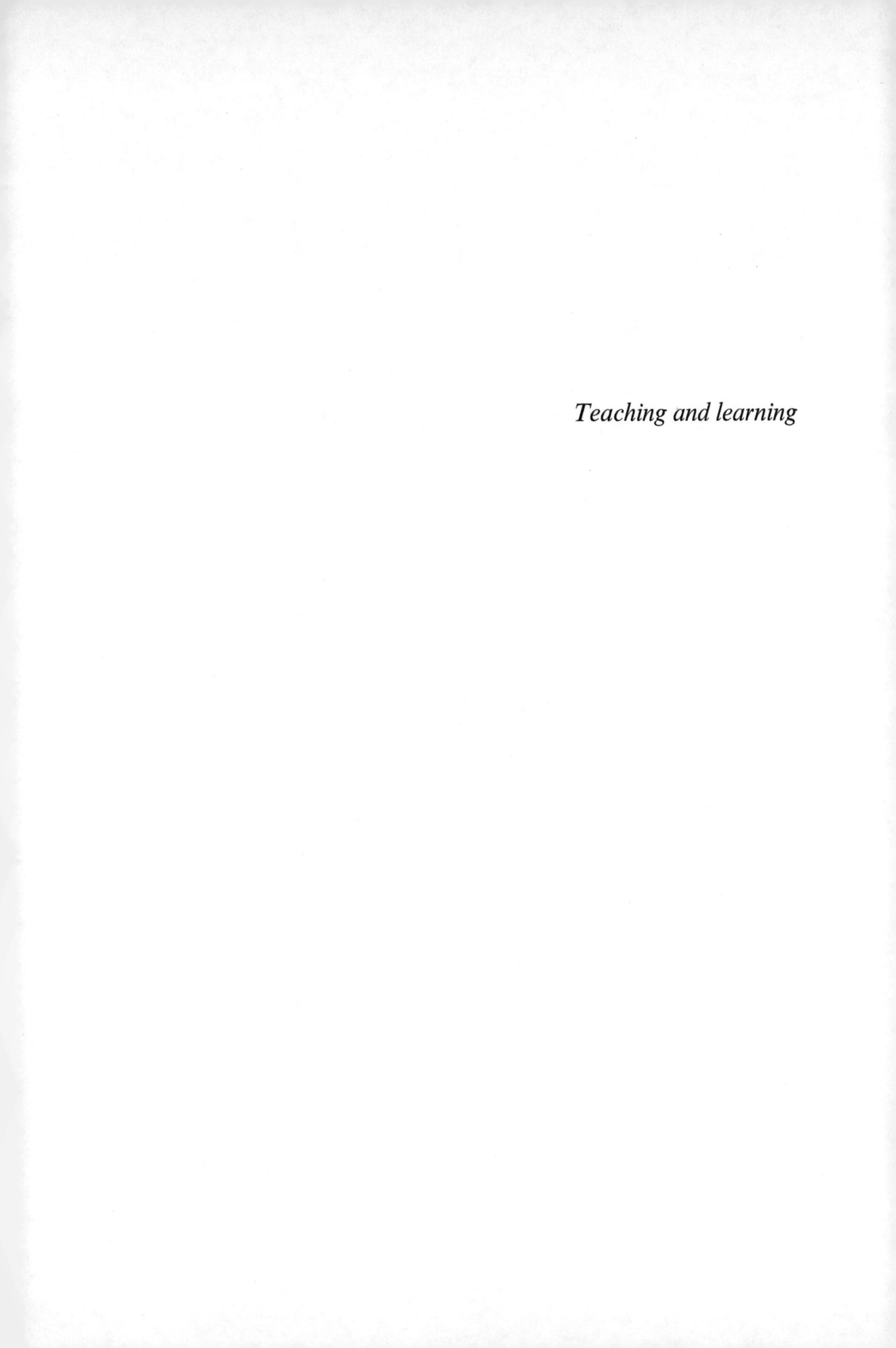

Teaching and learning

Borderless Education as a Challenge in the 5.0 Society – Abdullah, Adriany & Abdullah (eds)
© 2021 Taylor & Francis Group, London, ISBN 978-0-367-61960-2

Existence of Salafiyyah Islamic Boarding School in the middle era of industrial revolution 4.0

S.E. Cipta
Universitas Pendidikan Indonesia, Bandung, Indonesia

ABSTRACT: This research is a study that examines how pesantren as an Islamic education institution in Indonesia is able to survive amid the current modernization of pesantren education that changes the education system from traditional to modern. In this study there are several formulations of the problem, namely: 1) how was the initial development of Al-Barokah pesantren education, 2) how was the strategy of using the applied learning methods and models, 3) how was the final evaluation process at the Al-Barokah Islamic boarding school? This research uses the content study method or literature study using several written studies and the use of qualitative methods in writing this article.

1 INTRODUCTION

Education is an effort to shape the personality of children and students. In Islam education becomes a condition of obligation that must be carried out because in reality education can shape both behaviour and morals, of course, blessed by GOD. At least, there is an education system that is applied in Islam. First is the school-based Islamic education system. Islamic schools are the same as schools in general while still referring to the curriculum. It's just that what is distinguished between public schools and Islamic schools is located in the contents of the curriculum, especially material and lessons with additions such as moral education, jurisprudence education, hadith education, and memorizing surah juz the 30th which is mandatory for students as graduation requirements both in the field of study and the next stage. Second, Islamic boarding school-based education system. Often people mention that the pesantren education system is a very and very traditional education system. It was seen that Islamic boarding school education was strongly emphasized in the figure of Kiai and Habib who were also leaders or founders of Islamic boarding schools. Regarding the methods of education applied by pesantren in addition to education, fiqh education, hadith education, and memorizing surah in pesantren-based Islamic education also always emphasizes Quran interpretation education, and Quran studies. This method is often referred to as the yellow book learning method. Islamic boarding schools also carry out several roles, besides being Islamic education institutions.

Regarding the education system applied in the Salafiyyah Islamic boarding school, it is very different from other formal education systems. Islamic boarding schools are mainly based on salafiyyah or the education system in traditional boarding schools is focused on teaching religious learning in general such as Fiqh, Arabic, Hadiths, Sufism, Tawhid, and Tafsir Qur'an. The learning methods applied by Bandungan, Sorogan and Wetonan at the Salafiyah Islamic boarding school. The Yellow Book learning is used as the main learning media in the Salafiyyah Islamic boarding school. The study of the Yellow Book is to study the legal provisions of Fiqh and Islamic Law. The use of instructional media in salafiyah Islamic boarding schools is also as simple as the use of chalk boards and chalk as the main learning media with the unsupported use of modern learning media such as LCD projectors, laboratories, and the use of internet networks as learning support media. According to Dhofier (2011) classifying boarding schools is divided into several groups of boarding schools including; The old type (classical), the learning is only focused on the Yellow Book which consists only of nahwu and sharaf. The new type of learning is teaching material outside of the subject matter of religious learning. In the new type of learning classes have begun to open classes such as elementary, junior high and high school like boarding schools such as schools.

In the Salafiyyah Islamic Boarding School, the kiai becomes a respected and respected figure of the santri. The charismatic figure of the kiai makes the santri to submit to and obey the kiai so that not a few santri violate orders or oppose the kiai. For santri, in addition to the kiai being a teacher, a kiai is likened to a waliyullah who will demand for his followers and followers. That is in the structural position of the Islamic boarding school, the position of the kiai is a leader and caretaker of the boarding school education system (Sukamto 1999).

In the early 1980s there were changes and developments in pesantren educational institutions. Many Islamic boarding schools later transformed and adopted the modern education system as implemented by madrassas or formal schools. This is in line with the number of Islamic boarding schools that have begun to open classes such as MI, MA, MTS, SITP/public high school, religious tertiary institutions and even non-religious tertiary institutions. These changes can lead to new educational curricula in Islamic boarding schools with the existence of new educational patterns to the implementation of curriculum management in Islamic boarding schools (Fatmawati 2015, Saifuddin 2015). Supported by the role of the government in dealing with curriculum issues so that it results in changes in Islamic education as a whole, causing many Islamic boarding schools to change towards the modern (khalafiyah) of the traditional education system (salafiyah) (Baidlawi 2006).

According to Nafi (2007), said that pesantren have a role as a religious, scientific, coaching, community development, and at the same time a cultural knot. The role is usually not formed directly through step by step. The success of santri in understanding both Islam and teachings relating to Budi Perkerti was also seen as the success of a Kiai in delivering material to his students. The learning process that is applied generally starts with the delivery of the original subject matter, often the lecture method is carried out by the scholars and teachers and dignitaries starting to implement an independent learning system with the creation of study groups, then presenting. Al Barokah Islamic Boarding School is a boarding school that uses the Salafiyah method approach. Founded by KH. Mumuh Abdul Muhyi requested permission from the tower owner in 1982 when the land from the Al Barokah boarding school was one land with the tower owner on one of the cellular operators. The santri consisted of various regions such as Brebes, Cianjur, Sukabumi, Garut, and parts of Central and East Java. However, the reality of the Al Barokah Islamic Boarding School is that students who study at the lodge are almost all students.

2 RESEARCH METHODS

2.1 *Type of the research*

The method used by the author is a content study method that is using a literature study approach from several sources by examining a number of findings in the form of collecting books as a reference source as well as comparing the sources in this discussion (Hardiansyah 2010, Sugiyono 2009). The author also conducted interviews in an effort to strengthen the data during the research process.

2.2 *Place and time of research*

2.2.1 *Place the research*

The research entitled "Student Education System in Al Barokah Islamic Boarding School" was held at the Al Barokah Islamic Boarding School located on Jalan Cilandak RT04/RW05, Sukasari District, Sukarasa Sub-District, Bandung, West Java.

2.2.2 *Time of research*
The research was conducted on September January, 2019, at 10.00 WIB - 14.00 WIB, but before the research was conducted the researchers visited the Al Barokah Islamic Boarding School in the framework of the pre-study on 12 February, 2019.

2.3 *Research subjects*

The subjects in this study were two head of Islamic boarding school and three students in Al-Barokah Islamic Boarding School. The total number of subjects interviewed totalled 5 people.

2.4 *Data collections technique*

To obtain the data needed in this study, the researcher collected data. The data collection techniques carried out include:

2.4.1 *Observation*
Observations carried out in this study were researchers who participated in teaching and learning activities but were not directly involved or only limited to observing. The observation activities in this study aimed to find out the implementation of learning in Al Barokah Islamic Boarding School.

2.4.2 *Interview*
In the interview activity the researcher used several research instruments in the form of a list of questions asked to the resource persons who had been prepared previously. The list of questions asked is then recorded using a voice note to record answers delivered by the interviewees.

3 RESULTS AND DISCUSSION

The increasing number of santri who want to study in the pesantren. The status of the students who study at the Al Barokah Islamic Boarding School varies, mainly from students and students. Both MA/SMA, SMP/MTs students, also in Higher Education.

In 2005, one of the tower operator owners asked KH. Mumuh Abdul Muhyi (Kang Sepuh), in the Sundanese tradition the mention of Kang Sepuh is the senior leader or the establishment of pondok pesantren. Then given the permission to build the tower, the owner of the tower always sent the infaq as well as being the main donor for the smooth running of the Al Barokah Islamic Boarding School. Since 2016 the Al Barokah Islamic Boarding School has at least 100 santri consisting of various regions such as Garut, Cianjur, Sukabumi, Brebes, and so on. To be part of the santri of the Al Barokah Islamic Boarding School the santri were not charged at all during the study period, but the santri were only charged per-person electricity costs of Rp. 35,000. This is due to the sincerity of the sincerity of the teacher and the Kiai in guiding the students. But all this was done without intending but Al Barokah Islamic Boarding School remained consistent in carrying out its duties and obligations in guiding children into the behaviour of karimah and of course as a provision in continuing the Islamic mission when the students had become alumni and returned to their respective regions.

3.1 *Differences between students who follow outside learning activities with student who only follow learning activities in Al Barokah Islamic Boarding School*

In general, students in the Al Barokah Islamic Boarding School are not distinguished from the santri in general. The term 'santri' comes from an Indian language meaning it means the teacher of the Koran. According to C. C Berg (in Dhofier, 2011), that the term santri comes from the term shastri which in Indian means people who know the sacred books of Hinduism. The word santri comes from the root word shastra which means holy books, religious books or books about science. So, the understanding of pesantren is etymologically the origin of the santri which means the place of the santri. It's just that students who take part in outside learning activities such as schools and universities are not required to take part that are implemented by Kiai in full. In other words, the Al Barokah Islamic Boarding School strongly emphasizes the principles of humanism applied to its santri. This is in line with the presentation of Kang Anom (interview with Haji Yayat, September 13, 2018), namely, "... student students are only required to take part in the lessons done by Kiai at night only, the students are allowed to take part in their college activities in the morning." 'Besides the principle of humanism that is applied, Al Barokah Islamic Boarding School also allows students to obtain knowledge in general without having to stick to religious teachings. This was done so that the santri had the provision of knowledge both for the sake of the hereafter and their worldly interests to make the santri as educated and willing to educate.

In the perspective of Islam seeking knowledge is something that must be sought for every Muslim as long as the knowledge is still oriented to Islamic values. Searching for knowledge is also the same as looking for Ridho Allah Subhanallah Wa Taala because the process of gaining knowledge is done in a serious manner. Then there is a term, 'Look for science to China', meaning that humans will in essence continue to seek knowledge as long as humans live in the world for the good of themselves and others, especially in the provision after humans die (Hielmy 2001).

3.2 *Learning materials at Al Barokah Islamic Boarding School*

As in general, Al Barokah Islamic Boarding School is one of the pesantren that adheres to the salafiyah style or Islamic boarding school which still maintains Islamic values according to the beginning of the development of Islam. As a salafiyah boarding school, Al Barokah Islamic Boarding School in learning only focuses on issues relating to religion. Especially those relating to fiqh, hadith, and also the history of the Prophet Muhammad Saallahu Alaihi Wa Salaam. Santri are also taught about Islamic values that hold on to the Book of Yellow. According to Riki as students studying at the Al Barokah Islamic Boarding School said.

" .... which is a yellow book is a book that studies the teachings of Islam but is presented in the form of bare Arabic letters, not only books with yellow paper but books that teach about Islamic values. Even though the paper is white, but if you see the contents of Islamic values, it is still referred to as Kitab Kuning. "(Interview with Riki. September 13, 2014).

Regarding the origin of the naming of the yellow book as revealed by Riki it is probable that the word yellow scripture refers to the cover or cover of the Qur'an which is golden yellow. The Yellow Book itself is the result of writing from the scholars who started from tens of years and even hundreds of years ago. So that Kitab Kuning is also the only archive that is well preserved and cared for by the salafiyah santri. The yellow book also has types including Safinah and Jurumiyah. The Safinah book is one type of yellow book which teaches about jurisprudence and the science that discusses Islamic law. In the book of Safinah there are also Islamic laws from various views of four high priests such as Hanafi, Maliki, Hambali, and Syafii. Judging from the manner of behaviour applied by the santri of the Al Barokah Islamic boarding school, it refers more to the views of Imam Syafii, especially in the Tauhid doctrine problem. But there are also a number of things that refer to Imam Maliki such as the problem of procedures in the form of other deeds. Nevertheless, the santri must also study the views of other priests in order to solve a solution to problems concerning Islam.

The Jurumiyah book is one type of yellow book where the Jurumiyah book contains nahwu and shorof. Nahwu is the rules of Arabic to know the shape of the word and its conditions when it is still one word (Mufrod) or when it is composed (Murokkab). This includes the discussion of Shorof. Because the Science of Shorof is part of the Science of Nahwu, which is emphasized in the discussion of the word form and its state when mufrod. Shorof is often known as the grammar Arabic letter. To be able to know the Yellow Book, of course, you must be able to first understand about nahwu and its shorof. Because in Arabic when we misread the form of the message, the meaning of what is delivered in the Book of Yellow will be different. It takes at least 3-6 years to master the contents of the yellow book if you understand correctly nahwu and shorof as the main rules and guidelines in learning Arabic. The length of time in studying the contents of the yellow book depends on the personality of the santri both from the level of intelligence and the level of willingness to learn a santri in understanding the contents of the Yellow Book.

3.3 *Teaching methods in Al Barokah Islamic Boarding School*

The teaching method in the pesantren is bandhongan (known in Javanese as wethonan) and sorogan. Bandhongan is done by reading Arabic texts (in other words the Yellow Books), then the Kiai or lecturer translates into local languages, and at the same time explains the meaning of the contents contained in the book. This method was carried out in order to fulfil the cognitive competencies of the santri and expand scientific references for them (Nafi 2007). Indeed, in the bandhongan method, there was hardly any discussion between the Kiai and his santri. Especially in salafiyah pesantren like Al Barokah. For them this is a form of respect for the Kiai, also a tradition of one-way learning known as the chalk and talk method. The learning media used are very simple consisting of chalk and blackboard. The absence of LCD projectors or the like as supporting and supporting learning. The use of the one-way method (lecture) already existed at the beginning of the development of the history of Islamic Civilization. There is a reason why this method is still maintained in Islamic boarding schools, especially those with salafiyah models, does not mean that when students want to ask Kiai in the learning process, the Kiai cannot answer the questions raised by the students but it is a form of respect for the Kiai and leads to more forms of ethics santri to Kyai. For the student's views of the Al Barokah Islamic Boarding School, all the material presented by Kiai already has the truth that has been taken. So that the students also knew very well that the Kiai in terms of delivering the material were not from the origin but had read a number of references relating to religious issues, especially considering that the Kiai had once also been part of the santri. So, it is not surprising that this bandhongan method was taught from generation to generation, even since the Kiai and administrators were still santri and received similar education by their teachers.

Not only is the bandhongan method used, as for the sorogan method. This method aims to make students demanded to be more active learning in understanding the books that have been taught before by Kiai. Kiai gave the santri a minute to discuss the intent or essence of the contents of the Yellow Book. After discussion then Kiai told the students to face him to convey what had been discussed. Then after delivering what the Kiai had discussed, strengthening the material that had been given before. Al Barokah Islamic Boarding School when viewed from its teaching staff only has three Kiai. As explained by Lilis, “the instructors of the Al Barokah Islamic boarding school consist of Kang Sepuh (KH. Mumuh Abdul Muhyi), Kang Anom (Haji Yayat Nurul Hidayat), and Kang Haji Usman.” (Interview with Lilis Siti Nurjannah September 13, 2018). So, the learning pattern is different from modern Islamic boarding schools (khalafiyah). The use of methods in modern boarding schools is almost the same as the use of various learning methods and approaches that exist in public schools.

Particularly for the Al Barokah Islamic Boarding School, the santri strongly instilled the values of character in the Kiai figure, and ethics were highly respected. Maintaining ethics as well as seeking blessings also looks for Ridhoan from Allah Subhanallah Wa Taala. If one of the students who does not maintain ethics with the Kiai, the knowledge gained during his time as a santri will not be blessed even if he is an alumnus. So, respect for the Kiai that when he met the Kiai the santri bowed his body as a form of respect to the Kiai.

3.4 *Learning evaluation applied*

The application of the curriculum applied refers to the Learning of the Yellow Book as teaching material in the Al-Barokah Islamic boarding school. On the other hand, local materials or traditions from the Al-Barokah Islamic Boarding School are still maintained, one of them with a one-way learning system.

Al Barokah Islamic Boarding School has a unique form of evaluation, namely the santri test in front of the Kiai individually and there are times in groups. The indicator of the Test is how to read the yellow book and the Qur'an properly, the test method is known as imtihan. The schedule of the test was carried out when there were certain celebrations, especially during the Birthday of the Prophet Saallahu Alaihi Wa Salaam. The test was conducted to determine the extent of understanding of the students in understanding the contents of the Book of Yellow. Another form of test is tablighan, in this test the student seemed to be required to become preachers in conveying a problem related to Islam. The indicator of the tablighan test is that students are required to know the arguments of the Qur'an and Al Hadith arguments when they will explain their lectures before the Kiai. In addition, students were also required to find and find out the substantive solutions of the lectures delivered.

The results of the two tests, both imitations and tablighan, were not in numbers or quantitative, such as report cards, but the results of the tests came from the blessing of the Kiai who judged them. A student named Hanan said, "There is no report card as a result of the evaluation of learning but rather the appreciation made by the Kiai towards the students." (Interview with Hanan September 13, 2018). This is because the Al Barokah Islamic Boarding School does not have a class system and only studies in a hall. Nor does it have a class level so it forms a hierarchical, Kiai's blessing is the main reference for the santri.

3.5 *Tradition maintained at Al Barokah Islamic Boarding School*

Tradition is a manifestation of the cultural results of society that are continuously inherited or carried out continuously. In the Al Barokah Islamic Boarding School, the traditions that continue to be carried out by the Kiai and boarding school officials include:

- Barzanzi, a tradition that explains the history of the life of Rasulallah Saallahu Alaihi Wa Salaam. The difference between Barzanzi and Nabawiyah Sirah. lies in bookkeeping. Barzanzi is usually delivered directly by Kiai but the implications are more on prayer. So, in essence, Barzanzi is a combination of the history of the prophet who was sung through prayer.
- Marhaba, is a tradition in Sufism, namely the process towards the level of worship or interaction between humans and Allah Subhanallah Wa Taala to reach ma'rifatullah or human beings have reached the culmination of worship between man and His God. The practice of marhaba is prayer to Allah Subhanallah Wa Taala and Prophet Muhammad Saallahu Alaihi Wa Salaam.
- Muludan, is a tradition in commemoration of the birthday of Prophet Muhammad Saallahu Alaihi Wa Salaam. In addition, this tradition also commemorates the services that have been given to Al Barokah Islamic Boarding School. The form of this tradition is the study of both the santri and the surrounding community regarding the history and struggle of the Prophet Muhammad Sallahu Alaihi Wa Salaam in spreading the cause of Islam.
- Habkah, is a tradition that aims to strengthen the relationship between alumni, santri, and Kiai Al Barokah Islamic Boarding School. Habkah itself is (Himpunan Aluumni Al Barokah). The program is held once every year. Regarding the study period at the Al Barokah Islamic boarding school it is not determined to be a maximum of several years, but depends on the maturity level of the santri in plunging into the community and determined by Kiai's blessing that he is to become an alumni or still studying at the Islamic boarding school (pesantren).

4 CONCLUSION

In the early 1980s there were changes and developments in pesantren educational institutions. Many Islamic boarding schools later transformed and adopted the modern education system as implemented by madrassas or formal schools. As a traditional education the Al-Barokah Islamic boarding school still maintains traditional teaching and learning traditions. This boarding school prioritizes the use of the yellow book as the main medium of learning. Al-Barokah Islamic boarding school only relies on the blessing of the kiai as an evaluation process in learning at Al-Barokah Islamic boarding school. Al-Barokah Islamic Boarding School has a unique tradition that has been maintained since this educational institution was established. The curriculum created is a curriculum that is applied directly by the kiai as mentors in learning in traditional pesantren.

REFERENCES

Baidlawi, H.M. 2006. Modernisasi Pendidikan Islam (Telaah Atas Pembaharuan Pendidikan di Pesantren). *TADRIS: Jurnal Pendidikan Islam* 1(2).

Dzofier, Z. 2011. *Tradisi pesantren: Studi pandangan Hidup Kyai dan Visinya Mengenai Masa Depan Indonesia*. Jakarta: LP3ES.

Fatmawati, E. 2015. *Manajemen Pengembangan Kurikulum Pesantren Mahasiswa*. Malang: UIN Malang.

Hardiansyah, H. 2010. *Metodologi Penelitian Kualitatif untuk Ilmu-ilmu Sosial*. Jakarta: Salemba Humanika.

Hielmy, I. 2001. *Pesan Moral dan Pesantren: Meningkatkan Kualitas Umat, Menjaga Ukhuwah*. Bandung: Nuansa.

Nafi, M.D. 2007. *Praksis Pembelajaran Pesantren*. Yogyakarta: Institute for Training and Development.

Saifuddin, A. 2015. Eksistensi Kurikulum Pesantren dan Kebijakan Pendidikan. *Jurnal Pendidikan Agama Islam (Journal of Islamic Education Studies)* 3(1): 207–234.

Sugiyono 2009. *Metode Penelitian Kuantitatif, Kualitatif dan R&D*. Bandung: Alfabeta.

Sukamto 1999. *Kepemimpinan Kiai Dalam Pesantren*. Jakarta: LP3ES.

Borderless Education as a Challenge in the 5.0 Society – Abdullah, Adriany & Abdullah (eds)

Analytical investigation of lower secondary science teachers' knowledge and teaching practices

L. Myint & T.N. Oo
Yangon University of Education, Yangon, Myanmar

ABSTRACT: This article sets out to study the knowledge levels and teaching practice levels of lower secondary science teachers in Yangon Region. Quantitative and qualitative research methods were used, and a questionnaire survey was conducted. The reliability coefficient (Cronbach α) was 0.87 for the science teaching practice questionnaire, which was developed based on literature. Respondents were 320 lower secondary science teachers, selected using the cluster sampling method. Descriptive statistics, One-Way ANOVA, Tukey HSD test, and Pearson-product moment correlation coefficient were utilized to analyse the collected data. The level of knowledge of lower secondary science teachers was found to be at a satisfactory level (mean = 10.42, SD = 2.17), and the level of science teaching practice was at a moderately high level (mean = 3.87, SD = 0.54). The correlation between science teachers' knowledge and their teaching practices was statistically significant ($r = 0.340$, $p < 0.01$). Qualitative findings also suggested that the higher the knowledge level of science teachers, the better their teaching practices.

1 INTRODUCTION

Teachers play a central role in schools, inspiring children, parents, and management. As human resources, they should be molded, valued, and empowered. Reeves (2004) stated that teachers need to be empowered to reflect on instructional practices and make instructional decisions in the classrooms aimed at increasing student achievement. Empowering teachers leads to empowering children and the school as a whole. The knowledge spread in the class depends upon the teacher, so the more enlightened the teacher is, the more enlightened the class will be.

According to Hanushek & Rivkin (2006), teachers are central to any attempt to improve education. In addition, Veladat (2011) explained that the core of an excellent education system is talented teachers. Reeves (2004) also stated, "If teachers systematically examine their professional practices and their impact on student achievement, the result of such reflective analysis will finally transform educational accountability from a destructive and unedifying mess to a constructive and transformative force in education." In efforts to facilitate the type of force Reeves described, teachers need to be empowered to reflect on instructional practices and make instructional decisions in the classroom. It is a universally accepted fact that the quality of teaching determines the quality of education in a nation. Therefore, an analytical investigation of lower secondary science teachers' knowledge and teaching practices is necessary in order to improve the quality of instruction.

1.1 *Significance of the research*

Zaw (2001) pointed out that no educational system can ever be better than its teachers. The development of the education system and the quality of education solely depends on the quality of teachers.

Moreover, today's science teachers play vital roles in educating, inspiring, and guiding students to become responsible, scientifically literate citizens (Lanier 2009). The Secondary Education Modernization Program of the Republic of Trinidad and Tobago described how, at the lower secondary level, students' experiences in science will lead them to have a conceptual understanding of the natural world, of humanity's place in it, and of our responsibility to maintain and preserve it (De Lisle 2012). Thus, teaching science at the lower secondary level is incredibly important, because this level allows connection between primary and secondary levels. If teachers can instill love of learning science at the lower secondary level, children will go on to further study of science with enthusiasm.

According to Mishra & Mishra (2011), teachers are the prime source of knowledge in the classroom, giving strength to students with their capabilities and competencies. If such teachers are empowered, no doubt they will make better efforts toward turning students into resourceful persons, and they will show better performances in taking on extra behavioral roles. In addition, science education developers have recognized that teachers' instructional practices must change in order to reach the goal of science literacy for all.

The main aim of this study is to investigate the knowledge and teaching practices of lower secondary science teachers in Yangon Region. The specific aims of this study are: to study the knowledge levels of lower secondary science teachers; to study their teaching practices; and to study the relationship between their knowledge levels and practices.

1.2 *Review of related literature*

1.2.1 *Science education*

A major goal of science education today is the fostering of students' intellectual competencies, such as independent learning, problem-solving, decision-making, and critical thinking (National Research Council 1996).

According to Russel & Harlen (1990), science is about understanding certain aspects of the physical world around us; it involves testing and changing ideas about how things work, whether natural or manmade. It is a way of thinking about and understanding the world. It is a human endeavor, a personal way of exploring and knowing (Carin & Sund 1985). Cain & Evans (1990) identify four components of science: content or product; process or method; attitude; and technology.

1.2.2 *Science process skills*

Science process skills ensure active student participation, help students develop a sense of responsibility for their own learning, increase the permanence of learning, and also aid students in acquiring research methods – that is, science process skills enable students to think and behave like scientists (Ostlund 1992).

SAPA (Science: A Process Approach) groups process skills into two types: basic and integrated. Basic (simpler) process skills provide a foundation for learning integrated (more complex) skills. According to Jinks (1997), integrated science process skills are more appropriate for children in grade four and above.

According to Padilla (1990), the basic process skills are (1) observing, (2) inferring, (3) measuring, (4) communicating, (5) classifying, and (6) predicting; integrated process skills consist of (1) controling variables, (2) defining operationally, (3) formulating hypotheses, (4) interpreting data, (5) experimenting, and (6) formulating models. Students should be introduced to these skills early in their school experiences, because so much of their success in subsequent guided studies requires a sound understanding and appropriate use of them. Therefore, science teachers need to be expert in these process skills, and they also need to be expert in their teaching (Ango 2002).

1.2.3 *Teaching practices in science teaching*

Teachers' beliefs, practices, and attitudes are important for understanding and improving educational processes. In "How Students Learn: History, Mathematics, and Science in the Classroom" by the National Research Council (2005), four best pedagogical practices are identified as: supporting metacognition and student self-regulation; cooperative learning; engaging resilient preconceptions; and organizing knowledge around core concepts.

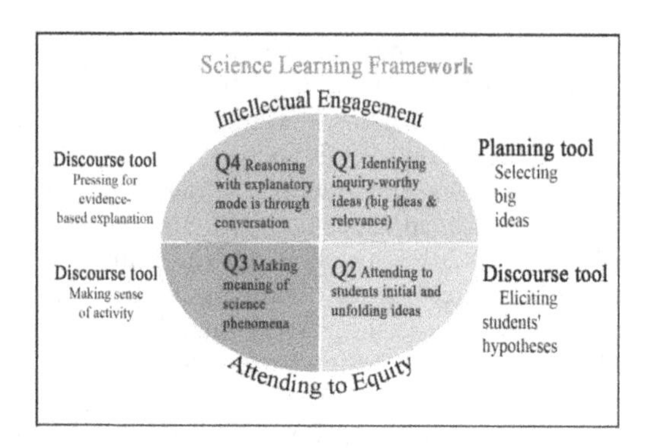

Figure 1. Science learning framework.

The National Science Foundation at the University of Washington identified several core instructional strategies that support science teaching. These "high-leverage" practices make up the Science Learning Framework (see Figure 1) and have been selected based on extensive research on how young people learn science, authentic forms of science activity, and how teachers learn to appropriate new practices.

According to Audunson (2007), science is an important discipline. It is through science that people develop an understanding of the world around them, make discoveries, and produce technological advancement. The teaching of science can be challenging because of the abstract and theoretical nature of the subject. To improve students' understanding of the complex processes involved, teachers should incorporate certain elements into their lessons: namely, metacognition; inquiry-based learning; and the learning cycle.

In addition to this, Mechling (2003) also suggested ten best teaching practices in science education: (1) hands-on activities; (2) constructivism and the learning cycle; (3) simulations; (4) simple, readily available materials; (5) combining science with children's literature; (6) standards-based instruction; (7) setting high student expectations; (8) cooperative learning; (9) effective questioning strategies; and (10) authentic assessment.

1.3 *Instructional strategies for science*

According to Cain & Evans (1984), teaching techniques refers to the methods of instruction used to aid students in attaining specified objectives. According to Bruner, instructional strategy can be categorized as action, imagery, and language modes (Lawrence 1969). These three modes of presentation can be used separately; however, they are likely more effective in combination for providing a format for science activities.

All daily science instruction should involve four elements. First comes attention-getting and motivating techniques. To begin a lesson, the teacher must first get the students' attention and then motivate them so they feel a need to continue with the instruction. Also useful is a verbal discussion that prepares the students for instruction by giving a preview of what is to come. Problem-solving or conflicts of opinion are two other effective introductions. They both tend to create situations that result in students using alternative means to find solutions.

The next portion of the lesson involves the learner in data gathering. Laboratory activities provide an excellent means of involving students with first-hand experience in data gathering. These activities can take place inside the classroom or outside as part of a field trip. The use of models, pictures, films, charts, or other visual aids provides students with the opportunity to examine data they cannot experience directly. Secondary sources (such as reading and lectures) can also provide information, but overuse of these sources can result in a content- or product-oriented science approach.

Once the data are collected, the teacher uses data-processing techniques to involve the students in organizing and analysing the information. Guided discussion can promote and encourage interaction among the learners. Engaging the students in both physical manipulation (making charts, graphs, and so on) and mental manipulation (formulating hypotheses, making inferences, making predictions) of the data is essential to all science instruction.

Developing techniques that help students reach closure is the final phase of instruction. A teacher monologue that summarizes the lesson or a question-and-answer format, in which students participate, can help learners put the lesson into proper perspective. Therefore, the following theoretical framework will lead this study.

The knowledge level of lower secondary science teachers was investigated using science process skills as outlined in Padilla (1990). These skills, as mentioned earlier, consist of basic process skills (observing, inferring, measuring, communicating, classifying, predicting) and integrated process skills (controling variables, defining operationally, formulating hypotheses, interpreting data, experimenting, formulating models). All twelve are important individually as well as in combination. As these process skills are the tools students use to investigate the world around them and to construct scientific concepts, it is essential for teachers to understand them well (Padilla 1990). Science teachers should also display competency in lesson preparation (LP), instructional strategies (IS), using teaching aids (UTA), and evaluation and feedback (E & F).

Regarding lesson preparation, the first instructional decision science teachers must make is determination of the learning objective. Science teachers should use a variety of teaching strategies, activities, and teaching aids to enhance students' learning. Projects, field trips, and science fairs all have a definite place in science instruction. The inquiry approach requires a skilled teacher who can arrange the learning environment so that students can make discoveries. Basic process skills are discovery processes; integrated process skills are inquiry processes. Teachers should play a major role in developing and maintaining a well-disciplined laboratory environment.

Science teachers can use positive and negative reinforcement to foster desired behaviors. To meet evaluation objectives, the teacher should become familiar with three major types of evaluation approaches: diagnostic, formative, and summative. Giving feedback helps students understand the reasons for the results received and shows them ways to do better next time.

2 RESEARCH METHODS

There are two phases in this study: (1) a quantitative study of science teachers' knowledge level and their teaching practices; and (2) a qualitative study (observation, interview, and documentation) of science teachers' teaching practices at the lower secondary level. The reliability coefficient (Cronbach α) was 0.87 for the practice questionnaire, which was developed based on literature. Three hundred and twenty lower secondary science teachers were selected from the Basic Education schools of Yangon Region using the cluster sampling method. Descriptive statistics was used to analyse the collected data.

The levels of knowledge and teaching practice perceived were determined as the item percent correct, mean value, and standard deviations. Moreover, One-Way ANOVA, Tukey HSD test, and Pearson-product moment correlation coefficient were utilized.

2.1 *Definition of key terms*

- Important terms were carefully defined in explaining the concepts underlying the development of the investigation.
- Teaching practice refers to something teachers do often, especially a particular way of teaching something.
- Science process skills refers to the skills that ensure active student participation, help students take responsibility for their own learning, increase the permanence of learning, and enable students to acquire research methods – in other words, ensuring they think and behave like scientists (Ostlund, 1992).

3 RESULTS AND DISCUSSION

The analysis of collected data as research findings will be discussed into two phases.

3.1 *Phase I – quantitative research findings*

Data collected in research is discussed in this phase – the knowledge level of teachers in terms of basic science process skills, integrated process skills, and science teaching methods were investigated, as well as varieties of teachers' knowledge and practices in terms of personal and school-related factors. In addition, the intercorrelation of knowledge and science teaching practices were presented in this phase.

3.1.1 *Data screening*

The data were screened for univariate outliers. The out-of-range values, due to administrative errors, were examined and recorded with the original responses from the questionnaire.

3.1.2 *Investigating science teachers' knowledge for science process skills*

Knowledge of science process skills (N = 320) was investigated with multiple-choice items. Table 1 shows item percent correct (IPC) values for knowledge for science process skills in multiple-choice items.

Table 1. Item percent correct (IPC) values showing lower secondary science teachers' knowledge of science process skills (multiple-choice items).

No.	Knowledge of science process skills items	No. of correct teachers	IPC
1	Giving students the opportunity to place pieces of paper on the playing drum and to press it with their hands helps to develop: Observing Skill* Measuring Skill Classifying Skill	266	83.1%
2	Letting students take body and classroom temperatures with the use of thermometer helps to develop: Classifying Skill Measuring Skill* Predicting Skill	289	90.3%
3	By using words or graphic symbols to describe an action, object, or event, students can develop: Predicting Skill* Experimenting Skill Communicating Skill	83	25.9%
4	By categorizing things according to similarities and differences, students can develop: Communicating Skill Classifying Skill* Operational Defining Skill	184	57.5%
5	After studying how a sound wave is produced by a vibrating object in different media, predicting whether sound traveling through a solid medium will be louder than that traveling through a liquid medium helps develop: Operational Defining Skill Inferring Skill Hypothesizing Skill*	58	18.1%

(*Continued*)

Table 1. (*Continued*)

No.	Knowledge of science process skills items	No. of correct teachers	IPC
6	By describing tools or units used in experiments, students can develop: Observing Skill Operational Defining Skill* Classifying Skill	172	53.8%
7	By identifying variables that can affect an experimental outcome, keeping the rest constant while manipulating only the independent variable, students can develop: Variable Controling Skill* Operational Defining Skill Experimenting Skill	114	35.6%
8	By stating the outcome of a future event based on a pattern of evidence, students can develop: Predicting Skill* Interpreting Skill Inferring Skill	28	8.8%
9	By making an "educated guess" about an object or event based on previously gathered data or information, students can develop: Predicting Skill Interpreting Skill Inferring Skill*	195	60.9%
10	Creating a mental or physical model of a process or event or a concept map of a lesson helps develop: Interpreting Skill Hypothesizing Skill Model Formulating Skill*	120	37.5%
11	Using a round-bottom flask and small bell to show that sound needs a medium to travel, students can develop: Variable Controlling Skill Experimenting Skill (a) and (b)*	68	21.3%
12	Answering which wax drop will melt first among three wax drops placed two inches apart on the left end of iron bar before the right end of that bar is heated, students can develop: Interpreting Skill Predicting Skill* Model Formulating Skill	192	60%
13	Stating the expected outcome of an experiment develops: Predicting Skill Hypothesizing Skill* Model Formulating Skill	96	30%
14	The most suitable teaching method to explain to students how to separate insolvent particles in a liquid is: Lecture Method Discussion Method Problem-Solving Method*	161	50.3%
15	Teaching about sounds in relation to a description of their rates is the teaching of: Facts* Science Concepts Problem-Solving Skill	49	15.3%
16	The most suitable method to teach students how to conduct experiments to collect heat energy from the sun by using convex lens and cotton wool is:	282	88.1%

(*Continued*)

Table 1. (*Continued*)

No.	Knowledge of science process skills items	No. of correct teachers	IPC
	Lecture-Cum-Demonstration Method Discussion and Question Method Problem Solving and Discovery Method*		
17	Of (1) induction (2) observation (3) project-based learning, and (4) problem-solving, the most suitable methods by which to teach why we should protect and maintain the natural environment are: (1), (2), and (3) (2), (3), and (4)* None of these	140	43.8%
18	The most suitable teaching method to teach how to make magnets by using electricity is: Problem-Solving Method* Field-Trip Method Lecture-Cum-Demonstration Method	271	84.7%
19	The most suitable teaching methods for teaching about plants and animals in the school and home environments is: Problem-Solving Method Field-Trip Method* Lecture-Cum-Demonstration Method	274	85.6%
20	In teaching lower secondary science, teachers should place great emphasis on: Making students memorize science concepts and facts in textbooks Using and developing students' science process skills* Teaching students to get good grades in the examination result	292	91.3%

* Correct Answer

Table 2. Number and percentages of participants showing the level of knowledge of science process skills among teachers.

Scoring range	No. of teachers (%)	Remark
< 50%	111 (35%)	Below satisfactory level
50%–74%	200 (62%)	Satisfactory level
≥ 75%	9 (3%)	Above satisfactory level

The IPC values for items 1, 2, 16, 18, 19, and 20 are greater than 75%. The IPC values for items 4, 6, 9, 12, and 14 are greater than 50%. However, the IPC values for items 3, 5, 7, 8, 10, 11, 13, 15, and 17 are less than 50%.

In scoring these items, a mark was given for each correct answer. If a participant teacher gave correct answers for 10 out of 20 items, her or her score would be 10, and the average score percent would be 50%. Teachers' levels of knowledge were then assessed according to the range of average score percent, as shown in Table 2.

As shown in Table 2, 111 (35%) of respondents scored at below satisfactory level; 200 (62%) scored at a satisfactory level; and 9 (3%) scored better-than-satisfactory.

The 20-item knowledge questionnaire comprised three parts: knowledge items for basic process skills (1, 2, 3, 4, 8, 9, 12); items for integrated process skills (5, 6, 7, 10, 11, 13); and items for science teaching methods (14, 15, 16, 17, 18, 19, 20). Science teachers' knowledge in these three parts was found as shown in Table 3.

Table 3. Number and percentages of participant teachers in science teachers' knowledge.

No.	Parts	No. of items	IPC
1	Basic process skills	7	55.2 %
2	Integrated process skills	6	32.7 %
3	Teaching methods	7	65.6 %
	Overall knowledge	20	52.1 %

In Table 3, according to the item percent correct, 55.2% of participant teachers were correct in items concerned with basic process skills; 32.7% were correct in items concerned with integrated process skills; and 65.6% were correct in items concerned with science teaching methods. Overall, 52.1% were correct in this knowledge questionnaire.

3.1.3 *Mean values for perceptions of lower secondary science teachers on knowledge grouped by district*

Mean values for perceptions grouped by district are shown in Table 4.

As shown in Table 4, all four groups of teachers were perceived to have satisfactory levels of knowledge. As Table 5 shows, there was no significant variation in perceptions of knowledge when grouped by district.

3.1.4 *Investigating lower secondary science teachers' teaching practices*

Mean values for perceptions of lower secondary science teachers on teaching practices grouped by district are shown in Table 6. All four groups of teachers were perceived as having moderately high-level teaching practices mentioned in this study.

Table 4. Mean values showing perceptions of lower secondary science teachers on knowledge grouped by district.

		District A	District B	District C	District D
No.	Variable	Mean (SD)	Mean (SD)	Mean (SD)	Mean (SD)
1	Science teachers' knowledge	10.33 (2.34)	10.65 (1.98)	10.44 (2.33)	10.29 (2.01)

Scoring direction: below 8.25 = below satisfactory level; 8.25–12.6 = satisfactory level; above 12.6 = above satisfactory level

Table 5. One-way ANOVA result showing perceptions of lower secondary science teachers on knowledge grouped by district.

Variable		Sum of squares	*df*	Mean square	*F*	*P*
Science teachers' knowledge	Between groups	6.38	3	2.13	0.45	n.s.
	Within groups	1493.83	316	4.73		
	Total	1500.20	319			

n.s. = no significance

Table 6. Mean values showing perceptions of lower secondary science teachers on teaching practices grouped by district.

No.	Variable	Mean (SD)			
		District A	District B	District C	District D
1.	Science teachers' teaching practices	3.97 (0.57)	3.82 (0.56)	3.89 (0.52)	3.81 (0.51)

As shown in Table 7, there was significant variation in perceptions of lower secondary science teachers on teaching practices among the teachers grouped by level of knowledge ($df = 2$, $F = 4.74$, $P < 0.01$).

As shown in Table 8, the Tukey test shows Group A was significantly different from Group B ($p < 0.01$, d = -0.33) in perceptions of lower secondary science teachers on teaching practices grouped by level of knowledge.

3.1.5 *Relationship between perceived science teachers' knowledge and their teaching practices*
The Pearson-product moment correlation coefficient was utilized to examine the relationship between a teacher's knowledge and his or her teaching practices. Table 9 shows that the correlation between knowledge and teaching practices was statistically significant because the "sig" is less than 0.01. There is an association between science teachers' knowledge and their teaching practices ($r = .340$, $p < 0.01$).

Table 7. One-way ANOVA result showing perceptions of lower secondary science teachers on teaching practices grouped by level of knowledge.

Variable		Sum of squares	*df*	Mean square	*F*	*P*
Science teachers' teaching practices	Between groups	2.72	2	1.36	4.74	0.009**
	Within groups	90.80	317	0 .29		
	Total	93.51	319			

**The mean difference is significant at the 0.01 level

Table 8. Tukey HSD result showing multiple comparison for the perceptions of lower secondary science teachers on teaching practice grouped by level of knowledge.

Variable	(I) Knowledge level	(J) Knowledge level	Mean difference (I-J)	*p*
Science teachers' teaching practices	Group A	Group B	-0.18*	0.011**
		Group C	-0.31	n.s.

Group A = below satisfactory level; Group B = satisfactory level; Group C = above satisfactory level
** The mean difference is significant at the 0.01 level.
n.s. = no significance

Table 9. Correlations between perceived science teachers' knowledge and their teaching practices.

Two groups	Science teachers' knowledge	Science teachers' teaching practices
Science teachers' knowledge	1	0.340**
Science teachers' teaching practices	0.340**	1

**Correlation is significant at the 0.01 level (2-tailed)

3.2 *Phase II – qualitative research findings*

This phase presents the results of classroom observation of science teaching practices in the lower secondary level of eight selected Basic Education Schools; the results of document analysis; and the results of interviews with teachers with below-satisfactory knowledge levels (Group A) and satisfactory knowledge levels (Group B).

3.2.1 *The result of classroom observation*

Classroom observation of lower secondary science classes was conducted using an observation checklist in four parts: lesson preparation (LP); instructional strategies (IS), using teaching aids (UTA), and evaluation and feedback (E & F).

Eight teachers were observed for at least two periods. Based on observation results, most of the Group A teachers placed more emphasis on the reading aloud of paragraphs individually, in groups, and by the whole class, rather than teaching science process skills: they tended to prefer teaching with a one-way lecture method. Group B teachers, on the other hand, tended to incorporate science process skills such as observation, inferring, communication, classification, prediction, and interpretation skills, by using teaching aids, especially real objects. Group B teachers were also more active and interested in teaching and were more likely to ask questions that aroused student to think and actively participate in the lesson than did Group A teachers.

3.2.2 *The result of documentation*

To investigate implementation of science teaching in schools, ten types of records were analysed: (1) student observation; (2) suggestions and discussion concerned with science teaching; (3) lesson plans; (4) students' participation in science activities; (5) students' science progress; (6) remedial aspects; (7) science-related support books; (8) special time allocated to discussion with other science teachersl (9) arrangements for guest expert lectures; and (10) lesson studies.

Based on documentation results, Group B had more complete documents to learn how to implement science teaching in their schools than Group A. Keeping student observation records ($n = 3$) and arrangements for guest expert lectures ($n = 4$) were not carried out in most schools for Group B; similar results, however, were found for Group A schools as well ($n = 3$ and $n = 3$ respectively). Overall, Group B teachers were able to implement science teaching more effectively than Group A teachers.

3.2.3 *The result of interview*

Based on the questionnaire survey instruments, an 11-item interview form was developed.

According to interview data, it was found that Group B teachers were more active, more interested in their students, and tried more to improve their own abilities and their student's achievement as compared to those in Group A. Similarly, Group B displayed commitment to teaching each student as much as they could and showed confidence in doing the best they could. Most Group B teachers were satisfied in their professional lives and tried to instil science process skills by using the resources available to them.

To sum up, according to both quantitative and qualitative research findings, it can be concluded that science teachers' teaching practices vary in terms of teacher knowledge levels.

Tony (2009) highlighted the notion that the nature of work worldwide dictates that outcomes or outputs of organizations largely depend on the emphases or attentions put to their inputs. Survey findings indicated that 111 (35%) of participant teachers had below-satisfactory scores, 200 (62%) achieved satisfactory levels, and 9 (3%) achieved greater than satisfactory scores. Overall, most participants achieved a satisfactory level of knowledge in science teaching. Science: A Process Approach (SAPA, cited in Padilla 1990) states that the basic (simpler) process skills provide a foundation for learning the integrated (more complex) skills. Similar conclusions can be drawn from this study. In addition, it was found that 55.2% of participant teachers were correct in items concerning basic process skills, 32.7% of the participant teachers were correct in items concerning integrated process skills, and 65.6% of the participant teachers were correct in items concerning science teaching methods. Overall, 52.1% of participant teachers were correct in the questionnaire; they showed more knowledge of basic process skills than of integrated process skills. Findings also suggested significant differences in teaching practices depending on level of knowledge: the higher the teacher knowledge level of science teachers, the better the teaching practices of them.

Regarding teaching practices, Collette & Chiappetta (1989) indicated that effective science teachers use a variety of instructional strategies and teaching skills within a given lesson. Moreover, they ask questions that require students to think, and they encourage all students to participate in the lesson. Similar conclusions can be drawn from this study. The results of a triangulation study also indicated that Group A teachers tended to place more emphasis on reading aloud rather than teaching science process skills to students; Group A preferred a one-way lecture method instead. Group B on the other hand were seen to place more emphasis on teaching process skills such as observation, inferring, communication, classification, prediction, and interpretation skills, especially by using teaching aids, including real objects. Group B teachers were more active, interested in teaching, and taught by asking questions that aroused students to think and actively participate in the lesson than did Group A teachers.

Group B teachers participated more in professional development activities such as reading, watching TV programs related to science, preparing lessons before teaching, and discussing with colleagues about lessons than did Group A teachers. Group B teachers also conducted better preparation for lessons, applied various instructional strategies, and used all available resources to develop students' science process skills; gave more emphasis on evaluating individual or group performance; and tried to give constructive feedback to emphasize ongoing learning, more so than Group A teachers. Therefore, science teachers need to be knowledgeable about their subject to feel more empowered.

4 CONCLUSION

4.1 *Recommendation*

The following recommendations are based on analysis of the research findings. School principals should make efforts to support teaching aids and instructional materials and to monitor teachers. Teachers should be supported via professional development activities, programs, trainings, and allowances to develop their science knowledge, skills, and attitude in order to improve the teaching of science. In self-directed teacher professional development, teachers should involve themselves in initiating and designing their own professional development and share materials and ideas as well as discuss challenges and solutions with each other. Teachers should be provided with workshops and training to develop their knowledge of science process skills, especially integrated process skills. Teachers should be supported with science materials, facilities, and resources to better implement science teaching. Teachers should use available resources as teaching aids in teaching science lesson effectively. A classroom should be used as a storage space for science materials and teaching aids, and teachers should use and maintain them with a logbook. Non-instructional workloads of teachers should be eliminated, to allocate more time to teaching science. Not only teacher–pupil ratios but also teacher–classroom ratios should be considered in order to provide students with inquiry-based learning opportunities. Formative and diagnosis assessment should be emphasized in teaching science in order to allow continuous reflection on learning science and identify students' weakness and misconceptions as to the teaching of science. An activity-based assessment system for science achievement should be used to match with how science should be taught.

4.2 *Need for further research*

Studies in all subject areas need to be conducted to investigate the knowledge and teaching practices of teachers in different subjects.

ACKNOWLEDGMENT

We would like to offer our respectful gratitude to Dr. Aye Aye Myint (retired rector, Yangon University of Education), Dr. Pyone Pyone Aung (pro-rector, Yangon University of Education), Dr. Kay Thwe Hlaing (pro-rector, Yangon University of Education), Dr. Aye Aye Cho (retired professor, head of department, Department of Educational Theory, Yangon University of Education), Dr. Daw Htay Khin (retired professor, head of department, Department of Educational Theory, Yangon University of Education), Dr. Daw Khin Mar Yee (retired professor, head of department, Department of Educational Theory, Sagaing University of Education), and Dr. San San Hla (retired lecturer, Department of Educational Theory, Yangon University of Education) for their valuable expertise, support, constructive suggestions, and criticism to complete our work, as well as members of the Dissertation Committee of Yangon University of Education for allowing us to pursue this dissertation.

REFERENCES

Ango, M. L. 2002. Mastery of science process S\skills and their effective use in the teaching of science. *Instructional Journal of Educology* 16(1): 11–30.

Audunson, R. 2007. Library and information science education – discipline, profession, vocation? *Journal of Education for Library and Information Science*: 94–107.

Cain, S. E., & Evans, J. M. 1984. *Sciencing* (2nd ed.). Columbus, OH: Merrill.

Carin, A. A., & Sund, R. B. 1985. *Teaching Modern Science* (4th ed.). Columbus, OH: Merrill.

Collette, A. T., & Chiappetta, E. L. 1989. *Science Instruction in the Middle and Secondary Schools* (2nd). Columbus, OH: Merrill.

De Lisle, J. 2012. Explaining Whole System Reform in Small States: The Case of Trinidad and Tobago Secondary Education Modernization Program. *Current Issues in Comparative Education*. 15(1): 64–82.

Hanushek, E. A., & Rivkin, S. G. 2006. Teacher quality. *Handbook of the Economics of Education* 2: 1051–1078.

Jinks, J. 1997. *The Science Processes*. Normal: Illinois State University.

Lanier, S. K. 2009. *Principal Instructional Leadership: How Does it Influence an Elementary Science Program amidst Contradictory Messages of Reform and Change?* Tallahassee, FL: DigiNole.

Lawrence, G. D. 1969. Bruner instructional theory or curriculum theory?. *Theory into Practice*, 8(1): 18–24.

Mechling, K. 2003. *Best Practices in Science Education: What Works?* [Online]. http://www.maisk-6scienceinquiry.org/practices.htm.

Mishra, U., & Mishra, Y. 2011. Teachers' empowerment: A need for making excellence in academics. *International Refereed Research Journal*, 17(1): 34–35.

National Research Council. 1996. *National Science Education Standards*. Washington, DC: National Academy Press.

National Research Council. 2005. *The Authentic Best Practices of Science Teaching*. [Online]. http://www.phy.ilstu.edu/pte/311content/effective/best_practice.html.

Ostlund, K. L. 1992. *Science Process Skills: Assessing Hands-on Student Performance*. New York, NY: Addison-Wesley.

Padilla, J. M. 1990. *The Science Process Skills. Research Matters – to the Science Teacher*. Reston, VA: National Association for Research in Science Teaching (NARST).

Reeves, D. B. 2004. *Accountability for Learning: How Teachers and School Leaders Can Take Charge*. Alexandria, VA: ASCD.

Russell, T., & Harlen, W. 1990. *Assessing Science in the Primary Classroom; Practical Tasks*. London: Paul Chapman Publishing.

Tony, M. C. 2009. *Perceptions on Teacher Empowerment and the Quality of Teaching in the Universal Primary Education Schools in Kampala Central Division*. Kampala: Makerere University.

Veladat, F., & Navehebrahim, A. 2011. Designing a model for managing talents of students in elementary school: A qualitative study based on grounded theory. *Procedia-Social and Behavioral Sciences* 29: 1052–1060.

Zaw, K. 2001. *DS 6: Theoretical Pedagogy (II) Ethic, PhD course*. Yangon: Yangon Institute of Education.

 ISBN 978-0-367-61960-2

Effectiveness of C-R-E-A-T-E model on building student creativity in making natural voltaic cells

W. Wahyu & R. Oktiani
Universitas Pendidikan Indonesia, Bandung, Indonesia

K. Komalia
SMA Laboratirium Percontohan, Bandung, Indonesia

ABSTRACT: This study was aimed to investigate the effectiveness of the Connecting-Restructuring-Elaborating-Applying-Tasking-Evaluating (C-R-E-A-T-E) model for building student creativity in making voltaic cells made from natural materials. This research used descriptive method. Data collection was carried out using research instruments consisting of assessment rubric of student worksheet, assessment rubric of voltaic cell products, student testimonial sheets, and observer testimonial sheets. Based on the results of assessment of the student worksheet, the creative product made by students, and students' and observers' testimonies, the C-R-E-A-T-E model was very effective to build student creativity. This model facilitated students to construct their knowledge and then make a product based on that knowledge and finally discuss and evaluate the product. It is suggested that the C-R-E-A-T-E model has the potential to be applied in the learning process to stimulate student creativity.

1 INTRODUCTION

Creativity is the process of generating ideas and combinations that are not the same as the others (Allen et al. 2017). Creativity is required because this is one of the four skills of the 21st century (Geisinger 2016). One way to build creativity in students is through the learning process (Burleson 2005).

The learning model has an important impact in building creativity through learning process. Researches on the application of learning models to build student creativity had been carried out successfully. Its learning model were problem-based learning (Amelia 2014, Awang & Ramly 2008, Nurhayati et al. 2013, Rudibyani 2019); project-based learning (Isabekov & Sadyrova 2018); and Search, Solve, Create, and Share or SSCS (Utami 2011).

Researches on the development of creativity in chemistry had been successfully carried out, such as in stoichiometry (Rudibyani 2019), petroleum (Nurhayati et al. 2013), and salt hydrolysis (Amelia 2014). According to Yahya (in Nurhayati et al. 2013), creativity can be seen from the ability to make new products from existing materials around the environment. Some research results showed that voltaic cells (batteries) could be made from natural materials. Some simple voltaic cells from natural materials had been successfully made, namely the orange juice clock (Castro-Acuña et al. 1996), lemon-powered clock (Letcher & Sonemann 1992), lemon cell battery (Muske et al. 2007), and lemon-powered calculator (Swartling & Morgan 1998). In addition, in its learning process, the Connecting-Restructuring-Elaborating-Applying-Evaluating (C-R-E-A-T-E) model could be applied to build student creativity. Students could create products to stimulate their creativity.

The purpose of this study was to investigate the effectiveness of the C-R-E-A-T-E model to develop student creativity in learning chemistry via making a voltaic cell from natural materials. The research method was descriptive model. The C-R-E-A-T-E model was effectively used to build students creativity in the making of voltaic cells from natural materials because in the C-R-E-A-T-E model students were directed to compile the knowledge needed to create the products then examined and evaluated the product. It showed that the C-R-E-A-T-E model could build student creativity through learning process effectively.

2 RESEARCH METHODS

This study used a descriptive method, describing building student creativity using the C-R-E-A-T-E model in the making of voltaic cell. The subjects of this study were 20 students of Class XII in one of the high schools in Bandung. During the learning process, the students were divided into five heterogeneous groups based on their cognitive abilities. The research instruments used were an assessment rubric of student worksheets, an assessment rubric of voltaic cell products, students' testimonial sheets, and observers' testimonial sheets.

Data collection of this research was through the scores of each the instruments. To assess effectiveness of the model through the student worksheet, the differences in the score of students were attributed to the effectiveness of the C-R-E-A-T-E model. After learning through the C-R-E-A-T-E model, student creativity was judges based on the average of percentage of the scores. To assess effectiveness of model through creative products made by the students, the differences in scores were used to describe students' creativity. After learning through the C-R-E-A-T-E model, student creativity was judged based on the calculation using the percentage of scores. To assess effectiveness of model through satisfaction of the student and observer, the differences in the score of students and observer were used to describe their satisfaction. After learning through the C-R-E-A-T-E model, student satisfaction increased, based on the calculation using the percentage of scores.

Data analysis techniques and analytical approaches were used, based on the scores of the instruments that measured their percentages. Based on the percentages, results were categorized through standard categorization: very effective, effective, or ineffective. Further, the description based on state of the finding and the result during the study were given.

3 RESULTS AND DISCUSSION

3.1 *Effectiveness of C-R-E-A-T-E model based on student worksheets*

The level of creativity of students in this study was based on the assessment rubric of student worksheets. The differences in the scores of students were used to describe effectiveness of the C-R-E-A-T-E model. Based on the average percentage of the scores, it is known that the students' creativity generally increased, as shown in Figure 1.

Based on the results of Figure 1, indicator 'originality' is highest; indicator 'fluency' is lowest, because the students are not familiar with this model. Thus, it can be concluded that syntax of the C-R-E-A-T-E model through students' worksheet is feasible. This is aligned with Isabekov & Sadyrova (2018) which states that good instruction through a student worksheet is able to facilitate students to build their creativity. Syntax of the C-R-E-A-T-E appeared in the student

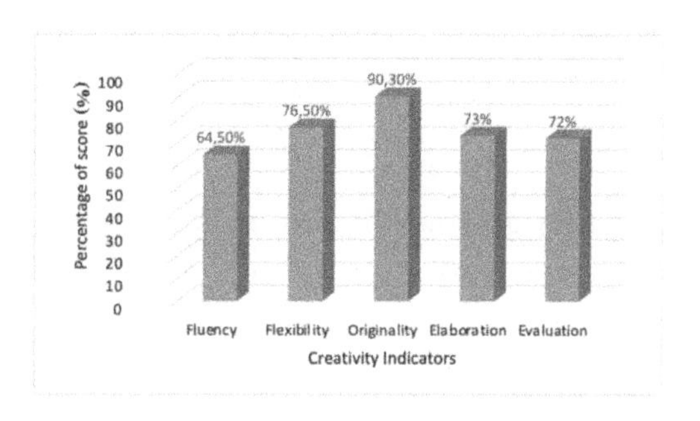

Figure 1. The average of scores of student's worksheets based on Williams' creativity indicators.

worksheet explicitly. The model provides a positive impact on increasing student motivation (Rudibyani 2019).

3.2 *Effectiveness of C-R-E-A-T-E model based on voltaic cell products made by the students*

The creativity of the students in terms of their voltaic cell products was measured using an assessment rubric of voltaic cell products. The differences in the scores of students are used to describe students' creativity. After learning through the C-R-E-A-T-E model students' creativity increased. Based on the calculation using the percentage of scores it is known that the students' creativity generally increased, as shown in Figure 2.

Based on the results of Figure 2, indicator 'originality' is highest; indicator 'evaluation' is lowest, because there are the students not making appropriate decisions when making creative products (in this case, the natural voltaic cells). Thus, it can be concluded that the creative products made by the students are very original. This is aligned with Isabekov & Sadyrova (2018) which states that an original creative product made by the students is able to build their creativities (Rudibyani 2019). Besides that, the sequence of activities contained in making natural voltaic cells in the C-R-E-A-T-E model also plays a role in building students' creativity (Isabekov & Sadyrova 2018).

3.3 *Effectiveness of C-R-E-A-T-E model based on students and observers' testimonies*

The opinions of the students about the effectiveness of the C-R-E-A-T-E model to support their voltaic cell products were measured using student testimonial sheets, and observer testimonial sheets. The syntax of the C-R-E-A-T-E model has greater percentage score in comparison with that of the introduction and closing greater the percentage of scores than introduction and closing of the lesson. This means implementation of the C-R-E-A-T-E model is more effective. The students' experience when making voltaic cells from natural ingredients is shown as very good. Average of the scores of students and observers' testimonies can be seen in Table 1.

Table 1 shows that total averages of percentage of each steps of the C-R-E-A-T-E model are very high (97% and 94%). It is mean that the students and observers felt satisfied during the instruction of this model. Finally, based on the results of Figure 1, Figure 2, and Table 1, it is shown that the implementation of the C-R-E-A-T-E model is more effective. Thus, it can be concluded that the C-R-E-A-T-E model is one of effective instruction. This is aligned with

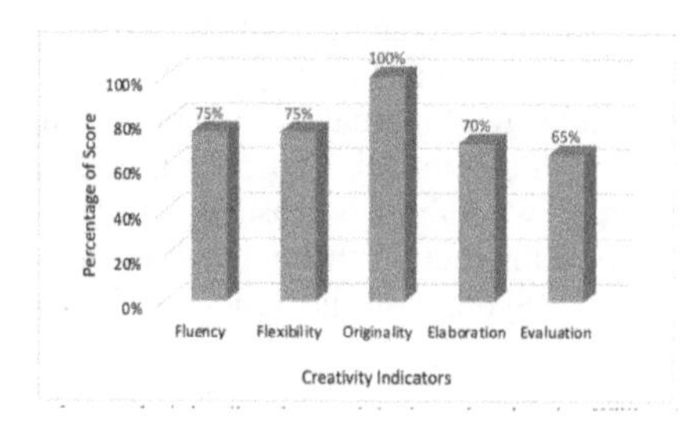

Figure 2. The average of scores of voltaic cell products made by the students based on Williams' creativity indicators.

Table 1. Average of the scores of students and observers' testimonies.

Steps of C-R-E-A-T-E	Average of students' testimonies (%)	Average of observers' testimonies (%)
Introduction	96	90
Syntax of model	98	98
Closing	97	94
Total of averages	97	94

Isabekov & Sadyrova (2018) which states that a good learning approach is able to motivate students to understand the content of the material they are studying. C-R-E-A-T-E is a learning model that provides a positive impact on increasing student motivation (Rudibyani 2019). Besides that, the sequence of activities contained in the C-R-E-A-T-E model also plays a role in building students' creativity (Isabekov & Sadyrova 2018). Students activities during implementation of the C-R-E-A-T-E process include connecting, restructuring, elaborating, applying, tasking, evaluating various sources of information, assessing others' opinions from different perspectives, applying abstract concepts to real situations, and finding solutions of problems in groups: this has made students more active (Swartling & Morgan 1998). Furthermore, Isabekov & Sadyrova (2018) suggest that students' activeness in learning by the activities (including the C-R-E-A-T-E model) provide a great opportunity for these students to maintain their knowledge over the long term and builds students' creativity.

4 CONCLUSION

Based on results and discussion, we concluded that the C-R-E-A-T-E model was very effective to build student creativity based on results of assessments of student worksheets, creative products made by students, and students and observers' testimonies. Hence, students' creativity was successfully built through learning using the C-R-E-A-T-E model. It showed the C-R-E-A-T-E model's potential to be applied in the learning process to stimulate student creativity.

ACKNOWLEDGMENT

We would like to express our gratitude to PDS (Penugasan Dosen di Sekolah) Programme at FKIP Universitas Pendidikan Indonesia 2019/2020 for granting research findings–based public service fund. Through the service, we are able to write this article as part of the services output.

REFERENCES

Allen, C., Lewis, M. A., & Fleming, D. 2017. Paint this picture: Infusing creativity through divergent thinking across all content areas. *National Youth-At-Risk Conference Savannah* 35.

Amelia, F. R. 2014. *The implementation of problem-based learning with multimedia based on computer to foster the students creativity and increase the students achievement in salt hydrolysis topic.* Doctoral dissertation. Medan: UNIMED.

Awang, H., & Ramly, I. 2008. Creative thinking skill approach through problem-based learning: Pedagogy and practice in the engineering classroom. *International Journal of Human and Social Sciences* 3(1): 18_23.

Burleson, W. (2005). Developing creativity, motivation, and self-actualization with learning systems. *International Journal of Human-Computer Studies* 63(4–5): 436–451.

Castro-Acuña, C. M., Kelter, P. B., Carr, J. D., & Johnson, T. 1996. The chemical and educational appeal of the orange juice clock. *Journal of Chemical Education*, 73(12): 1123.

Geisinger, K. F. 2016. 21st century skills: What are they and how do we assess them? *Applied Measurement in Education*, 29(4): 245–249.

Isabekov, A., & Sadyrova, G. 2018. Project-based learning to develop creative abilities in students. *Vocational Teacher Education in Central Asia*: 43–49.

Letcher, T. M., & Sonemann, A.W. 1992. A lemon-powered clock. *Journal of chemical education* 69(2): 157.

Muske, K. R., Nigh, C. W. & Weinstein, R. D. 2007. A lemon cell battery for high-power applications. *Journal of ChEmical education* 84(4): 635.

Nurhayati, L., Martini, K. S., & Redjeki, T. 2013. Peningkatan Kreativitas dan Prestasi Belajar Pada Materi Minyak Bumi Melalui Penerapan Model Pembelajaran Problem Based Learning (PBL) Dengan Media Crossword. *Jurnal Pendidikan Kimia* 2(4): 151–158.

Rudibyani, R. B. 2019. Improving students' creative thinking ability through problem based learning models on stoichiometric materials. *Journal of Physics: Conference Series* 1155(1): 012049.

Swartling, D. J., & Morgan, C. 1998. Lemon cells revisited: The lemon-powered calculator. *Journal of Chemical Education* 75(2): 181.

Utami, R. P. 2011. Pengaruh Model Pembelajaran Search, Solve, Create, and Share (SSCS) AND Problem Based Instruction (PBI) Terhadap Prestasi Belajar dan Kreativitas Siswa. *Bioedukasi* 4(2): 57–71.

Borderless Education as a Challenge in the 5.0 Society – Abdullah, Adriany & Abdullah (eds)
© 2021 Taylor & Francis Group, London, ISBN 978-0-367-61960-2

Personal safety assessment instruments for children with intellectual disabilities

O.S. Homdijah & A.S. Rizky
Universitas Pendidikan Indonesia, Bandung, Indonesia

ABSTRACT: This research is motivated by the unavailability of personal safety skills assessment instruments in schools, so schools cannot conduct personal safety skills assessments for children with intellectual disabilities. This study aims to obtain an instrument for assessing personal safety skills for children with intellectual disabilities, as well as obtaining data on testing the instrument for assessing personal safety skills. This research uses a research and development model and is carried out in two stages: preliminary research and instrument development. In the preliminary study, the data collection techniques used were interviews with three teachers. and at the stage of developing the instruments, the data collection techniques used were questionnaires. Data analysis was performed by calculating the validity score given by three experts and calculating the reliability score based on the results of the trial of the instrument of personal safety skills assessment. Based on the results of the trial, it can be concluded that the personal safety skills assessments of children with intellectual disabilities are appropriate for use by school teachers.

1 INTRODUCTION

Every child needs protection and affection, and the right to be protected from violence. In fact, there are many dangers that stake out children, one of which is the phenomenon of child abuse or violence that occurs everywhere. Child abuse is defined by Richard J. Gelles as a deliberate act that causes harm to children physically or emotionally. Examples of child abuse include sexual abuse, child exploitation, and physical violence (Huraerah 2012).

Child abuse occurs in all parts of the world, including Indonesia. Personal safety refers to the freedom from physical harm and threat of physical harm, and freedom from hostility, aggression, harassment, and devaluation by members of the community. Safety includes worry about being victimized as well as actual incidents. Another personal safety definition refers to freedom from sexual violence, child exploitation, and physical abuse. Yet another definition put forward by the Department of Public Safety of East Tennessee State University mentions personal safety as general recognition and avoidance of possible harmful situations or persons in your surroundings.

Child abuse can occur to anyone, including children with intellectual disabilities. Some studies have revealed children with intellectual disabilities who experience abuse. Sullivan & Knutson (2000) investigated the incidence of child abuse among an entire school-based population that included all 50,278 children during the 1994/95 school year in Omaha, Nebraska. They collected child abuse registry records, foster care records, law enforcement records, and school records to obtain evidence of child abuse and information about disability status. The results showed that children with intellectual disabilities were about 4.0 times more likely to be the victims of child abuse than their peers without disabilities. In particular, these children were 4.0 times as likely to be sexually abused, 3.8 times as likely to be physically abused, 3.8 times as likely to be emotionally abused, and 3.7 times as likely to be neglected as children without disabilities.

The AAIDD definition of intellectual disability specifies significant limitations in both intellectual functioning and adaptive functioning and stipulates that the disability originates before 18 years of age (Heward et al. 2017). The Diagnostic and Statistical Manual of Mental Disorders, 5th edition (DSM-5) defines intellectual disability (intellectual developmental disorder) as a disorder with onset during the developmental period that includes both intellectual and adaptive functioning deficits in conceptual, social, and practical domains. The following three criteria must be met: (1) Deficits in intellectual functions, such as reasoning, problem solving, planning, abstract thinking, judgment, academic learning, and learning from experience, confirmed by both clinical assessment and individualized, standardized intelligence testing. (2) Deficits in adaptive functioning that result in failure to meet developmental and sociocultural standards for personal independence and social responsibility. Without ongoing support, the adaptive deficits limit functioning in one or more activities of daily life, such as communication, social participation, and independent living, across multiple environments, such as home, school, work, and community. (3) Onset of intellectual and adaptive deficits during the developmental period (American Psychiatric Association 2013).

Personality is not stable and easily intimidated resulting in children with intellectual disabilities more easily becoming victims of violence or bullying, therefore children with intellectual disabilities must get lessons about personal safety. Teaching personal safety to children with intellectual disabilities must be in accordance with their learning needs.

2 RESEARCH METHODS

Participants in this study totaled thirteen people: ten children with mild intellectual disabilities and three classroom teachers. The research instruments used were interview and questionnaire guidelines. Data collection techniques used interview and questionnaire techniques.

Interviewing is the process of obtaining information for research purposes by means of questions and answers face-to-face between the interviewer and the person being interviewed, with or without using interview guidelines (Bungin 2007). In this study the researchers interviewed three classroom teachers to determine their understanding of personal safety skills, the implementation of personal safety learning, the objective conditions of the personal safety skills assessment instrument, and the implementation of personal safety skills assessments in three study locations, where the results of this interview were used as initial research and became a background in research.

The data analysis technique used to test the assessment instruments was the validity and reliability test. Validity tests check the compatibility between test items made with indicators, material, or learning objectives that have been set (Susetyo 2015). The formulas used are as follows:

$$\mathrm{CVR} = \frac{2Mp}{M} - 1 \tag{1}$$

Reliability instrument assessment was tested using the following formula:

$$\rho_{\alpha} = \frac{N}{N-1}\left(1 - \frac{\sum \sigma_{\mathrm{i}}^{2}}{\sigma_{\mathrm{A}}^{2}}\right) \tag{2}$$

3 RESULTS

In this study, researchers conducted the preparation of personal safety skills assessment instruments. The components of personal safety skills that will be developed into an assessment instrument consist of (1) knowledge of self-identity; (2) knowledge of correct genital limb names; (3) assertive skills; (4) opening to adults about events experienced; (5) distinguishing touches; and (6) travel with adult permission. These components are based on the ability that must be mastered by children to improve the personal safety skills presented by the California Childcare Health Program (CCHP). The CCHP is a program from the Department of Family Health Care Nursing, University of California–San Francisco School of Nursing (UCSF). After determining the components, the instrument lines are then compiled and contain components, sub-components, assessment techniques, and indicators, which are then developed into instrument items.

The assessment instrument that was made was then tested for content validity by three experts. In the first validation, it is known that 14 out of 28 questions were declared invalid and not suitable enough so that the second stage of validation was needed. In the second stage of validation, all instruments received a CVR value > 0, which is 1, which means that all experts stated that the instrument was valid and worth testing.

Trial of the assessment instrument was carried out on 10 children with intellectual disabilities. The results of testing the assessment instrument can be seen in Figure 1 below:

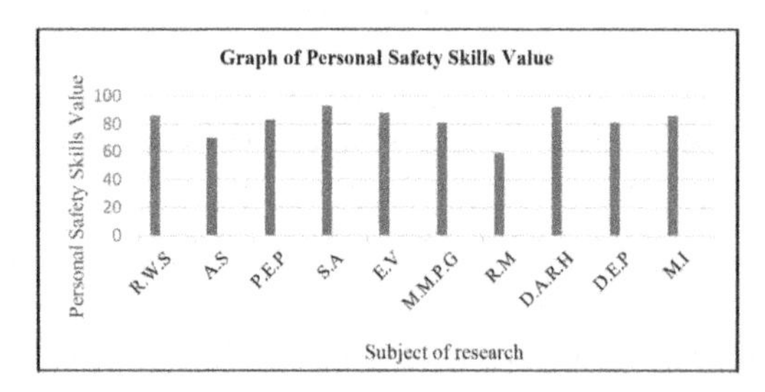

Figure 1. Graph of personal safety skills value.

Reliability is related to the level of trust. The assessment instrument can be trusted if the test results do not change or are relatively the same if the tests are repeated (Susetyo 2015). The results of calculating the reliability coefficient using the alpha Cronbach formula note that the instrument for assessing the personal safety skills of children with intelligence barriers has a value of 0.93, so it can be concluded that the assessment instrument has a fairly high reliability, so that when the instrument is used for reassessment, the test results are not likely to change and are relatively the same.

The results of the data analysis showed that the instrument of personal safety skills of children with intelligence barriers was valid and feasible to be used to assess the personal safety skills of children with intellectual disabilities.

4 DISCUSSION

The knowledge and attitudes of people with an intellectual disability about sexuality has relevance to their overall wellbeing and quality of life (Galea et al. 2004),, but children with intellectual disabilities receive less sex education both at school and at home (Levy & Packman 2004, McCabe 1999, Murphy & O'Callaghan 2004, Schaafsma et al. 2015). Children with intellectual disabilities must have personal safety skills, and they must receive sex education in order to have personal safety skills. The level of knowledge about sexuality of children with intellectual barriers must be known before learning about sex education begins. This was obtained from the results of the assessment.

As a result of the interviews, researchers found that both SLB Centra PK-PLK in Cimahi City, SLB Sukagalih, and SLB Pancaran Iman had not carried out personal safety skills assessments due to the unavailability of personal safety skills assessment instruments for children with impaired intelligence. Assessment is the basis for designing a child's learning planning. The results of the assessment analysis can determine a child's learning needs. Sunardi and Sunaryo (in Soendari & Nani 2011) suggest that, in general, the assessment aims to (1) obtain relevant, objective, accurate, and comprehensive data on the child's current condition; (2) know the full profile of the children, especially the problems and learning constraints they face, their potential, their special needs, and the environmental carrying capacity they need; and (3) determine the services needed in order to meet their special needs and monitor progress.

Children with intelligence barriers have a non-stable personality, they can be influenced by other people and intimidated, so they might also more easily fall victim to bullying. This means they need to be taught how to protect themselves from the threats that exist in their environment, so that they do not become victims of violence. To learn this, an assessment instrument is needed to find out the strengths, weaknesses, and learning needs of children, for teachers to start learning.

5 CONCLUSION

Based on the results of instrument testing, it is known that 100% of instructions for preparation of assessment, instructions for evaluating, instructions for assessing assessment results, as well for instructions for making personal safety skill profiles are easily understood by teachers, so teachers at school can assess the personal safety skills of children with intellectual disabilities. Instrument items and drawing illustrations are also suitable and do not need to be revised. The personal safety assessment instrument for children with intellectual disabilities is reliable, as is indicated by the results of the calculation of reliability, which is at 0.93, so the reliability table shows a high level of reliability. Based on the results of the trials and the calculation of reliability, it can be concluded that the instrument to assess the personal safety skills of children with intellectual disabilities that has been prepared can be used by teachers in schools.

REFERENCES

American Psychiatric Association 2013. *Diagnostic and Statistical Manual of Mental Disorders* (5th edition). Arlington, VA: American Psychiatric Association.

Bungin, B. 2007. *Penelitian kualitatif* (Edisi Kedua). Jakarta: Kencana.

Galea, J., Butler, J., Iacono, T., & Leighton, D. 2004. The assessment of sexual knowledge in people with intellectual disability. *Journal of Intellectual and Developmental Disability* 29(4): 350–365.

Heward, W. L., Morgan, S. R. A., and Konrad, M. 2017. *Exceptional Children: An Introduction to Special Education* (11th). Pearson.

Huraerah, A. 2012. *Kekerasan terhadap anak*. Bandung: Nuansa Cendekia.

Levy, H., & Packman, W. 2004. Sexual abuse prevention for individuals with mental retardation: Considerations for genetic counselors. *Journal of Genetic Counseling* 13(3): 189_205.

McCabe, M. P. 1999. Sexual knowledge, experience and feelings among people with disability. *Sexuality and Disability* 17(2): 157–170.

Murphy, G. H., & O'callaghan, A. 2004. Capacity of adults with intellectual disabilities to consent to sexual relationships. *Psychological Medicine* 34(7): 1347–1357.

Schaafsma, D., Kok, G., Stoffelen, J. M., & Curfs, L. M. 2015. Identifying effective methods for teaching sex education to individuals with intellectual disabilities: A systematic review. *Journal of Sex Research* 52(4): 412–432.

Soendari, T., & Nani, E. 2011. *Asesmen dalam pendidikan anak berkebutuhan khusus*. Bandung: Amanah Offset.

Stewart C. C. 2012. Beyond the call: Mothers of children with developmental disabilities responding to sexual abuse. *Journal of Child Sexual Abuse* 21: 701–727.

Susetyo, B. 2015. *Prosedur penyusunan & analisis tes untuk penilaian hasil belajar bidang kognitif*. Bandung: Refika Aditama.

Borderless Education as a Challenge in the 5.0 Society – Abdullah, Adriany & Abdullah (eds)
 ISBN 978-0-367-61960-2

Adaptation of mathematics learning programs for students with mathematical difficulties in elementary school

T. Soendari & I.D. Aprilia
Universitas Pendidikan Indonesia, Bandung, Indonesia

ABSTRACT: Students with mathematics difficulties are experience obstacles in reasoning and mathematical calculations. Mathematical learning problems faced by students at the level of basic education must be anticipated by appropriate interventions through an adaptive learning program design, thus, the difficulties faced can be eliminated or removed which can, in turn, improve students' mathematical abilities optimally. This research aims to find out adaptive learning programs for improving the mathematical abilities of students with mathematics difficulties in elementary school. To achieve this goal, this research uses a descriptive qualitative approach with research subjects of six students in third grade and two teachers from two elementary schools in Bandung-Indonesia. The research produces (1) the formulation of an adaptive mathematics learning program for students with mathematical difficulties in elementary school, and (2) the resulting program becomes a reference and can be disseminated wider. This research has implications for elementary school teachers and for education policy makers: (1) Implications for teachers that the adaptation of this learning program can be used as a guide in teaching students with learning difficulties, especially in mathematics. (2) Implications for government policy, that the adaptation of learning programs for students with learning difficulties must be considered in the preparation of the national curriculum, especially the education curriculum at the elementary school level.

1 INTRODUCTION

Whenever political and education leaders speak or write about how the economic success of nations depends on the academic success of its students, the first example mentioned is often mathematics (American Institutes for Research [AIR] 2006, National Research Council 2004). The USA is falling behind in the world, often told that Asian and European peers are surpassing the US in mathematics, therefore, soon they will surpass the US in ability to manage and profit from new technologies, increasingly complex science, and advanced engineering. Those all depend on a solid core of skill in mathematics. If mathematics is essential to success in all quantitative endeavors and occupations, then success in basic mathematics is of particular importance. In elementary school, students learn the basic mathematical ideas and operations. They learn math is fun and worthwhile, or that it is tedious and unrewarding, so that sometimes they think that they are either "good at math" or "not good at math." These early learnings of students can have long-term consequences.

Students with mathematics difficulties, are likely to perform poorly in basic math. Believing they are not good at math, many avoid math activities and, ultimately, professions that require mathematical proficiency (Barton & Coley 2010, Ganley & Lubienski 2016). In light of the importance of elementary mathematics and the social and economic implications of ethnicity and social class, it is clear that substantial investments in improving elementary mathematics are justified. Carnine et al. (1997) stated that individuals who experience learning difficulties, including in mathematics, not have intellectual disabilities or impairments, but also are suffering from ineffective learning. Therefore, one of the things that needs to be improved to increase students' mathematical abilities is the quality of learning delivered by the teacher, because the teacher is the main source of student knowledge in the early grades. Students with mathematical difficulties in elementary school need appropriate education services that are adjusted to the level of difficulty and needs of children to obtain appropriate educational achievements. Yet which programs and practices are most likely to increase achievement and reduce gaps? It is undeniable that the availability of a good program can change the direction of the learning process to achieve the expected goals. There are many kinds of learning programs that have been developed by experts and education practitioners, such as: *Systematic Design of Instruction* from Dick et al. (2001); the program design model from Kemp; the IDI Model developed by the University Consortium for Instructional Development and Technology (UCIDT); Banathy (1968); and Briggs. However, the types of programs that have been widely developed there are not learning programs based on student learning needs. Adaptation of learning programs is one of the assessment-based learning programs that emphasizes the abilities, difficulties, and learning needs of students. Blackwell & Rossetti (2014) mentioned "limited research on the ways in which assessment information is utilized." The implication is that additional research is needed to learn more about the ways in which assessment practices and data are actually utilized to develop adaptation of learning programs, particularly in the area of assistive technology considerations. Adaptation of learning programs should meet the unique needs of every child with disabilities. The primary tool for enabling schools to provide this required level of support to students with disabilities is the adaptation of a learning program. Most adaptation mathematics learning program goals are not developed with powerful opportunities for students to showcase their reasoning, rather, goals generally focus on remediating gaps in knowledge (Tan 2015) or developing fluency in calculations (Barnes et al. 2015). Adaptation of learning programs serves as a roadmap for special education (SE) services (Conroy et al. 2008, Diliberto & Brewer 2012, Gartin & Murdick 2005). This research aims to find adaptive learning programs to improve the mathematics abilities of students with mathematical difficulties in elementary school. To achieve these objectives, researchers need data about the profile of learning needs of students with mathematical difficulties and the profile of learning programs conducted in elementary schools.

2 RESEARCH METHODS

The expected goal of this research is the formulation of an adaptive learning program for improving the mathematical abilities of students with mathematical difficulties in elementary school. To achieve this goal, this research uses a descriptive qualitative approach. Retrieval of data sources was done purposively: participants were six students in third grade and two teachers from two elementary schools. The data analysis technique used in this research is the inductive technique (analytical induction). More operationally, this research is designed to strengthen the scientific field based on field issues related to the implementation of inclusive education in the context of learning practices in primary schools. Broadly speaking, this study consists of four stages.

The initial or first stage to develop a mathematics learning program in primary schools in an effort to overcome student learning difficulties can be illustrated through data collection on students with mathematical difficulties profiles and existing mathematics learning programs. This initial stage is expected to reveal the learning difficulties faced by students and the mathematics learning program used by teachers in overcoming these learning difficulties. From the practice of implementing learning there will also be teachers' obstacles in developing and implementing mathematics learning programs for students who have difficulty learning internally and externally. The second stage is analysing and evaluating field data. The characteristics that emerge from the class teacher in compiling and implementing learning programs to their students are expected to bring up ideas that can enrich and color alternative hypothetical programs supported by relevant theories. The third stage is formulating a draft mathematics learning program. In order for ideas or principles generated in the program to be more meaningful, practitioner validation process is a must. Including suggestions and ideas from special education experts, practitioners (teachers), and parents will strengthen the feasibility of the program. The fourth stage is the test of program implementation. This activity is carried out through training for classroom teachers. The training material contains the steps of development and academic assessment, development of assessment instruments, preparation of individualized learning programs, and implementation.

This learning program is expected to be able to improve the skills of teachers in developing and implementing the adaptive learning programs that can increase the mathematical achievements of students who experience mathematics learning difficulties. An adaptive mathematics learning program can also develop the role of schools with various devices in it (teachers and curriculum) as well as other system support such as a student-friendly learning environment, with the result that students can optimally actualize their potential.

3 RESULTS AND DISCUSSION

3.1 *Learning needs profile of students with mathematical difficulties and mathematical learning process conducted in elementary school*

Profile of students with mathematical difficulties and mathematical learning process conducted in elementary school can bee seen in Table 1 and Table 2.

Table 1. Profile of students with mathematics difficulties grade 3 elementary school.

Students	Difficulties	Learning Needs
AN	• Determine the place values of tens and units • Adding 2-digit numbers • Solving problems involving the addition of 2-digit numbers	• Place values for tens and units • Addition of 2-digit numbers • Expressing daily problems associate with adding 2-digit numbers
BA	• Determine the place values of tens and units • Reducing 2-digit numbers • Solving problems that involve adding and subtracting 2-digit numbers	• Place values for tens and units • Addition and subtraction of 2-digit numbers • Expressing daily problems related to addition and subtraction of 2-digit numbers.
BM	• Determine the place values of tens and units • Addition and subtraction of 2-digit numbers • Solving problems that involve adding and subtracting 2-digit numbers	• Place values for tens and units • Addition and subtraction of 2-digit numbers • Expressing daily problems related to addition and subtraction of 2-digit numbers
MA	• Determine the place values of tens and units • Addition and subtraction 2-digit numbers • Solving the problems that involve adding and subtracting 2-digit numbers	• Place values for tens and units • Addition and subtraction of 2-digit numbers • Expressing daily problems related to addition and subtraction of 2-digit numbers
NM	• Determine the place values of tens and units • Addition and subtraction 2-digit numbers • Solving problems that involve adding and subtracting 2-digit numbers	• Place values for tens and units • Addition and subtraction of 2-digit numbers • Expressing daily problems related to addition and subtraction of 2-digit numbers
NB	• Determine the place values of tens and units • Addition and subtraction 2-digit numbers • Solving problems that involve adding and subtracting 2-digit numbers	• Place values for tens and units • Addition and subtraction of 2-digit numbers • Expressing daily problems related to addition and subtraction of 2-digit numbers

Table 2. Profile of mathematics learning process in elementary school.

Aspect	Condition	Needs
Teacher Mindset	Mathematics is considered a difficult subject to understand, not only by elementary school students but even by the college students, because it is abstract	The changes in the mindset of teachers about learning process of mathematics
Curriculum	Using the 2013 National curriculum	Modification curriculum
Assessment	No assessment, no assessment tools	Mathematical assessment tool
Planning	Using the same syllabus, semester program, and learning implementation plan for all students in grade 3	Special programs that fit each students' needs
Implementation	Performed classically using the same material, methods, and media for all students in grade 3	Individualized learning
Evaluation	Use the same problem in accordance with material provided	Conformity with learning objectives
Follow-up	Sometimes do the conclusions and give homework	Variation of tasks according to students' abilities

3.2 *The draft of an adaptive learning program design in improving the mathematical abilities of students with mathematics difficulties in elementary school*

Adaptation of mathematics learning programs that is able to facilitate and accommodate the abilities and learning needs of students with learning difficulties, so that their potential can develop optimally. Program components consist of student, school, and subject identity; student learning needs, learning objectives, learning indicators, descriptions of learning services (learning activities, materials, methods, learning media, and learning resources); evaluation and follow-up. Flowchart of adaptive learning program design for improving mathematical ability of students in elementary school with mathematics difficulties can be seen in Figure 1.

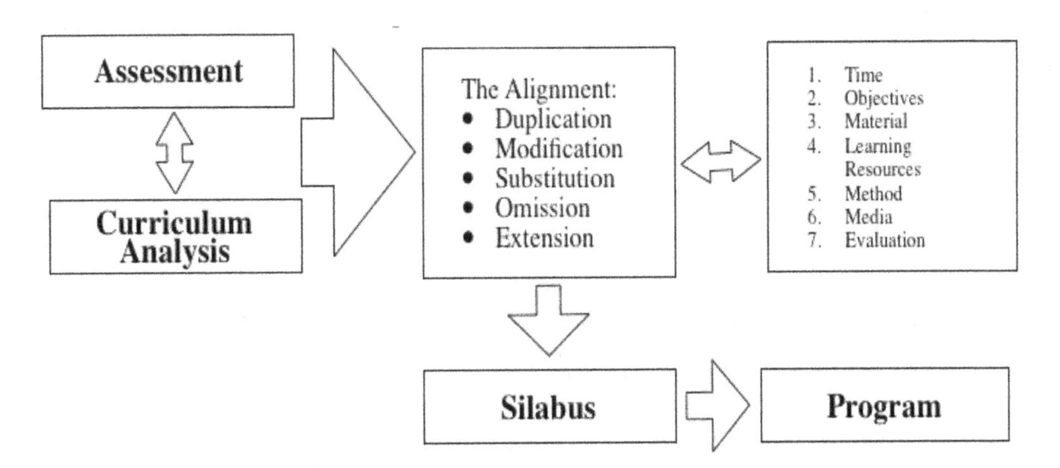

Figure 1. Flowchart of adaptive learning program design for improving mathematical ability of students in elementary school with mathematics difficulties.

3.3 *The result of an adaptive learning program design implementation for improving the mathematical abilities of students with mathematics difficulties in elementary school*

- An adaptive program can increase the mathematics abilities of students with difficulties in elementary school; the six students or subjects tend to be able to determine the value of tens and units, able to do the sum of 2-digit numbers; able to solve everyday problems that involve addition or subtraction of 2-digit numbers.
- The language used in the program is easily understood by the teacher.
- Descriptions of each aspects are quite operational.
- The program prepared can be a very useful guide for schools in order to improve the mathematical abilities of students with learning difficulties in elementary school and can support teachers for improving the quality of the learning.

3.4 *Discussion*

The results of research show that teacher said that mathematics was considered as a difficult science to understand because it was abstract, not only by elementary school students but even college students. If this assumption is left unchecked, children will be less interested in learning mathematics. Math will continue to be something frightening for children. In this case, Krech (1962) argues: "Man acts upon his ideas. His irrational acts no less than his rational acts are guided by what he thanks, what he believes, what he anticipates." This means that people's actions, both rational and irrational, are guided by what they think, what they believe and what they anticipate. Thus, the application of a learning is motivated by what the teacher understands about the topic, therefpre the teacher's improper understanding of mathematics learning needs to be straightened out, so that learning mathematics is not something scary, in fact it becomes something fun.

The results of the research indicate that teachers in elementary schools currently do not understand the learning programs that are appropriate for dealing with students with mathematics difficulties. The condition of the teachers in these elementary schools is very reasonable because they have no special educational background (Rahardja 2017). This condition has implications for the need to develop learning programs that are in accordance with student learning needs, namely through adaptation of learning programs, especially in learning mathematics. The adaptation of the mathematics learning program emphasizes student profiles as a result of assessment analysis. The facts show that regular elementary school teachers do not conduct assessments. This is because in addition to the teacher not understanding the concept of assessment, there are also no mathematical assessment tools available for children with mathematical difficulties in regular primary school. The assessment is seen as a systematic effort to find out the abilities, difficulties, and needs of students, especially students with learning difficulty in certain fields including mathematics. The results of the assessment data are used as a foundation in the preparation of the learning program for the students concerned, so what will be taught to them is in accordance with what they need. As Soendari (2011) states: "Assessment is a comprehensive appraisal through the process of gathering systematic information about an individual to find out his abilities, the difficulties that student experienced, and find out the background of why obstacles or difficulties arise, so that, the student's learning needs can be identified as a material for consideration and decision in the preparation of the intervention/learning programs for the individual involved."

Through the teacher assessment activities, teachers can obtain relevant, objective, and accurate data about the abilities of students, learning constraints faced, and student learning needs, so that the planning of learning programs can be arranged in accordance with the difficulties and materials needed by students. The importance of assessment in the planning of learning programs for students with mathematical difficulties cannot be doubted anymore, because with the results of the assessment the teacher can determine the learning objectives, the material to be provided, the methods and media used that are appropriate to the learning needs of students, and determine the evaluation to know whether the needs of students are fulfilled or not.

In relation to the implementation of the program, it shows that the adaptive program prepared is very useful, and will be able to become a guide for schools in improving the mathematical abilities of students with mathematical difficulties in elementary school and support teachers in improving the quality of learning. In this case, Sutikno (2009), asserted that "planning is one of the absolute requirements for every management activity. Without planning, the implementation of an activity will face any difficulty and even failure in achieving the expected goals." Furthermore Majid (2008) suggests several benefits of a learning planning program, as follow:

- Direction of activities in achieving the objectives;
- Basic pattern in regulating the duties and authority for each element involved in the activity;
- Work guidelines for each element, both teacher and student elements;
- Measuring instrument is effective or not, so, at any time know the accuracy and delay in work;
- To data preparation material for work balance; and
- To save time, energy, costs, and equipment.

Realizing the benefits above, an adaptive learning planning program is essential for teachers in elementary schools, so that implementation of mathematics learning for students with mathematics difficulties runs effectively and efficiently. The results of this research also indicate that the adaptation of learning programs for students with mathematical difficulties must be considered in the preparation of the national curriculum, especially the mathematics curriculum for elementary education level.

4 CONCLUSION

Four important things were found in this research: (1) The profile of students with mathematical difficulties shows they need special learning services to improve their mathematical abilities. (2) In terms of the profile of learning mathematics process in elementary school, professional teachers were highlighted in terms of understanding the whole learning plan, both theoretically and practically. The teachers are expected to not be confined by the status quo satisfied with the existing knowledge, but must be more active in developing abilities in their field, both in the delivery and mastery of the material. In other words, the teachers must like novelty and try to change the mindset of only relying fully on their own experience. (3) The formulation of the mathematics adaptation program as a learning program able to facilitate and accommodate the abilities and learning needs of students with mathematical difficulties. The program adaptation consists of seven program components, beginning with the formation of an assessment team consisting of general education teachers, special needs education teachers, and parents of students with mathematical difficulties. (4) The trial results of the program show that this adaptive program can improve students' mathematical abilities to continuously learn in elementary school. This program can be implemented by providing various benefits both for teachers and students who are continuously learning mathematics. Thus, it is recommended to teachers that this learning program can be adapted as a guide in teaching students with learning difficulties, especially in mathematics. Based on the conditions in the field, namely that the current curriculum is still classical, because it is also recommended to government policy makers that the adaptation of learning programs for students with mathematics difficulties should be considered in the preparation of the national curriculum, especially the mathematics curriculum of education at the elementary level.

REFERENCES

American Institutes for Research 2006. *A Review of the Literature in Adult Numeracy: Research and Conceptual Issues*. Washington, DC: American Institutes for Research.

Banathy, B. H. 1968, *Instructional System*. Belmont, CA: Fearon Publisher.

Barnes, B., Agness, J., & Craig, K. 2015. *Developing Mathematics IEP Goals and Objectives that Work!* Solution Tree Blog. http://www.solution-tree.com/blog/developing-mathematics-iep-goals-and-objectives-that-work/

Barton, P. E., & Coley, R. J. 2010. *The Black-White Achievement Gap: When Progress Stopped. Policy Information Report*. Princeton, NJ: Educational Testing Service.

Blackwell, W. H., & Rossetti, Z. S. 2014. The development of individualized education programs: Where have we been and where should we go now? *Sage Open* 4(2).

Carnine, D., Jitendra, A. K., & Silbert, J. 1997. A descriptive analysis of mathematics curricular materials from a pedagogical perspective: A case study of fractions. *Remedial and Special education* 18(2): 66–81.

Conroy, T., Yell, M.L. & Katsiyannis, A. 2008. Schaffer v. Weast: The Supreme Court on the burden of persuasion when challenging IEPs. *Remedial and Special Education* 29(2): 108–117.

Dick, W., Carey, L., & Carey, J.O. 2001. *The systematic design of instruction* (6th ed). New York: Longmann.

Diliberto, J. A., & Brewer, D. 2012. Six tips for successful IEP meetings. *Teaching Exceptional Children* 44(4): 30–37.

Ganley, C. M., & Lubienski, S. T. 2016. Mathematics confidence, interest, and performance: Examining gender patterns and reciprocal relations. *Learning and Individual Differences* 47: 182–193.

Gartin, B. C., & Murdick, N.L. 2005. Using functional behavioral assessment with individuals with mental retardation/developmental disabilities. *Assessment for Effective intervention* 30(4): 25–32.

Krech, D., Crutchfield, R. S., & Ballachey, E. L. 1962. *Individual in Society: A Textbook of Social Psychology*. USA: McGraw-Hill Book Company.

Majid, A. (2008). *Perencanaan Pembelajaran Mengembangkan Standar Kompetensi Guru*. Bandung: PT Remaja Rosdakarya.

National Research Council 2004. *On Evaluating Curricular Effectiveness: Judging the Quality of K-12 Mathematics Evaluations*. Washington, DC: The National Academies Press.

Rahardja, D. 2017. Understanding of special teachers in teaching children with special needs at inclusive school. *International Conference on Special Education In Southeast Asia Region (ICSAR) 7th* 1(1): 13–17.

Soendari, T. 2011. *Asesmen dalam Pendidikan Anak Berkebutuhan Khusus*. Bandung: Amanah Offset.

Sutikno, M. S. 2009. *Pengelolaan Pendidikan: Tinjauan Umum dan Konsep Islami*. Bandung: Prospect.

Tan, A. G. 2015. Teaching mathematics creatively. *The Routledge International Handbook of Research and Teaching Thinking*: 411–423.

Borderless Education as a Challenge in the 5.0 Society – Abdullah, Adriany & Abdullah (eds)
© 2021 Taylor & Francis Group, London, ISBN 978-0-367-61960-2

Management capacity improvement by group-dynamic-based training in Rumah Pintar Nurul Falah Kelurahan Sukaluyu Kecamatan Regol Kota Bandung

J.S. Ardiwinata, O. Komar & C. Sukmana
Universitas Pendidikan Indonesia, Bandung, Indonesia

ABSTRACT: Strengthening the Nurul Falah Smart House as the centre of various activities in the Regol area of Bandung. Not yet optimal management performance and not yet maximum efforts in developing centres of each existing centre, are the main problems of this institution. The purpose of the study is to describe the steps to strengthen the capacity of the manager of the Nurul Falah Smart Home Craft Center. This research method uses a qualitative approach with a case method. The data collection techniques used were interview, observation, literature study, documentation study. The results of the study indicate that the steps to strengthen the capacity of Nurul Falah smart home craft centre managers are a) organizing training based on group dynamics. b) mentoring after training. Strengthening the capacity of managers has a positive impact on managers in implementing smart home programs.

1 INTRODUCTION

National education in Indonesia still faces three big complex challenges. The first challenge, as a result of the economic crisis, the world of education is required to be able to maintain the results of educational development that has been achieved. Secondly, to anticipate the global era of education the world is required to prepare competent human resources to be able to compete in the global job market. Third, in line with the implementation of regional autonomy, it is necessary to make changes and adjustments to the national education system so that it can realize a more democratic education process, pay attention to the diversity of needs/conditions of regions and students, and encourage increased community participation. In addition, national education is still faced with several prominent issues, namely:

- The low level of equity in obtaining education;
- The quality and relevance of education is still low; and
- The weak management of education, aside from not yet achieving independence and scientific and technological excellence in academics.

Non-Formal Educational Institutions Units such as Smart Houses, PKBM, learning centres, Islamic boarding schools, padepokan, and other educational activity organizers such as foundations and others have the capacity to develop non-formal education which is an important part of education development programs and overall community development. Non-formal education is a deliberate conscious effort to help the community so that they can change their attitudes and behaviour and can use these attitudes and behaviours to improve their standard of living. Rumah Pintar is a "House of Education" for the community that aims to:

- Increase interest in reading, develop children's intelligence potential, technology introduction through learning in 4 centres.
- Develop and empower community skills based on local potential through craft centres.

- Encouraging the creativity of the community to maintain and preserve local culture.
- Developing entrepreneurial skills based on local potential.
- Improve family living standards.

Community empowerment can be chosen to support the community of Sukaluyu subdistrict Regol Bandung city in training. One form of community empowerment activities is through the provision of group dynamics-based training in Nurul Falah smart home institutions. Training is empowering in the field of education so that the community is able to explore traditional wisdom (indigenous-technology) and easily adopt innovations that benefit the lives of families and their communities (Mardikanto & Soebiato 2012). Community empowerment is considered a national investment so it is expected to improve human skills and community knowledge (Muljono 2011). The results of the interviews conducted at the preliminary study stage revealed that there had never been a training based on group dynamics. Group dynamics-based training activities for managers of smart home institutions can be designed by considering aspects such as (Emawati et al. 2012):

- Based on identification results learning needs of local managers; and
- Pay attention to the suitability between the needs or types of skills with the potential of existing resources

Sentra Kriya, a centre for the provision and service of skills and other life skills in accordance with community needs. Activities at the craft centre are a means of community empowerment focused on providing participants with life and vocational skills in working while working so as to foster independent entrepreneurial attitudes. According to Komar (2018) strengthening the capacity of institutional managers must be based on needs5, this is in accordance with the existence of the home smart is very much needed as a home for education with the aim of increasing interest in reading, technology literacy and increasing skills activities as well as information centres, partnership networks and potential development events and is a concrete manifestation of human empowerment with the motto "From, By, For Society" based on the determination "Towards Society Mandiri Through Optimization of Potential and Self-Reliance". The era of globalization and the development of information technology that has an influence on communication patterns, Rumah Pintar Nurul Falah as a media of facilitation for rural communities to be able to have skills and utilize information sources to improve their standard of living have an urgent role in efforts to increase public awareness about the importance of education. The lack of available resources spurs smart houses to develop and contribute as much as possible to the progress of society with the work of the community can build independence.

2 RESEARCH METHODS

The approach used in this research is a qualitative study, which prioritizes the emic view, which is concerned with the views of informants without coercion from researchers. Data collection was carried out with interview guidelines and participant observation guidelines (Moleong 2007). The informants of this research are smart home managers and tutors. The end result is the process of strengthening the capacity of nurul falah smart home craft centre managers through group dynamics-based training.

3 DISCUSSION

3.1 *Steps to strengthen the capacity of the Nurul Falah craft centre manager smart house Nurul Falah*

Steps to strengthen the capacity of Nurul Falah smart home management centres are: 1) Group dynamics-based training. 2) Assistance; Capacity development is an approach that is now widely used in community development.

This term has been used since the 1990s by donor countries to improve the capacity of partner countries (countries that receive assistance). Capacity building efforts are an important part of various aspects of life. It is important for government officials to improve the performance of the apparatus in carrying out their duties as servants of the state, carrying out regulations and deregulation of government policies, so that in the context of overall development capacity building efforts are an inseparable part. That is, it is impossible for a development process to occur in any case without capacity building efforts for actors as well as the system that governs it. Among them is through capacity building, a series of strategies aimed at increasing efficiency, effectiveness and responsiveness of performance. That is, as the ability of an organization or company to create value where the ability is obtained from various types of resources owned by the company.

The purpose of Capacity Building is: In general, to embody the sustainability of a system. Specifically, to achieve better performance seen from:

- Time efficiency and resources needed to achieve an outcome.
- Effectiveness in the form of appropriateness of business carried out for the desired results.
- Responsiveness is how to synchronize between the needs and capabilities of the purpose.
- Learning that is indicated on the performance of individuals, groups, organizations and systems.

Therefore, capacity building will change systematically and consistently the work mechanism system of mind-set and culture-set organizations that have not been efficient, effective, productive, professional and bureaucrats who do not have the mindset of serving the community, have not achieved better performance (better performance), and have not been results oriented. Public services cannot accommodate the interests of all strata of society and do not fulfill the basic rights of citizens/residents. The implementation of public services is not in accordance with the expectations of an increasingly nation move forward in increasingly fierce global competition Benefits of Capacity Building activities in human resource development, to:

- Reduce and eliminate poor performance.
- Increase productivity.
- Increase the flexibility of the workforce.
- Increase employee commitment.
- Reducing turn over and absenteeism.

The researcher describes the steps in strengthening the manager of Nurul Falah Smart Home Craft Center as follows:

3.1.1 *Group dynamics-based training*

First stage is the recruitment of participants based on group dynamics. Recruitment of participants is key to determining the success of the next step in group dynamics-based training. This stage the organizers set several requirements that must be met by the participants of the group dynamics-based training. After recruitment, the management participant identified the learning needs, learning resources and possible obstacles. Identification of learning needs is the activity of searching, finding, recording, and processing data about learning needs desired or expected by the organization. To be able to find this learning need, various approaches can be used. Three sources can be used as a basis for identifying learning needs, namely managers/human resources in smart homes, potentials and problems in society as a whole. Furthermore, the organizer determines and formulates training objectives based on group dynamics. The objectives of the group dynamics-based training that were formulated will guide the implementation of the training from the beginning to the end of the activity, from making program plans to evaluating program results. Therefore, the formulation of goals must be done carefully. The training objectives generally contain the things that must be achieved by group dynamics-based training. These general objectives are translated into more specific goals. To facilitate the organizers, the formulation of objectives must be formulated concretely and clearly about what must be achieved with group dynamics-based training.

After developing general objectives, the organizer prepares the initial evaluation and final evaluation tools. The initial evaluation of a group dynamics-based training program is intended to find out the "entry-level behavioural" of a group dynamics-based training participant. In addition to determining the learning material and methods correctly, this search is also intended to group and place the training participants proportionally. Final evaluation is intended to measure the level of material acceptance by trainees based on group dynamics. In addition, to find out materials that need to be deepened and improved. Furthermore, the manager arranges the sequence of training activities based on group dynamics. At this stage the organizer of the training based on group dynamics determines learning material, selects and determines learning methods and techniques, and determines the media to be used. The sequence that must be arranged here is the whole set of activities from opening to closing. In arranging this sequence of activities, the factors that must be considered are the training participants, learning resources, time, available facilities, forms of training and training materials. Then the trainer must understand the training program as a whole. The sequence of activities, scope, training materials, methods used, and media used should be understood correctly by the trainer. In addition, the trainer must also understand the characteristics of the trainees and their needs. Therefore, orientation for the trainer is very important to do.

The next stage the organizer conducts an evaluation for the training participants based on group dynamics. The initial evaluation which is usually done with a pretest can be done verbally or in writing. This stage is the core of training activities, namely the process of educational interactions between learning resources and learning citizens in achieving the goals set. In this process various dynamics occur, all of which must be directed to the effectiveness of group dynamics-based training. All capabilities and all components must be combined so that the training process produces optimal output. The final stage the organizers conduct a final evaluation of the training program based on group dynamics. This stage is carried out to determine the success of learning. With this activity it is expected to know the absorption and acceptance of learning citizens towards various materials that have been delivered. That way the organizer can determine the follow-up steps that must be taken. Evaluation of training programs based on group dynamics. Evaluation of training programs based on group dynamics is an activity to assess all training activities from beginning to end, and the results are input for the development of further training. Kartika is of the opinion that the implementation of such a model can be said to be a standard step in every training event. With this activity, in addition to knowing the perfect factors that must be maintained, it is also expected to know the weak points in each component, every step, and every activity that has been carried out. In this activity, the results are not only assessed, but also the process that has been carried out (Komalasari 2012). Thus, a comprehensive and objective picture of the activities that have been carried out is obtained. The following Table 1 steps of training based on group dynamics to participants.

Basically, this training model is almost the same as the previous training model, there is the formulation of objectives and training evaluations. However, in this model the formulation of objectives is set and determined based on the needs of the trainees (Mujiman 2011). So before that a trainer must gather information relating to the trainees from the characteristics of the prospective trainees, their needs to what is expected by the participants after attending this training. This training model is usually applied when carrying out training based on community empowerment.

3.1.2 *Post training assistance*

Assistance in the training program can be interpreted as a follow-up of the training that has been given by the main instructor through continuous interaction between the instructor and the participant until the participant is deemed capable of compiling the work program of smart home institutions al barokah. Kamil argued that assistance was an

Table 1. Training steps based on participant group dynamics.

Aspects of Formation	The Dynamics of Cooperation	Learning Dynamics	Business Dynamics
Duties and Functions of The Group	• Divide tasks and functions according to the abilities or competencies of each farmer member • Coordinate between leaders and members, or fellow members • Building group participation with more planned activities, success and objectives can be measured	• Organizing activities that are able to provide constructive new information for the group • Each member understands the tasks that must be carried out to achieve the learning objectives	• Develop business strategies based on the results of group thought and deliberation • Invite members to participate in business activities
Group Coaching	• Maintain work practices and develop group participation • Utilize a means of activity to facilitate group aspirations • Supervision of every new member and supervisor of the group's performance is carried out • Give satisfaction to each component of group members by providing evidence of success and appreciation for those who excel	• Supervise information facilities in support of a conducive learning environment for participants • Maintaining a conducive learning environment, with the formation of a schedule of meetings that discuss new information	• Supervise the business activities and business sectors managed by participants. • Conducting consultations in the field of entrepreneurship, to measure marketing capabilities and open business opportunities
Group Atmosphere	• Build integrated action between group members • Build solidarity and a spirit of cooperation within the group	• Participatory approaches and variations in the methods used in building a group atmosphere • Build a climate of discussion among group members	• Build business cooperation for the progress of the group and personal participants • Joint deliberation in resolving various problems and obstacles experienced in the business sector
Group Goals	• Make a mapping of participants' problems and needs as a basis for the participant's objective needs • Discussing the main problems and objective needs of the participant • Formulating the participants' needs systematically, starting from the priority needs first.	• Formulate the problem based on identifying the participant's needs in achieving participant's learning needs. • Establish priorities of learning programs that are indicated as priority needs • Coordination between leaders and members in maintaining consistency in group goals	• Arrange participants' needs and classify them as products that have business opportunities • Designing a business design that is able to be carried out by farmer group members • Build commitment between members to achieve joint business goals

(Continued)

Table 1. (*Continued*)

Aspects of Formation	The Dynamics of Cooperation	Learning Dynamics	Business Dynamics
Group Structure	• Develop organizational structures, especially in the field of partnership • Compile clarity in building power relations • The division of tasks in accordance with the organizational structure • Building cooperation and communication between team members including schedule time categorized into routine activities and monthly activities	• Building a "learning organization", each member is given the responsibility to always seek information, continue to learn/learn about the organization in developing his group • Establish a conducive means of communication in providing information facilities to each group member	• The division of tasks in accordance with the ability of members • Build commitment between members to achieve business goals
Group Effectiveness	• Take into account more efficient production to clarification the various risks and its opportunities • Building effective communication by fostering personal relationships with the principle of "kinship". • Build and remind each other of the commitment of the group's vision and mission in each meeting or activity	• Creating an effective learning communication climate, through question and answer and intensive discussion • Always remind and motivate participants about the group's vision and mission • Give each farmer group roles and responsibilities to find the latest information, then discuss together	• Give assignments to each member to build a network and search for business networks for groups and members • Build creativity from ideas and ideas adopted by each member of the farmer group for business progress • Building teamwork in building a business

activity carried out by someone who was consultative, interactive, communicative, motivative, and negotiates. Consultative in question is creating a condition where the companion or the mentored person can consult together in solving problems together, interactively means that between the companion and the person being accompanied must be equally active, communicative means what is conveyed by the companion or the person being accompanied can be understood together, motivational means the companion must be able to foster self-confidence and be able to provide enthusiasm/motivation, and negotiation means that the companion and the person accompanied are easy to make adjustments (Kamil 2010).

Based on the results of the group dynamics-based training, a picture was obtained that in general the organization of the training based on group dynamics so far had been perceived by the informants quite well. The results of the interviews with the participants of this activity are very useful and important because the managers of smart home institutions have not received training activities so far using this model. The results of interviews with the head of the training institute are very useful for managers to determine the program to be implemented according to needs. Managers in deciding on programs must be based on an analysis of the program needs of the participants. Needs analysis is an effort to find out what exactly is needed by participants in the training. The results of interviews with trainees regarding the relevance of the material to the needs are very appropriate because what the manager will do in accordance with what was trained in the training activities.

The results of interviews with participants regarding the evaluation of training based on group dynamics are very important because every program that is carried out must have targets and results achieved in a program. The evaluation is carried out and adjusted to the needs of the training participants, namely the evaluation of the training program. This is as explained by Chaudhery that evaluations are carried out to help managers, employees, and professional human resources to make decisions about certain programs and methods" (Chaudhery 2014). Evaluations begin with a clear identification, objectives or expected results of the program training by focusing on goals and results.

4 CONCLUSION

The conclusions of the steps to strengthen the capacity of the manager of the smart home craft centres are:

- Organization of training based on group dynamics. as follows:
  - Recruitment of training participants based on group dynamics
  - identification of learning resource needs
  - Determine and formulate training objectives based on group dynamics
  - Develop evaluation tool
  - Arrange a sequence of training activities based on group dynamics
  - training for trainers
  - Conduct evaluation for training participants based on group dynamics
  - Implement group dynamics-based training
  - final evaluation of group dynamics-based training program
- Post training assistance: carried out consultatively. Consultative in question is creating a condition where the companion or the mentored person can consult together in solving problems together, especially the problem of craft dysentery which is the basis of entrepreneurial development, interactive meaning between the companion and the mentored must be equally active meaning there is a group dynamics, communicative the meaning is what delivered by the facilitator or the assisted can be understood together, motivative means that the companion must be able to foster self-confidence and be able to provide enthusiasm/motivation, and negotiation means the companion and the person being accompanied can easily make adjustments.

ACKNOWLEDGMENTS

Thank you to the Graduate School, Universitas Pendidikan Indonesia for funding this research.

REFERENCES

Chaudhery, U.A. 2014. A Critical Study of Evaluation Models for Training and Development. *Journal of Global Research Computer Science and Technology* 1(143).

Emawati, S., Lutojo, L., Irianto, H., Rahayu, E.T. & Sari, A.I. 2012. Efektivitas model pelatihan keterampilan berbasis usaha pertanian-peternakan terpadu pasca bencana erupsi Gunung Merapi di Kecamatan Selo, Kabupaten Boyolali. *Sains Peternakan: Jurnal Penelitian Ilmu Peternakan* 10(2): 85–92.

Kamil, M. 2010. *Model Pendidikan dan Pelatihan (Konsep dan Aplikasi)*. Bandung: Alfabeta.

Komalasari, K. 2012) The Living Values-Based Contextual Learning to Develop the Students' Character. *Journal of social sciences* 8(2): 246.

Komar, O. 2018. Penguatan Kapasitas Pengelola Rumah Pintar Al Barokah Melalui Pelatihan Berbasis Kebutuhan, *International Conferences Education and Science.*

Mardikanto, T. & Soebiato, P. 2012. *Pemberdayaan Masyarakat Dalam Perspektif Kebijakan Publik.* Bandung: Alfabeta.

Moleong, L.J. 2007. *Metodologi Penelitian Kualitatif*, Bandung: PT Remaja Rosdakarya Offset.

Mujiman, H. 2011. *Manajemen Pelatihan Berbasis Belajar Mandiri.* Yogyakarta: Pustaka Pelajar

Muljono, P. 2011. The model of family empowerment program for community development in West Java, Indonesia. *Journal of Agricultural Extension and Rural Development* 3(11): 193–201.

Borderless Education as a Challenge in the 5.0 Society – Abdullah, Adriany & Abdullah (eds)
 ISBN 978-0-367-61960-2

The mathematical content knowledge of elementary school pre-service teachers

A.D. Fitriani, S. Prabawanto & D. Darhim
Universitas Pendidikan Indonesia, Bandung, Indonesia

ABSTRACT: This research aims to illustrate mathematical content knowledge in elementary school, specifically in the topic of negative integers. The mastery of content knowledge as one of the aspects of subject matter knowledge that elementary school pre-service teachers must have. This ability includes the mastery of mathematical knowledge possessed by pre-service teachers and how they organized knowledge into a design of teaching materials. The research method used is qualitative research with case study techniques, with the subject of research being pre-service teachers for elementary school. Based on the results of the study, pre-service teachers' mathematical content knowledge is translated into two aspects, namely common content knowledge or minimal ability in terms of the concept of negative integers, and specialized content knowledge or mathematical knowledge used to teach mathematics. The two sub-abilities are already possessed by elementary school pre-service teachers, although this still needs to be improved.

1 INTRODUCTION

Regulation of the Minister of National Education of the Republic of Indonesia No. 16 of 2007 concerning Academic Qualification Standards and Teacher Competencies states that teachers must have pedagogical, personal, professional, and social competencies. One indicator of professional competence is mastery of extensive and in-depth content knowledge. The mastery of extensive and in-depth content knowledge can be demonstrated by the mastery of scientific substance to deepen the knowledge or material of a particular field of study. Some studies state that teachers must master the concept of what will be taught (Loewenberg Ball et al. 2008). A teacher who has insufficient knowledge is unlikely to be able to teach students to understand concepts.

Building and mastering mathematical concepts certainly requires a long time and process. Three main components of knowledge must be possessed by the teacher, namely subject matter knowledge, pedagogical content knowledge, and general pedagogical knowledge (Loewenberg Ball et al. 2008). Pre-service teachers and teachers must have subject matter knowledge. Attorps (2006) states that subject matter knowledge is material that contains other knowledge, namely: mathematical knowledge possessed by the teacher and how the teacher organizes that knowledge; knowledge of substantive structure, which is a way to organize facts, concepts, and principles in mathematics; knowledge of syntactic structure, which is an understanding of the rules of proof in mathematics; conceptual knowledge, which is computational ability; and procedural knowledge, which is the ability to identify components, algorithms, and mathematical definitions. Meanwhile, Hill et al. (2008) state that subject matter knowledge is divided into common content knowledge, specialized content knowledge, and knowledge at the mathematical horizon. Subject matter knowledge is the teachers' understanding of the material, or in other words, subject matter knowledge is the level of teacher's knowledge in mastering the material that will be taught. Understanding the content being taught is a condition for a teacher because the process of raising a problem, explaining the material, and the evaluation process is very dependent on the teacher's understanding of the topic. A teacher must have as much knowledge as possible to be able to provide knowledge to his students.

Content knowledge is knowledge about the topic to be taught, in the form of facts, theories, concepts, procedures, relationships between concepts, evidence, and content (Koehler et al. 2014). Mathematical content knowledge is divided into two main domains, namely common content knowledge and specialized content knowledge (Copur-Gencturk et al. 2019). In this research, mathematical content knowledge is described as the result of pre-service teachers' knowledge as illustrated in mastering a concept to be developed in the learning process, especially in the concept of numbers.

In the context of learning at the elementary school level, one topic that is considered quite tricky is integers, especially concerning negative numbers (Gallardo 2002). Students often see negative numbers as something absurd (Shanty 2016); most students have difficulty in conceptualizing negative numbers (Cetin 2019), representing negative numbers as mathematical objects, and using rules that are applied to arithmetic integers (Stephan & Akyuz 2012). The biggest problem faced with operations in negative numbers is the use of symbols '+' and '–', which also represent addition and subtraction operations (Hativa & Cohen 1995, Li and Smith 2007, Van de Walle 1998). Most of the studies that have emerged have highlighted students' mathematical content knowledge. From another point of view, students' difficulties or obstacles are very likely to arise from the teachers' mathematical content knowledge (or lack thereof).

The situation currently, is that there are still teachers and elementary school pre-service teachers who are lacking knowledge of mathematics, and lacking knowledge in the matter of how to teach mathematics. This situation can occur because when the teacher teaches mathematics in elementary school, the teacher has no other choice but to teach mathematics, even though the teachers' mathematical content knowledge is not enough to teach mathematics. For teachers to be ready to teach mathematics effectively, the educational process must prepare pre-service teachers' mathematical knowledge (Li & Smith 2007). This paper will examine the mathematical content of pre-service teachers to support the ability to design mathematics learning in the topic of numbers.

2 METHODS

This research used qualitative research with case study techniques. The data processing includes data exploration, data credibility, and data confirmability. Beginning the study, researchers examined information about mathematical content knowledge of pre-service teachers through observation of the subject. Furthermore, researchers build the credibility of the data through observation and triangulation of data. Confirmability is done by re-interviewing the results of learning design from different time intervals, and checking the suitability of the data obtained. The research subjects in this activity are elementary school pre-service teachers in the third year, assuming the research subjects will soon enter the profession.

3 RESULTS AND DISCUSSION

In a broad sense, mathematical content knowledge is divided into common content knowledge and specialized content knowledge. Common content knowledge emphasizes basic concepts to be taught in schools (Schilling 2007). Common content knowledge in this context is the minimum mathematical concepts that must be possessed by elementary school pre-service teachers, for example, when going to teach the concept of integers. Specifically, abilities related to common content knowledge include the ability of pre-service teachers to be able to detect incorrect mathematical statements and solve mathematical problems (Loewenberg Ball et al. 2008), though it does not require in-depth knowledge of a particular concept. Specialized content knowledge can be interpreted as mathematical knowledge that can be used to teach mathematics. Specialized content knowledge in this context is the ability of pre-service teachers to provide explanations for mathematical ideas, procedures, and algorithms, as well as how pre-service teachers learn to diagnose student misconceptions (Hill et al. 2008, Schilling 2007).

Based on results of the research conducted, the problem that arises at the beginning is that some subjects have difficulty in defining and identifying an activity or learning activity that represents negative integersbut has no difficulty with positive integers. The research subjects represented positive integers using objects around them that would later be easily understood by students.

The implementation of research is more focused on how to understand the research subject concept in introducing negative integers. Most research subjects choose the concept of debt to make it easy to introduce negative integers to students. Research subjects try to find what forms of learning are engaging for students when introducing negative integers. One of them is the game concept, for example, Monopoly games or hop-shop games (buy items with limited money, thus allowing players to borrow money). Some research subjects introduced negative numbers through the concept of temperature. However, there was a conflict between research subjects, and another subject queried how to show students the temperature of –2, something that would be difficult for students to understand. Yet another subject introduced negative integers through the concept of seawater depth, following the findings of other studies which state that to introduce negative integers can be through the concept of money and loans, thermometers, and the height of an area and water depth (Van de Walle 1998).

Furthermore, the subject is invited to build a learning flow design, illustrating possible approaches that can be designed for learning. The design arranged in the subject poured into the form of learning instructional trajectory and learning iceberg. One of the results obtained are as follows (see Figure 1).

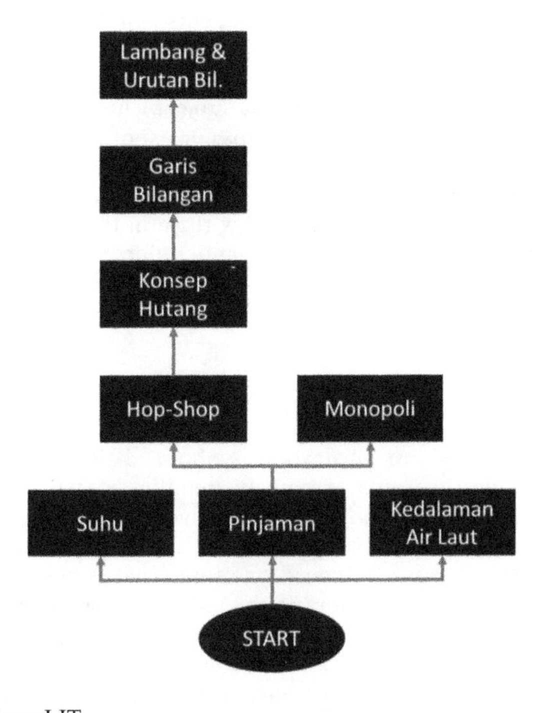

Figure 1. Negative integer LIT.

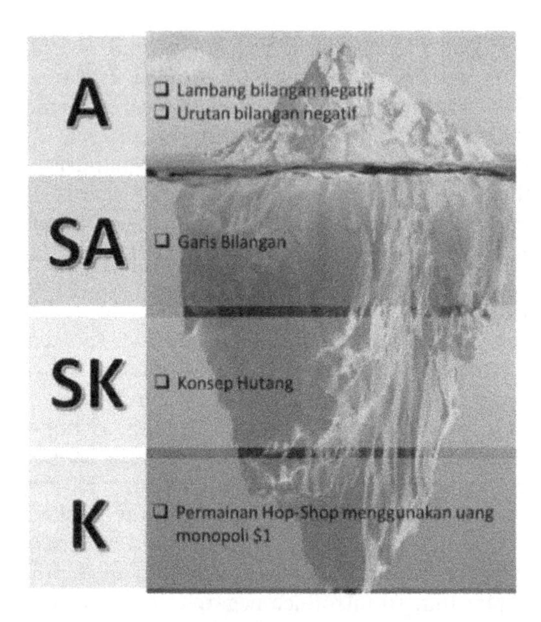

Figure 2. Negative integer iceberg.

Based on the LIT, the subject designs learning based on the students' thinking stages as put forward by Piaget's learning theory, namely through the concrete, semi-concrete, semi-abstract, and abstract stages. The steps taken by the subject are using loan representation through play (concrete stage). After playing the game, students are expected to recognize negative integers (semi-concrete stages), and then be able to draw them in the form of number lines (semi-abstract stages). After students can draw a number line, then students can sort and write negative symbols (abstract stage). Negative integer iceberg can be seen in Figure 2.

Another thing that arises is the opinion of one subject who stated that to recognize negative integers, one can use a bottle cap. Bottle caps consist of 2 sides, where the top side can be considered positive, and the bottom side can be considered negative. So that later students will know that positive numbers and negative numbers are numbers that have opposite values and make it easier for students to draw them in the form of a number line. The use of two approaches (bottle caps or coins and number lines) for learning like this together in teaching integers is seen as giving better results than using only one of them (Cetin 2019).

4 CONCLUSION

Based on the results obtained, an illustration is obtained that the mathematical content knowledge studied includes two aspects, namely common content knowledge and specialized content knowledge. Common content knowledge or the minimal ability of the concept, in this case, negative integers, shows that the knowledge of some subjects is still at a minimal level. Negative integers is indeed one topic that is considered difficult by pre-service teachers to represent in learning. Some of the approaches used by pre-service teachers are commonly used approaches. From the aspect of specialized content knowledge or mathematics knowledge that is used to teach mathematics, pre-service teachers have been able to design learning instructional trajectories that will later be developed in learning.

REFERENCES

Attorps, I. 2006. *Mathematics Teachers' Conceptions about Equations*. Helsinki: University of Helsinki.

Cetin, H. 2019. Explaining the Concept and Operations of Integer in Primary School Mathematics Teaching: Opposite Model Sample. *Universal Journal of Educational Research* 7(2): 365–370.

Copur-Gencturk, Y., Tolar, T., Jacobson, E., & Fan, W. 2019. An empirical study of the dimensionality of the mathematical knowledge for teaching construct. *Journal of Teacher Education* 70(5): 485–497.

Gallardo, A. 2002. The extension of the natural-number domain to the integers in the transition from arithmetic to algebra. *Educational Studies in Mathematics* 49(2): 171–192.

Hativa, N., & Cohen, D. 1995. Self-learning of negative number concepts by lower division elementary students through solving computer-provided numerical problems. *Educational Studies in Mathematics* 28(4): 401–431.

Hill, H. C., Ball, D. L., & Schilling, S. G. 2008. Unpacking pedagogical content knowledge: Conceptualizing and measuring teachers' topic-specific knowledge of students. *Journal for research in mathematics education*: 372–400.

Koehler, M. J., Mishra, P., Kereluik, K., Shin, T. S., & Graham, C. R. 2014. The technological pedagogical content knowledge framework. *Handbook of research on educational communications and technology*: 101–111.

Li, Y., & Smith, D. 2007. Prospective middle school teachers' knowledge in mathematics and pedagogy for teaching: The case of fraction division. *Proceedings of the 31st Conference of the International Group for the Psychology of Mathematics Education* 3: 185–192.

Loewenberg Ball, D., Thames, M. H., & Phelps, G. 2008. Content knowledge for teaching: What makes it special? *Journal of Teacher Education* 59(5): 389–407.

Schilling, S. G. 2007. The role of psychometric modelling in test validation: An application of multidimensional item response theory. *Measurement* 5(2–3): 93–106.

Shanty, N. O. 2016. Investigating students' development of learning integer concept and integer addition. *Journal on Mathematics Education* 7(2): 57–72.

Stephan, M., & Akyuz, D. 2012. A proposed instructional theory for integer addition and subtraction. *Journal for Research in Mathematics Education* 43(4): 428–464.

Van de Walle, J.A. 1998. *Elementary and middle school mathematics: Teaching developmentally*. Reading, MA: Addison-Wesley Longman, Inc.

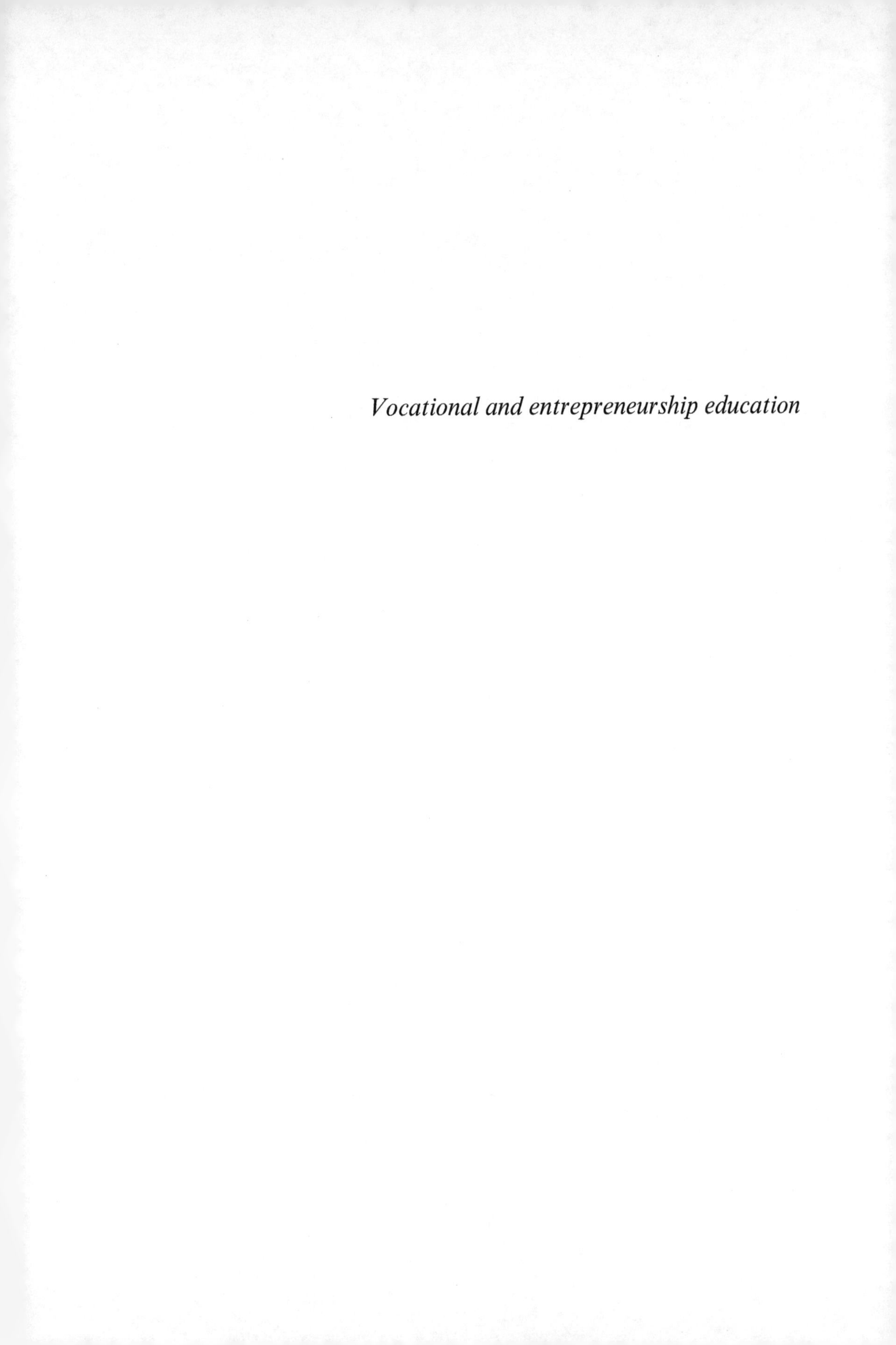

Vocational and entrepreneurship education

Borderless Education as a Challenge in the 5.0 Society – Abdullah, Adriany & Abdullah (eds)
© 2021 Taylor & Francis Group, London, ISBN 978-0-367-61960-2

Marketing training model of community learning center

Y. Shantini
Universitas Pendidikan Indonesia, Bandung, Indonesia

D. Hidayat
Universitas Singaperbangsa, Karawang, Indonesia

L. Oktiwanti
Universitas Siliwangi, Tasikmalaya, Indonesia

ABSTRACT: CLC's efforts in empowering the community through skills training encourage productivity to produce goods that are expected to be sold to increase the income of its members. The product of the learning outcomes is expected to be marketed both locally and more widely. Continuing strategies and training are needed for CLC members to be able to produce quality goods and market the products. This research tries to determine training models that can educate CLC members to be able to sell their products. The method used was the ADDIE model and data collection was carried out through interviews, focus group discussions, and observations on nine CLC in five regions in West Java. The results show that marketing by CLC currently tends to be traditional; only three out of nine CLCs are starting to market their products in modern style, professionally, and with broad scope. Marketing training can be used to focus on improvement and readiness to market products, including improving product quality, product standardization, product legality, product packaging, and product marketing. CLC marketing training to produce a product is divided into three parts: (1) scope of skills in addition to more focused on the learning process, (2) scope of skills as something that is compulsory that only can make a product for student's self, and (3) scope of skills as something that is compulsory for production and professional product marketing.

1 INTRODUCTION

Advances in technology mean almost all large and small companies around the world use technology to market their products: it is called digital marketing. Digital marketing is the use of technology to help marketing activities in order to improve customer knowledge by matching their needs (Chaffey & Smith 2017). This is because of the many benefits gained by marketing their products through technology. Broader market reach and cheap marketing costs make companies more expansive in marketing their products using technological assistance. Companies have realized the importance of digital marketing, and recognize that they will have to merge online with traditional methods to meet customer needs of more precisely (Parsons et al. 1998). Marketing products using technology are also very helpful for small and medium-sized businesses in Indonesia. Nowadays, they can sell their products using their social media, which is certainly very helpful and easy for them. In Indonesia, some companies – large or small, ranging from technology companies to manufacturers – require digital marketing and online shopping to lift marketing from the digital side (Suriansha 2017). According to Yannopoulus (2011), the Internet is the most powerful tool for businesses.

Utilization of technology for small and medium businesses is often done by people who have individual and group businesses, or by educational institutions and community empowerees who have obstacles in marketing their products. One such institution is the Centre for

Community Learning (CLC). Marketing products produced by CLC is not easy, because the price of the product is not in accordance with the average expenditure incurred by students. In addition, only one product with one package was marketed (Frandinata et al. 2019). There are several CLCs that have tried to market their products through social media accounts from the CLC itself, although the results have not been optimal. Constraints in marketing these CLC products are felt by many CLCs in Indonesia, due to various factors. Some of the causes include the lack of open knowledge of citizens to learn about technology; lack of knowledge about ITC (information and communication technology) as well as ways of using and utilizing good social media; being impatient to get immediate results; and lack of understanding to create interesting "posts" (Purwana et al. 2017). The resulting products still need to be fostered until the products are suitable for sale. According to Frandinata et al. (2019), products ofter do not even have a clear brand, sticker, or attractive design. People do not yet accept CLC products and the marketing strategies of these products are still not optimal. Improved product marketing management by improving labels, packaging design, marketing attributes, and business feasibility analysis for products are needed so that they are more attractive for consumers and can be expected to increase turnover.

Therefore, coaching needs to be done for students to be better prepared to do community-based business. Learning citizens must be taught as well as possible to increase their knowledge, attitudes, and skills in doing business. Coaching efforts can be carried out in various forms, one of which is through training. Training is one of the educational processes that is quite effective if carried out according to procedures and objectives to be achieved. Harrison (1993) states that training and development are fields related to organizational activities that aim to improve individual performance. The Constitution of the Republic of Indonesia No. 13/2003 on Manpower, states that "Training is an overall activity to provide, obtain, improve and develop work competence, productivity, discipline, behaviour, and work ethic at several competencies and skill levels related to the level of work and qualifications."

Training is an important learning process for dynamic development of organizations or institutions and human resources, which are an important factor in the country's economic development. De Cieri et al. (2003) stated that the function of training is to develop the knowledge of citizens learning about the culture of their organizations and competitors, to help learning citizens acquire the skills to work with new technologies, and to help citizens learn to understand how to work efficiently and effectively in teams to produce products and services of good quality. To conduct training, program design is required. Program design is a series of ongoing processes consisting of program planning, describing and organizing learning experiences to achieve the goals set through effective planning and implementation. In general, training is a learning activity that aims to improve workers' knowledge and skills, change attitudes, and teach new practices (Batten 1962). Meanwhile, according to Estrin (1987), program design is one of the activities to meet goals and changes. This involves three main management processes: planning, implementation, and evaluation. In managing the training pro-gram, organizational goals will be formed as a planning reference frame. Overfield (1989) has identified two important aspects of work as managing and involvement. He found that managing was the determining factor for achievement in work and planning was the process of selecting and organizing scheduled activities. Meanwhile, implementation is the process of working on activities that are scheduled to achieve the objectives identified, while evaluation is analyzing the results and effectiveness of the program. The results of the evaluation will be used to design future programs (Rogers 1970).

1.1 *The four level Kirkpatrick training evaluation model*

There are many ways to evaluate training or what is called a training evaluation method that can be used. Kirkpatrick (1998), stipulates:

- Level 1: Reaction Level measures how trainees or trainees react to training. It is important to measure reaction, because it can help to understand how well training was received by

the participants. It also helps improve training for potential trainees, including identifying important areas or topics that are missing from the training;
- Level 2: Learning Level measures what participants have learned. Planning a training session usually starts with a list of specific learning objectives, which can be a starting point for measurement. It is important to measure this level, because knowing what participants are learning and what they are not learning will help improve training in the future;
- Level 3: My Behavior Level evaluates how far participants have changed their behavior based on the training received. It is important to realize that behavior can only change if conditions are favorable, so, this stage is best measured after Levels 1 and 2 are completed. However, it is important to remember that just because behavior hasn't changed, it doesn't mean that the participants haven't learned anything;
- Level 4: Results Level will analyse the final results of the training. This includes results that have been determined by the company to be good for business, good for employees, or good for the bottom line.

This model was formed to help planners or trainers to plan, implement, and evaluate training programs more effectively (Pesson 1968). In designing and modelling, people really try to find ways to plan, implement, and evaluate training programs more systematically and follow certain rules. A model is important in building training programs because it shows certain actions or stages that must be followed so that the actual objectives will be achieved. Nadler (1983) once mentioned that the model was like a road map: before going to a place, one must go through the map first to avoid getting lost; having a complete map still does not ensure one's travel because there are many other things to consider. Based on the Nadler, the training model can determine the important components and processes involved in program planning and can also clearly show the results of training activities. Looking at the practice carried out by training program planners, it will be interesting to learn whether they are following what is recommended in the literature.

1.2 *Approaches and models in designing training programs*

The program design model aims to understand and improve training program design practices. It calculates all activities carried out by the planner or trainer of the training program. Good practice in design of training programs requires integration between multiple views or approaches. Various aspects have been identified in each stage or step in this approach:

- Analysing the training and recipient contexts;
- Identifying training requirements;
- Creating goals;
- Selecting program content;
- Choosing resources;
- Determine the budget and;
- Evaluating the program.

The classical view founded by Tyler (1949) emphasized continuous learning. According to Tyler, there are four important stages for creating a program design model. The stages are:

- Setting goals;
- Identifying learning experiences to achieve goals;
- Organizing the learning experience in a more effective way; and
- Evaluating whether the objectives were achieved.

2 RESEARCH METHODS

This study uses the ADDIE model, which can be used for various forms of product development such as learning strategies and methods, media, and teaching materials. The

ADDIE model can serve as a guideline in building training or learning program tools. Development style ADDIE is quite effective, dynamic, and supports the performance of a program. The ADDIE model consists of five interrelated components systematically structured. The stages are very simple when compared to other design models, so this model is easy to understand and apply. The stages are (1) Analysis, at this stage an analysis of product feasibility and development requirements is conducted. (2) Design, at this stage design systemic activities by setting goals, designing designs, and evaluation tools to be used. (3) Development, at this stage the product design is implemented or realized but not yet implemented in the market/market. (4) Implementation, at this stage the designs and methods that have been developed are applied to the actual conditions. (5) Evaluation, at this stage the evaluation is carried out at the stage of the specified period and on the whole activity. Data collection was carried out through interviews, focus group discussions, and observations of nine CLCs in five regions in West Java for 6 months.

The research procedure was carried out as follows: (1) the Research Team conducted a field study on the education unit and the regional technical implementation unit to obtain organizing data and an understanding of the conditions of the CLC. The field study was coordinated by each research member from representatives of the Bandung and surrounding areas, Tasikmalaya, and Karawang. The selection of sample areas was done by (a) appointing a CLC that produces quite good products but has problems in marketing and (b) taking several samples from different regions by looking at the opportunities for locality products of each region; (2) the Research Team and nine CLC Managers conducted focus group discussions (FGD) on conditions, obstacles, and solutions that must be pursued in the process of improving product quality and marketing; and (3) the Research Team examined relevant literature on the implementation of e-commerce marketing.

3 RESULTS

The training at CLC to build community businesses was carried out through various programs, including through the life skills program. Below is the condition of the life skills program in eight CLC and one Rumah Pintar in West Java.

3.1 *Programs organized by CLC to train skills*

Based on the results of interviews and focus group discussions at eight CLC institutions and one Smart House Institution in West Java, the results are presented in the following table:

Table 1 illustrates the life skills program in general at nine institutions, showing that the number of students who are involved in the production process is not too great: around 10 to 30 people depending on market demand. This also affects the relatively small number (2 or 3) of groups truly devoted to the production process. The life skill learning program source is referring to books or skills possessed by the tutor or also developing resources from the Internet. As for tutors, those from the nine institutions have been educated to the Bachelor level. Learning is facilitated by various types of reference books, learning media, and other supporting facilities. Eight out of nine institutions already have permanent buildings that belong to foundations or managers, while one institution is still semi-permanent and the status is still leased to the community. The most prominent learning motivation is to increase income. Citizens of learning hope for an increase in income after getting skills from the life skills program to improve their standard of living. The life skills program funding has been provided by the Government; two institutions have also partnered with companies to receive CSR assistance, while one institution still relies on the assistance of temporary donors.

Table 1. General description of life skill programs in CLC and Rumah Pintar.

Life Skill Program	Students	Learning Source	Instructor	Learning Facilitated by…	Learning Place	Group Learning	Learning Motivation	Learning Finance
Danis Jaya CLC	10	Books, Instructor	Bachelors	Books, Learning media, Other facilities	Permanent, belonging to foundation	2 groups	Increase income	Government
Ulul Albab CLC	10	Books, Instructor	Bachelors	Books, Learning media, Other facilities	Permanent, belonging to the manager	2 groups	Increase income	Government
Al Huda CLC	10	Books, Instructor	Bachelors	Books, Learning media, Other facilities	Permanent, belonging to foundation	2 groups	Increase income	Government
An-Nur Ibun CLC	25	Books, Instructor, Internet	Bachelors	Books, Learning media, Other facilities	Permanent, belonging to foundation	3 groups	Increase income	Government, CSR
Bina Mandiri Cipageran CLC	> 20	Books, Instructor, Internet	Bachelors	Books, Learning media, Library, Other facilities	Permanent, belonging to the manager	2 groups	Increase income	Government, Students
Rumah Pintar Al Barokah	> 20	Books, Instructor	Bachelors	Books, Learning media, Other facilities	Permanent, belonging to foundation	2 groups	Increase income	Temporary donors
Asholiyah CLC	30	Books, Instructor, Internet	Bachelors	Books, Learning media, Other facilities	Permanent, belonging to foundation	3 groups	Increase income	Government, CSR
Ummul Yatama CLC	10	Books, Instructor	Bachelors	Books, Learning media, Other facilities	Permanent, belonging to foundation	2 groups	Increase income	Government
Nurul Furqon CLC	10	Books, Instructor	Bachelors	Books, Learning media, Library, Other facilities	Semi-permanent, Rent	2 groups	Increase income	Government

From Table 2, it can be seen that the purpose of a life skills program at CLC and Rumah Pintar is to increase the income of learning citizens. This is people's main goal when joining the life skills program: other goals are to improve literacy and numeracy skills as well as to expand relationships with the community. The types of skills taught are very diverse, depending on the skills of the tutor and the condition of the environment around the institution. The types of skills taught at eight CLC and one Rumah Pintar in the West Java region include the skills to make a mukena with Tasikmalaya's distinctive embroidery motifs; making low-quality senda from well-known brands but that are durable at low prices; making banana sale with various types of flavors, making blazers from unique patterned sarongs; making milk soap of various types; making fish, shrimp paste, and shellfish crackers; and making hats. These are the upper-level skills

Table 2. Life skill programs in CLC and Rumah Pintar in Indonesia.

Life Skill Program	Goals	Skills Taught	Learning Evaluation	Impact
Danis Jaya CLC	Increase income	Making mukena (Muslim women's prayer dresses)	Emphasizing ability of community to learn	Improve the ability/skills of community
Ulul Albab CLC	Increase income	Making sandals	Stressing environmental needs	Increase income
Al Huda CLC	Increase income	Making smoked or dried banana fritters	Stressing environmental needs	Increase income
An-Nur Ibun CLC	Increase income	Making blazer from sarong	Stressing environmental needs	Improve the ability/skills of community
Bina Mandiri Cipageran CLC	Increase income	Making milk soaps	Emphasizing ability of community to learn	Open learning opportunities in formal education
Rumah Pintar Al Barokah	Increase income	Making Kere Jaer (fried jerky Mujaer fish meat)	Emphasizing ability of community to learn	Improve the ability/skills of community
Asholiyah CLC	Increase income	Making terasi (shrimp paste)	Stressing environmental needs	Increase income
Ummul Yatama CLC	Increase income	Making clam crackers	Stressing environmental needs	Change learning attitude
Nurul Furqon CLC	Increase income	Making hats	Emphasizing ability of community to learn	Increase income

taught in each institution, although there are still many other skills taught in these institutions, such as making brooches, processing rice, making ornaments from acrylic, making knitted goods, etc. Five of nine institutions emphasize the ability of citizens to learn certain skills, while the other four institutions emphasize environmental needs adapted to the natural and human resources around the institution. There are two impacts of the life skills program: increasing the ability to learn and increasing the income of learning citizens, as per people's initial goal in following the life skills program at the CLC or Rumah Pintar institutions.

3.2 *Achievement of community change through the life skills program*

The results of the data in the field show that the community changes through life skill programs are changes in knowledge about entrepreneurship, increased learning skills of citizens, as well as the motivation of citizens to start doing businesses because they are trained for entrepreneurship while participating in programs in the CLC or the Smart House. Several other studies show that there are achievements obtained by learning citizens after participating in life skill programs, including changes in the ability to work and the spirit of entrepreneurial independence, the ability to solve problems in entrepreneurial activities, the ability to think creatively and create innovative businesses in determining the type of business innovative and in accordance with market developments, the ability to work meticulously and productively, the ability to market the products owned by learning residents (still at an early stage), and the ability of citizens to learn in earning income (even learning citizens who have not been able to manage their businesses).

3.3 *Difficulties faced in learning and empowering communities*

Below is a table of SWOT analysis conducted at the CLC and Rumah Pintar institutions in West Java:

The SWOT analysis results in Table 3 show the strength of each upper-level product produced by the institution. Danis Jaya CLC sees the strength of the product produced as the

Table 3. SWOT life skill program analysis.

Internal	Strength	Opportunities	External	Weakness	Treats
Danis Jaya CLC	Mukena with Tasikmalaya embroidery motifs at low prices	Many competitors of similar products	Danis Jaya CLC	Human Resources Marketing has not been maximized, Limited tools	Market interest is still low
Ulul Albab CLC	Original model sandals, fake goods, low price, quite durable	Imported raw materials, Human Resources, Work partners	Ulul Albab CLC	Human Resources Marketing has not been maximized, Limited tools	Market interest is still low
Al Huda CLC	Smoked or dried banana fritters, many flavors, cheap	Many competitors of similar products, The weather, Low learning motivation	Al Huda CLC	Marketing has not been maximized	Market interest is still low
An-Nur Ibun CLC	Blazer sarong, unique, varied models, neat stitches, competitive prices	Innovation in the model and quality of human resources	An-Nur Ibun CLC	Students are not focused and give up quickly when there is a problem, Product packaging and marketing still depend on the CLC manager	The difficulty of increasing students' confidence
Bina Mandiri Cipageran CLC	Milk soap, more milk content, unique shape, competitive price	Don't have BPOM permission yet	Bina Mandiri Cipageran CLC	Product packaging and marketing still depend on the CLC manager	Market interest is still low
Rumah Pintar Al Barokah	Kere Jaer (fried jerky Mujaer fish meat), unique, delicious Kere, cheap	Producing its own raw materials (Mujaer cultivation)	Rumah Pintar Al Barokah	Production is not well scheduled, Limited tools, Marketing has not been maximized	Raw materials are difficult to obtain
Asholiyah CLC	Shrimp paste, shrimp content more rebon shrimp, competitive prices, neat packaging	Marketing expansion supermarkets	Asholiyah CLC	Marketing has not been maximized	Market interest is still low
Ummul Yatama CLC	Shellfish/clam crackers, unique, cheap	Packaging innovation and flavor variants	Ummul Yatama CLC	Not yet focused, because the tutor has other activities, Activities are not carried out routinely	Do not have permanent customers, consumers are not yet reliable
Nurul Furqon CLC	Hats, neat stitches, cheap	Marketing expansion and hat model innovation	Nurul Furqon CLC	Production is not well scheduled, Limited tools, Marketing has not been maximized	Do not have permanent customers, consumers are not yet reliable

characteristics of the region but what distinguishes it from other products is the affordable price. Therefore, the challenge is the number of competitors with similar products. The weaknesses of production are the lack of human resources who want to be involved and the ability to sew and the drill is still limited, marketing is not optimal, and sewing and embroidery tools are quite expensive for the institution. The threat from mukena production is that market interest is still low on the mukena products it makes because of the many rivals with similar products.

Ulul Albab CLC produces sandals, and its strength is that the model is similar to the models of famous brands but modified and sold at low prices. They use the ATM principle (observe, imitate, modify). The challenges for production are that raw materials are still imported as they have not found local raw materials of the same quality, human resources are still not good in terms of quality or quantity; and there are not many partnerships with other organizations. Disadvantages for this CLC's sandal production are human resources, marketing, and the limitations of the means of production. The threat is that market interest is still low.

CLC Al Huda sells banana fritters produced with many different flavors, including original flavor, ginger, cinnamon, alley chili, and others. The challenges in the production of this banana product are the number of competitors with similar products and that the motivation of citizens to continuing to improve their skills is still low. The disadvantage of this banana sale is that marketing is not yet extensive or maximized. The threat is that market interest is still quite low.

An-Nur Ibun CLC has a superior product in the form of a blazer made of sarongs with a unique motif and provides various models, which is an advantage of its products. The challenge of its production is that it must continue to develop its products and continue to improve the quality of its human resources (learning citizens). The weaknesses of the production are that (a) the learning community is still not focused on running it and the motivation is low so that it is easy to give up when exposed to problems, and (b) packaging and marketing of the product still cannot be independent, rather still depends on the CLC manager. The threat is that it is difficult to increase the confidence of citizens learning to do business.

Cipageran milk soap produced by CLC Bina Mandiri Cipageran is one of the excellent products produced by the CLC's learning citizens. The advantage of this product is greater and fresher milk content, formed in a unique format and at an affordable price for the community. The challenge is that up to now there is no BPOM permit. The weakness of this milk soap production is that packaging and marketing still depend on the CLC manager. The threat is that market interest is still low on milk soap because the price is slightly higher than ordinary soap.

Tilapia fish that is used by Al Barokah Smart House as one of its superior products, "Kere Jaer," has advantages in that it is rarely found so it seems unique, it tastes good and distinctive, and has a low price. The challenge in the production of this commodity is that the Smart House must try to produce tilapia fish or cultivate it by itself so that the raw material is more easily available and the price is cheaper, than currently. The disadvantage is that the production of Jean's Kere has not been well-scheduled, there is limited production equipment, and marketing has not been maximized. The threat is that raw materials are still difficult to obtain.

The coastal area produces a lot of sea fish and shrimp, thus CLC Asholahiyah exploits this potential by processing rebon shrimp to make shrimp paste, which is one of its superior products in addition to the Karawang rice that it produces. The advantages of the CLC's shrimp paste are the shrimp content is more pronounced, the packaging is neat, and it is sold at a low price. The challenge is to continue to expand marketing, especially in more exclusive supermarkets. The disadvantage is that marketing is still not optimal. The challenge is that market interest is still low in products that do not have well-known brands in the community.

One of CLC Ummul Yatama's superior products is clam crackers, which have the advantages of beig rarely found so it seems unique, tastes good, and the price is cheap. The challenge of the production of these shellfish crackers is that they must continue to innovate both their packaging and the flavor variants or shapes of crackers to make them more attractive to customers. The disadvantage of the production of shellfish crackers is that the tutor or the learning community is still not focused on running the production of these shellfish crackers because they have other activities besides CLC, so the shellfish cracker production schedule is still not done routinely. The threat from the production of shellfish crackers is that they do not have permanent customers, so consumers of these products are not yet reliable.

CLC Nurul Furqon produces hats as one of its superior products. The advantage of the hat that it manufactures is that it has neat stitches and is sold cheaply. The challenge in this production is that it must continue to expand market reach and continue to innovate to make attractive cap models. The weakness of this production is that it has not been well-scheduled, the limited means of production, and that marketing has not been maximized. The threat is that they do not have permanent customers so consumers are not reliable.

4 DISCUSSION

4.1 *Programs organized by CLC to train skills*

According to Brolin (1997), life skills education can provide skills so that a person can live independently, through education in daily skills, personal skills, and proficiency for work. Factors that influence community participation are individual characteristics such as age, level of formal education, non-formal education, attitudes toward life skills, motivation toward life skills, level of knowledge about life skills, functional skills, and previous life skills experience (Asnamawati et al. 2014). Cohen & Uphoff (1977) divided the type of participation into four, namely: (1) participation in decision making (planning), (2) participation and implementation, (3) participation in evaluation, and (4) enjoying the results. The life skills program at nine CLC shows that, in general, the learners are people who are involved or participate in the production process or in implementation program, and a relatively small number devoted to the production process. Characteristics of life skills learning are (1) the process of identifying learning needs; (2) the process of awareness raising for joint learning; (3) the alignment of developmental learning activities, study, independent businesses, joint ventures; (4) the process of mastering personal, social, vocational, academic, managerial, and entrepreneurial skills; (5) the process of providing experience in doing work properly, producing quality products; (6) the process of interaction of mutual learning from experts; (7) the competency assessment process; and (8) the occurrence of technical assistance to work and form joint ventures (Anwar 2006). The life skill learning program source and supplemental sources are available in various formats. The most prominent learning motivation for following the life skills program is to increase income. Citizens of learning hope for an increase in income to improve their standard of living after getting skills from the life skills program. According to Brolin (1997), life skills are the interaction of various knowledge and skills that are very important so someone can live independently. Brolin groups life skills into three skill groups, namely: (1) daily living skills, (2) personal/social skills, and (3) occupational skills. Life skills programs in nine of the CLCs have as objective of the program to give occupational skills. MONE (2004) defined occupational skills as including: skills to choose work, work planning, preparation of work skills, skills training, mastery of competencies, running a profession, awareness to master various skills, ability to master and apply technology, design and implement work processes, and produce goods and services. Life skills programs have been funded by the Government, companies that receive CSR assistance, or temporary donors. The evaluation conducted on the life skills programs emphasizes the ability of citizens to learn certain skills and the environmental needs that are adapted to the natural and human resources around the CLC. There are two impacts of the life skills programs, namely increasing the ability of citizens to learn and increasing the income of learning citizens, often the initial purpose for following the life skills program at the CLC.

4.2 *Achievement of community change through the life skill program*

Life skill programs lead to community changes in knowledge about entrepreneurship, increased learning skills of citizens, as well as the motivation of citizens to start doing business because they are trained for entrepreneurship while participating in programs in the CLC or the Rumah Pintar. Life skills education is directed at the development of people who are noble, smart, skilled, healthy, independent, and have high productivity and work ethic. The implementation of life skills education in non-formal education units and programs, especially in the context of poverty alleviation and unemployment reduction is more emphasized on learning efforts that can provide income (learning and earnings) (Baruwadi 2008). The purpose of the life skills program is to provide basic provisions, training, and skills to students that can be used for solve problems that exist in society and can improve standard of living (Tumiyati 2013). According to Anwar (2006), the purpose of life skills education is actualizing the potential of students to solve the problem at hand. Several other studies show that there are achievements obtained by learning citizens after participating in life skill programs, including changes in the ability to work and the spirit of entrepreneurial independence, the ability to solve problems in entrepreneurial activities, the ability to think creatively and create innovative businesses in determining the type of business innovative and in accordance with market developments, the ability to work meticulously and productively. As well, the ability to market the products owned by learning residents is still at a basic stage, as is the ability of citizens to learn in earning income, even those learning citizens who have not been able to manage their businesses. The implementation of life skills education is directed at poverty alleviation efforts and efforts to solve the problem of unemployment, therefore the selection of skills to be learned by residents is based on community needs, local potential, and market needs, so that it is expected to provide positive benefits for learning citizens, surrounding communities, and government (MONE 2004).

4.3 *Difficulties faced in learning and empowering communities*

The SWOT analysis results show the strength of each superior product produced by the institution. The strength of the product produced in nine CLCs include that it embodies characteristics of the region but what distinguishes it from other products is the affordable price; the model is similar to the models of famous brands but modified and sold at low prices because they use the ATM principle (observe, imitate, and modify); the product has many variants offered, is a unique product and provides various types of models; is affordable price for the community; or the product is rarely found so it seems unique, tastes good and is distinctive at a low price. The challenge of these products are the number of competitors with similar products; raw materials still imported; local raw materials of the same quality not found; human resources still not good in terms of quality or quantity; and other challenges such as not many partnerships with other organizations; the learning motivation of citizens to learn is still low in continuing to improve their skills; the cbusiness must continue to improve quality; production not well scheduled,; limited production equipment; and marketing has not maximized to expand marketing (especially in more exclusive supermarkets).

The weaknesses of production are the lack of human resources who want to be involved and the ability to produce products; marketing is not optimal; tools that are quite expensive for the institution; need to continue developing products and human resources (students); students still not focused on running the business and motivation is low so that it is easy to give up when exposed to problems; and packaging and marketing of the product still cannot be independent and depends on the CLC manager. The threats are that market interest is still low; marketing is not yet extensive and maximized; it is difficult to increase student confidence to do business; the price is slightly higher than ordinary products in the market; and products do not have well-known brands in the community.

4.4 *CLC marketing training model*

The marketing of CLC business products has basically been done through direct marketing, both face-to-face or through chat media. There are also a number of CLCs that carry out consignment/entrusted sales systems in the markets or souvenir shops in the city centre. However, in marketing, not all institutions are able to meet market needs in a short time, due to constraints of raw materials and human resources. As for the classification of marketing products, CLC businesses can be divided into three types: (1) marketing for the scope of additional skills, which means that students in CLCs are more focused on activities to increase knowledge not on skills, so that the skills taught are just additional learning, which causes residents to learn not to

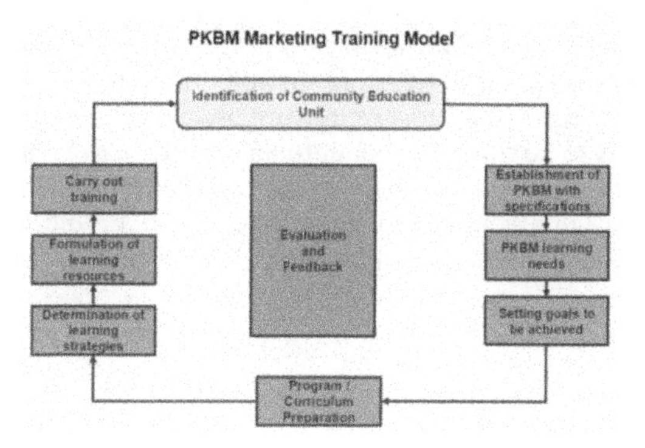

Figure 1. Training model.

focus on making business products, so the results of their skills are only for themselves or can be sold directly within the scope of a small market/around the house; (2) marketing for the scope of compulsory skills, means that skills learning activities are the main learning so that students can produce a product that is relatively good and can be sold to the market in a fairly broad environment; and (3) professional marketing of CLC business products, meaning that the learning community is indeed directed to focus on entrepreneurship, product quality, product standards, legality, packaging and product labels, as well as extensive and quite massive marketing. CLC readiness to be able to use these patterns.

Training models to educate learning citizens to be able to market CLC products by looking at marketing conditions have been carried out by CLC. The form of marketing training for CLC products uses certain marketing patterns that are conceptually formulated as follows (see Figure 1).

The training model conducted by CLC and Rumah Pintar is teaching the community to understand how to produce goods, package, legalize products, and market products by identifying community learning needs by community education units. The training specifications are prioritized and then adjusted to the identified learning needs of the community. After formulating this, the objective is to determine the objectives to be discussed, which are then elaborated in the detailed program/curriculum preparation, including the determination of learning strategies, formulation of learning resources, and execution of training implementation. This continues in a cycle if other types of training are needed, and evaluation and feedback are carried out.

This training is divided into three opportunities that depend on the CLC concession in carrying out community businesses. So, the training specifications refer to the CLC business product classification. If the training makes the skills tangential and not the main focus, the training is merely providing basic knowledge about product marketing in theory. If CLC training makes skills learning mandatory so that products are possible to sell to the market or the community, the training must pay more attention to competence in marketing products and ensure the market is neither too broad nor only around the community near the CLC or surrounding communities learning residents (family, neighbors, friends, etc.). If the CLC makes skill learning one of its focuses, then training must be done in stages, so it is necessary to practice or simulate in the field with guidance from the CLC or other parties who are experts in the field of marketing.

5 CONCLUSION

The model of product marketing training for CLC and Smart House is by identifying community education units, setting CLC with specifications and community learning needs, setting goals to be achieved, preparing curriculum, determining learning strategies, formulating learning units, and conducting training by continuing to evaluate and feedback from the cycle.

In CLC training method, the trainees are not only taught basic marketing skills, but also more advanced ones.

The type of training adapted to the classification of CLC conditions in marketing products, namely CLC which makes the scope of skills only as additions only required training by providing competence or basic knowledge about marketing, in CLC which makes the scope of skills as an obligation, Training needs to be started in depth to achieve mid-level marketing competence with a market scope that is not too broad. To produce and market products professionally, training is needed in stages to achieve product marketing competencies in a professional manner. The training program has to do simulations in the field with guidance from the CLC or marketing experts.

REFERENCES

Anwar, M. 2006. *Pendidikan Kecakapan Hidup (Life Skills Education)*. Bandung: Alfabeta.

Asnamawati, L., Purnaningsih, N., & Hatmodjosoewito, S. J. 2014. Tingkat Partisipasi dalam Kegiatan Pendidikan Kecakapan Hidup. *Jurnal Penyuluhan* 10(2).

Baruwadi, D. 2008. Penyelenggaraan Pendidikan Kecakapan Hidup dalam Peningkatan Kemandirian Pemuda. *Jurnal Pendidikan Luar Sekolah* 8(1).

Batten, C. W. 1962. *Training for Community Development: Getting Agriculture Moving*. New York: The Agriculture Department Council, Inc.

Brolin, D. E. 1997. *Life Centered Career Education: A Competency-Based Approach*. Reston, VA: The Council for Exceptional Children.

Chaffey, D., & Smith, P. R. 2017. *Digital Marketing Excellence: Planning, Optimizing and Integrating Online Marketing*. Taylor & Francis.

Cohen, J. M., & Uphoff, N. T. 1977. Rural development participation: concepts and measures for project design, implementation and evaluation. *Rural Development Participation: Concepts and Measures for Project Design, Implementation and Evaluation* (2).

De Cieri, H. L., Kramer, R., Noe, R. A., Hollenbeck, J. R., Gerhart, B., & Wright, P. M. 2003. *Human Resource Management in Australia. Strategy-People-Performance*. McGraw-Hill Education.

Estrin, J. 1987. What do contract trainers need to know to be successful? *Training and Development Journal* 41(4).

Frandinata, J., Renaldy, R., Gunawan, H., Angel, V., & Kuntjoro, A. P. 2019. Economy Development of Merpati Bangsa Community Learning Centre through Snack Bussiness. *Jurnal Pemberdayaan Masyarakat* 1(1).

Harrison, R. 1993. Developing people for whose bottom line. *Human Resource Management: Issues and Strategies*.

Kirkpatrick, D .L. 1998. The four levels of evaluation. In *Evaluating Corporate Training: Models and Issues*. Springer, Dordrecht.

MONE (Ministry of National Education) 2004. *Guidelines for Implementing Non Formal Education Life Skills Programs*. Jakarta: Directorate General of Non-Formal Education and Youth Ministry of National Education.

Nadler, L. 1983. *Designing Training Programs: The Critical Event Model*. Reading: Addison-Wesley.

Overfield, K. 1989. Program development for the real world. *Training & Development Journal* 43(11): 66–72.

Parsons, A., Zeisser, M., & Waitman, R. 1998. Organizing today for the digital marketing of tomorrow. *Journal of Interactive Marketing* 12(1): 31–46.

Pesson, L. L. 1968. *Principles of Extension Teaching*. Kuala Lumpur: College of Agriculture.

Purwana, D., Rahmi, R., & Aditya, S. 2017. Pemanfaatan Digital Marketing Bagi Usaha Mikro, Kecil, Dan Menengah (UMKM) Di Kelurahan Malaka Sari, Duren Sawit. *Jurnal Pemberdayaan Masyarakat Madani (JPMM)* 1(1): 1–17.

Rogers, T. G. 1970. *The Recruitment and Training of Graduates*. London: Institute of Personnel Management.

Suriansha, R. 2017. Digital Marketing and Online Shopping in Indonesia. *International Journal of Scientific Research in Science and Technology (IJSRST)* 3(7): 136–140.

Tumiyati. 2013. *The Implementation of Class B Package Learning Based on Life Skills in Bhakti Persada CLC*, Yogyakarta: Universitas Negeri Yogyakarta.

Tyler, R. W. 1949. *Basic Principles of Curriculum and Instruction*. Chicago: The University of Chicago Press.

Yannopoulos, P. 2011. Impact of the Internet on marketing strategy formulation. *International Journal of Business and Social Science* 2(18).

Borderless Education as a Challenge in the 5.0 Society – Abdullah, Adriany & Abdullah (eds)
 ISBN 978-0-367-61960-2

Entrepreneurship learning based on literacy conservation in lifelong educational perspectives

I. Hatimah
Universitas Pendidikan Indonesia, Bandung, Indonesia

ABSTRACT: Lifelong education (PSH) as a concept that underlies the community education profession is intended for all people who need it in the context of community empowerment in accordance with their needs and potential. The program for maintaining adult literacy is one of the efforts to achieve the target of increasing population literacy that has an effect on improving the education index as part of the Human Development Index (HDI) component. The purpose of this study is to (1) maintain and develop the literacy of students who have participated in and/or attain basic literacy competencies, (2) increase the ability of independent businesses to develop and realize the various potentials possessed by students, and (3) build business nodes independently oriented toward improving the standard of living of the community. This research used a descriptive method with quantitative and qualitative approaches. The participants of this study were 113 people, consisting of 13 PKBM managers and 100 participants or learning residents. The research locations were six groups of PKBM in Indramayu Regency, three districts of Cirebon Regency PKBM, and three groups of Subang Regency PKBM. Data collection techniques used were interviews, observation, documentation studies, and questionnaires as supporting data collectors. The findings of this study are: (1) There has been an increase in people's ability to read, write, and count, due to the accuracy of the use of learning methods, learning materials according to the needs of the community, high motivation of participants to participate in entrepreneurship programs, and the role of speakers in the delivery of material; (2) Citizens learn to have knowledge, attitudes, and skills about entrepreneurship. In the aspect of knowledge an increase in understanding of entrepreneurial material, ways to make culinary products, simple bookkeeping methods, and ways of marketing. In the aspect of attitude, people learn to have an attitude of confidence, dare to start a business, be more open to the opinions of others, motivate each other between groups, and show increasing willingness to learn and try. In the aspect of skills, the residents learned skilfully to make fish-bone crackers and shredded fish; (3) The business nodes that were developed were making fish-bone crackers and shredded fish.

1 INTRODUCTION

Lifelong education (PSH) as a concept that underlies public education, leads to learning activities in the community. Lifelong education is one of the principles of proper education for people who live in a world of transformation and in a society where people influence one another, as in the current era of globalization when the global institutionalist theory emphasizes strong putative links between education, individual development, and national progress (Meyer et al. 1998). Prior cross-national studies of education identify global patterns of educational institutionalization and conclude that education for development has become a world-wide taken-for-granted norm (Ramirez & Boli 1982). The need for lifelong education, especially for literacy programs, is felt internationally. This is evidenced by a joint agreement signed by 179 ministers of education in Jomtien Thailand called Education for All (EFA). This conference was sponsored by four organizations: UNICEF,

UNESCO, the World Bank, and the United Nations Development Programme (UNDP) (Chabbott 1998). The continuation of this agreement was reformulated in Dakkar Senegal in 2000which is famous for the Dakar Declaration. The Dakkar Declaration agreed to fight for six educational action frameworks for all: (1) expanding and improving the overall care and education of young children comprehensively, especially those who are very vulnerable and neglected, (2) gender equality in education, (3) life-skills programs for youth and adults, (4) eradicating illiteracy, (5) compulsory education, and (6) improving the quality of education.

Education is fundamental to development and growth. The Education Sector Strategy 2020 lays out the World Bank Group's agenda for achieving "Learning for All" in the developing world over the next decade (World Bank 2011). Through lifelong education, it is expected that people who are literate can continue to maintain their literacy, so they can develop their abilities through various skills that can be used to improve their lives. One effort to maintain adult literacy is the sustainability of the program through the economic sector in the form of business. This is important to exploring the potentials that exist in the community, so that people have the ability to develop themselves independently but can still maintain their literacy abilities.

In the framework of lifelong education, the literacy program is one of the programs that continues to be developed, because those who are illiterate or literate at the basic stage must continue to be helped to maintain their literacy skills, so as not to be illiterate again.

One effort to maintain literacy is through business development. This is very important to do, in the context of community empowerment in the economic field, by (1) developing independence through strategic economic development based on local potentials, (2) developing people-based economic development models based on self-reliance and independence, and (3) strengthening institutions and community economic institutions in order to grow the collective economic system and strengthen capital dependently and independently.

2 LITERATURE REVIEW

2.1 *Lifelong education*

Lifelong education focuses on the motivation of a person or group to obtain learning experiences on an ongoing basis. According to Djuju (2004), the presence of lifelong education is caused by the emergence of learning needs and educational needs that continue to grow and develop during the course of human life. The three main characteristics in interpreting lifelong education according to Cropely (in Abdulhak 2000), are:

- Vertical education integration, meaning learning occurs normally and naturally throughout life, in accordance with the personal and physical continuous development of a person from infancy to old age;
- Horizontal integration, meaning the relationship between education and life, in the sense that social skills are widely learned through interactions with the environment from day to day, from small things to complex ones, from observing to practicing so that people become more present, independent, and creative as individuals, group members, and community members; and
- Prerequisites for learning-equipment for lifelong education, in the sense that education throughout life can be done if you have sufficient skills to be involved in it.

Literacy education has to do with lifelong education, because people have the opportunity to continue to attend education in accordance with their needs. Literacy education based on lifelong education is oriented to the process of changing the attitudes and behaviors of students toward maturity. Adult people develop their potential and strive to achieve self-satisfaction in a good life that is meaningful for themselves and their environment. People mature trying to meet new needs or to solve problems encountered through the learning process. These new needs may be to deepen, enhance, or expand the capabilities they already have. Adult people carry out activities dynamically, namely from efforts to meet a need to the next effort to meet needs that arise later.

Based on the concept of lifelong education, education in the community includes equivalency education which provides basic education for citizens who are unable to receive formal education. Lifelong education is needed for each individual, and is based on the idea that learning for someone happens throughout the life, even though in different ways and processes according to the needs of individuals. The concept of lifelong education and lifelong learning are related. As expressed by Knapper & Cropley (2000), "Lifelong education has been defined as a set of organizational, administrative, methodological and procedural measures" as well as "lifelong learning described the habit of continuously learning throughout life, a made of behaviour.

2.2 *Entrepreneurship*

According to economists, an entrepreneur is a person who combines resources, labor, materials and other equipment to increase value higher than before, and also those who introduce changes, innovations, and other improvements in production (Alma 2009).

Entrepreneurship also has an important role in society, particularly for regional economic development, through the initiation of small-middle-micro business in order to reduce unemployment levels in society. Unemployment is actually a serious issue for developing countries. To reduce the problem and to further improve prosperity of society, one effort is by giving social entrepreneurship education at the earliest stage possible. According to United Nation and World Bank (in Bhuiyan & Ivlevs 2019) entrepreneurship has been long hailed as an important tool to reduce poverty and promote economic growth in developing countries.

According to Suryana (2006), entrepreneurial activities are an effort to develop the ability to think creatively and innovatively in business activities to create new ideas that are different from before as the basis for utilizing opportunities for success in business. Furthermore, according to Thomas, Takamichi, and Shuichi (1997), entrepreneurship has character and values that can provide attitudes and unique characteristics that support and differentiate them from other lives in the community.

According to Alma (2009), characteristics of entrepreneurs are:

- Confidence,
- Oriented in tasks and results,
- Daring to take risks,
- Leadership,
- Indifference, and
- Oriented to the future.

According to Suryana (2006), characteristics and indicators of entrepreneurs are:

- Full of confidence: the indicators are full of confidence, optimistic, committed, disciplined, responsible,
- Having initiative: the indicator is full of energy, agile in acting, and active,
- Has a motive for achievement: the indicator is results-oriented and looking to the future,
- Having a leadership spirit: the indicator is daring to be different, trustworthy, and strong in acting,
- Dare to take calculated risks, and like challenges.

Entrepreneurship is the start. No matter how small an entrepreneur enterprise is, if it is developed with economic principles and carried out professionally, it will foster the spirit of entrepreneurship. We can learn from the entrepreneurship experience, such as the way to organize and run a business well and the way to evaluate the business to develop it into a larger business. And entrepreneurship relates to fulfillment and well-being, which is of utmost importance (Wiklund et al. 2019).

Entrepreneurs and social enterprise, in recent years, have become prominent issues mainly because of their ability to address pressing global concerns (Agarwal et al. 2018, Sakarya et al. 2012, Stephan et al. 2016, Viswanathan & Rosa 2010). Entrepreneurship can be viewed from three main elements, namely: motivation, organization, and society (Durieux & Stebbins 2010).

2.3 *Literacy*

One of the main contents of the Dakkar Agreement is the achievement of a 50% improvement in adult literacy rates by 2015, especially for women, and equitable access to basic and continuing education for all adults. Literacy is the soul of an educational and cultural program that provides a set of values that are useful for making wise choices. Bhola (1990) says, "Literacy can be defined in instrumental terms as the ability to read and write in the mother tongue or in national language. ... Numeracy the ability to deal with numbers at a primary level is typically considered part of literacy. Based on this definition, literacy implies that charity as an instrument is closely related to human civilization in the form of literacy as the main language used by every nation in the world. Literacy ability (literacy) is closely related to cultural development, including the interaction of all the factors that support literacy."

According to UNESCO (2006), "Literacy is the ability to identify, understand, interpret, create, communicate and compute, using printed and written materials associated with varying contexts. Literacy involves a continuum of learning in enabling individuals to achieve their goals, to develop their knowledge and potential, and to participate fully in their community and wider society."

According to John Hunter (in Kusnadi 2005), there are three basic categories of literacy definitions in people's lives, namely: (1) Literacy as set of basic skills, abilities, or competencies, (2) Literacy as the necessary foundation for higher quality of life, and (3) Literacy as a reflection of political and structural realities.

3 METHODOLOGY

This research used a descriptive method, with mixed-methods approach. Mixed-methods research combines quantitative methods and qualitative methods to be used together in a research activity, in order to obtain more comprehensive, valid, reliable, and objective data. Data collection techniques used are interviews, observation, documentation, and questionnaires.

Participants as research subjects totalled 113 people, namely 3 managers, 10 tutors, and 100 participants or learning citizens. The research was conducted at PKBM Famili in Indramayu Regency, PKBM Bina Kreatif Bahari Cirebon District, and PKBM Bima Sakti Subang Regency.

4 RESULTS AND DISCUSSION

4.1 *Conservation of literacy*

This study looked at three abilities, namely, reading, writing and arithmetic. Based on the results of data processing conducted, the reading ability of 63% of the students was very high, 27% were in the high category and 10% the medium category. For writing skills, 35% of students rated very high, then 22% high, 17% were still low, and 7% were still very low. The ability to count showed that as many as 28% of the students had very high ability, 22% high ability, 19% medium ability, 18% low ability, and 13% were still very low.

Based on these results, there were variations among the study participants. This is reasonable because the participants are adults who have complex issues in following their learning. However, these results indicate an increase in the ability to read, write, and count. This happens because they have the motivation to continue learning throughout their lives so that they become knowledgeable communities. This is in accordance with the opinion of Jarvis (2007) who says that a knowledgeable society is a source of development that can increase individuals and communities toward the better, and furthermore be able to change their culture so that they can respond and adjust to the social, political, and economic changes that occur in their lives.

Post-literacy programs are inseparable from the fact that many new literates are again illiterate. This happens if the ability to read and write is not utilized for a long time. To avoid this, there need to be opportunities to assist them in maintaining their fluency continuously through various follow-up programs so these skills last a lifetime. Lifelong education provides direction to help students develop themselves throughout adulthood. Humans who mature always try to find satisfaction through self-actualization; lifelong education focuses on the motivation of a person or group to obtain learning experiences on an ongoing basis. The function of lifelong education is to motivate students so that they can carry out learning activities based on self-directed encouragement (self-directed learning) by way of thinking and doing in the world. According to,ju (2004), the application of the principle of lifelong education in public education has three general characteristics, namely: (1) providing educational opportunities for everyone according to their interests, age and learning needs. This opportunity can be obtained in programs of group learning activities, individual learning activities, and learning activities through mass media; (2) in organizing education, always involve students, starting from planning activities, implementation, processes, results, and to the effects of learning activities carried out, and (3) having a purpose in accordance with the needs of individuals carried out in the education process. This goal is spelled out in the process of learning activities that lead to efforts to foster a democratic atmosphere of life, improve the living standards and lives of students and the community, and develop student behavior toward adulthood.

Associated with literacy programs, according to UNESCO (1993), the main purpose of post-literacy programs is to develop learning citizens who have learning styles:

- Respect for objectivity,
- Ability to interpret complex patterns,
- Tolerant of ambiguity,
- Broadening views,
- Desire for complexity,
- Awareness of socio-economic problems, and
- Sense of interdependence.

Another review of an increase in reading, writing, and arithmetic skills is associated with the efforts of methods and content of learning conducted based on business practices. This was felt by the participants to be very useful to implement in their lives, making it easier to digest the material in their learning.

4.2 *Entrepreneurship learning*

Based on the findings, after the respondents participated in entrepreneurial learning, (1) they showed very high self-confidence (27%), (2) they had the desire to practice entrepreneurship training (as much as 26%), (3) the emergence of new ideas and ideas for doing business was quite high (24%), (4) the benefits obtained are still low (15%), and (5) the desire to remain in training is low (8%).

Based on the results of the pre- and posttest, entrepreneurship learning shows that there is an increase in knowledge and skills by 9.3%. This happens because of the strong motivation of students to have knowledge and skills of certain types of businesses. The learning process occurs when the stimulus situation together with memory affects students in such a way that their actions change from the time before the situation. This result of the study is related to Mueller's (2011) finding that entrepreneurship education influences changes in attitudes, perceptions, and intentions of the community toward entrepreneurship.

The results indicated that the material provided in entrepreneurship learning showed a positive trend that is based on the opinions of students, that the material presented was something new according to the needs of learning citizens. Learning needs of learning citizens must be the basis for determining programs on community education. This is a benchmark in meeting learning needs, so that it can be meaningful for citizens to learn and it can be useful in their lives. With reference to the opinion of Ahmadi (2007), an individual's attitude is formed by the information he receives and changes in an individual's attitude in relation to various objects that exist within him or outside him can cause the attitude to be stronger or vice versa, depending on the individual's experience. In line with the opinion of Lee and Stearns (2012), the implementation of entrepreneurial activities cannot be separated from the psychological factors that influence the personality of the entrepreneurs involved.

4.3 *Business nodes*

The business nodes pioneered by the learning community are the manufacture of fish-bone crackers and shredded fish. The business node is based on available local potential. Utilizing local potential brought the business node many advantages, namely making it easier to supply raw materials. Referring to the opinion of Hatimah (2016), local potential can be a force that gives the authority of community members to utilize it in their lives. Local potential can be a carrying capacity for human activities. In utilizing local potential, it depends on the ability of its human resources, because human resources have an important role in maintaining the sustainability of their resources.

Local potential can be used as a tool to defend people in the social, economic, political, and spiritual fields. In the social field, it means that utilizing local potential can increase knowledge and skills for people who are powerless. In the economic field, local potential can build social economic institutions that are able to provide opportunities for the community to get decent employment and income, dignity, and personal subsistence. In the political field, local potential can create an open and democratic political climate. In utilizing local potential based on spiritual belief, every human being will have a very high awareness of various potentials that exist in the environment because it is related to the creator.

5 CONCLUSION

Viewed from the perspective of lifelong education, literacy maintenance programs have a positive impact in helping illiterate students to continue learning. Conclusions from the results of the study are:

- There has been an increase in people's ability to read, write, and count due to the accuracy of the use of learning methods, learning materials according to the needs of the community, high motivation of participants to participate in entrepreneurship programs, and the role of speakers in the delivery of material.
- Citizens learn to have knowledge, attitudes, and skills about entrepreneurship. In the aspect of knowledge an increase in understanding of entrepreneurial material, ways to make culinary products, simple bookkeeping methods, and ways of marketing. In the aspect of attitude, people learn to have an attitude of confidence, dare to start a business, be more open to the opinions of others, motivate each other between groups, increasing willingness to learn and try. In the aspect of skills, residents learned skilfully to make fish-bone crackers and shredded fish.
- Business nodes that are developed are the manufacture of fishbone crackers and shredded fish.

REFERENCES

Abdulhak, I. 2000. Metodologi *Pembelajaran Orang Dewasa*. Bandung: Andira.

Agarwal, N., Chakrabarti, R., Brem, A., & Bocken, N. 2018. Market driving at Bottom of the Pyramid (BoP): An analysis of social enterprises from the healthcare sector. *Journal of Business Research* 86: 234–244.

Ahmadi, A. 2007. *Psikologi Sosial*. Jakarta: Rineka Cipta.

Alma, B. 2009. *Manajemen Pemasaran dan Pemasaran Jasa*. Bandung: Alvabeta.

Bhola, H. S. 1990. Literature on adult literacy: New directions in the 1980s. *Comparative education review* 34(1): 139–144.

Bhuiyan, M. F., & Ivlevs, A. 2019. Micro-entrepreneurship and subjective well-being: Evidence from rural Bangladesh. *Journal of Business Venturing* 34(4): 625–645.

Chabbott, C. 1998. Constructing Educational Consensus: International Development Professionals and The World Conference on Education For All. *International Journal of Educational Development* 18(3): 207–218.

Djuju, S. 2004. *Pendidikan Non Formal: Wawasan, Sejarah Perkembangan, Filsafat dan Teori Pendukung, serta Asas*. Bandung: Falah Production.

Durieux, M., & Stebbins, R. 2010. *Social entrepreneurship for dummies*. Indianapolis: Wiley Publishing, Inc.

Hatimah, I. 2016. *Pendidikan Berbasis Masyarakat*. Bandung: Rizqi Press.

Jarvis, P. 2007. *Globalization, Lifelong Learning and the Learning Society: Sociological Perspective*. London and New York: Routledge.

Knapper, C., & Cropley, A.J. 2000. *Lifelong learning in higher education*. Psychology Press.

Kusnadi 2005. *Pendidikan Keaksaraan: Filosofis, Strategi, dan Implementasi*. Jakarta: Departemen Pendidikan Nasional.

Lee, S. S., & Stearns, T. M. 2012. Critical success factors in the performance of female-owned businesses: A study of female entrepreneurs in Korea. *International journal of management* 29(1): 3.

Meyer, J. W., Boli, J., Thomas, G. M., and Ramirez, F. 1998. Theories of culture: Institutional vs. actor-centered approaches. *American Journal of Sociology* 103(1): 144–181.

Mueller, S. 2011. Increasing entrepreneurial intention: Effective entrepreneurship course characteristics. *International Journal of Entrepreneurship and Small Business* 13(1): 55–74.

Ramirez, F. O., & Boli, J. 1982. *Global patterns of educational institutionalization. Comparative Education*. New York. Macmill.

Sakarya, S., Bodur, M., Yildirim-Öktem, Ö., & Selekler-Göksen, N. 2012. Social alliances: Business and social enterprise collaboration for social transformation. *Journal of Business Research* 65(12): 1710–1720.

Stephan, U., Patterson, M., Kelly, C. & Mair, J. 2016. Organizations driving positive social change: A review and an integrative framework of change processes. *Journal of Management* 42(5): 1250–1281.

Suryana, D., 2006. *Kewirausahaan: Pedoman Praktis (Kiat dan proses menuju sukses)*. Jakarta: Salemba Empat.

Thomas, J.E., Takamichi, U. & Shuichi, S. 1997. New lifelong learning law in Japan: promise or threat? *International Journal of Lifelong Education* 16(2): 132–140.

UNESCO 1993. *World Education Report*. Paris: UNESCO Digital Library.

UNESCO 2006. *Education for All Global Monitoring Report*. Paris: UNESCO Digital Library.

Viswanathan, M., & Rosa, J. A. 2010. Understanding subsistence marketplaces: Toward sustainable consumption and commerce for a better world. *Journal of Business Research* 63(6): 535–537.

Wiklund, J., Nikolaev, B., Shir, N., Foo, M.,D., & Bradley, S. 2019. Entrepreneurship and well-being: Past, present, and future. *Journal of Business Venturing Journal*.

World Bank 2011. *Learning for All: Investing in People's Knowledge and Skills to Promote Development. World Bank Group Education Strategy 2020*. Washington, DC: The World Bank.

Vocational education

Borderless Education as a Challenge in the 5.0 Society – Abdullah, Adriany & Abdullah (eds)
© 2021 Taylor & Francis Group, London, ISBN 978-0-367-61960-2

The effectiveness of implementation of industrial electricity installation trainers based on "training within industry" in the subject of industrial electricity installation

E. Elfizon, A. Asnil & M. Muskhir
Universitas Negeri Padang, Padang, Indonesia

ABSTRACT: The purpose of the paper is to determine and evaluate the effectiveness of the implementation industrial electricity installation based on training within industry in the subject of industrial electricity installation in the Department of Electrical Engineering, Faculty of Engineering, Universitas Negeri Padang (UNP). The Quasy Experimental is applied as a method to develop the research by taking 16 students which are divided into two groups from Department of Electrical Engineering as the subject and the instrument used are step performances and results of the product. The research results show that the average value of performance is 82,3 % meanwhile the result of product assessment obtained is 88,75 %. The achievement of student learning outcomes with the implementation of this product can be classified into two categories, high category (46%) and very high category (54%). Therefore, it can be concluded that the implementation of industrial electrical installation trainers based on "training withing industry" is effective to improve the student learning outcomes in the Department of Electrical Engineering, Faculty of Engineering, UNP.

1 INTRODUCTION

Universities as one of the organizers of vocational education are required to be able to follow technological developments so as to produce graduates who are competent cognitively, psychomotorically, and affective. The introduction of new technology must be done in the lecture process so that students are able to become individuals who are ready to face the challenges of the world in the technological era. Vocational education graduates should be able to compete and adapt to the changes and technological advancements that are developing in the business and industrial world. Changes and technological advances make the structure of the type of work in the world of work/industry also change. Various types of jobs also require new competencies so that industrial production with new technology can be of economic value to the nation and state. Higher education should be able to produce graduates who have high competence in entering the world of work, competitiveness in the field of work that is acted on, responsive and anticipatory to changes and technological advancements, live in harmony with the work environment and can achieve decent career goals in his life (Azhar 2011).

Padang State University as one of the institutions of vocational education providers in Indonesia always strives to realize the function of national education which is also listed in the National Education System Law No. 20 of 2003, which is to develop capabilities and shape the dignified character and civilization of the nation in order to educate the nation's life. Sharing efforts has been made by UNP in realizing qualified, superior and competent graduates who are able to compete and adapt to changes and technological advances that develop in the business and industrial world. One of the real efforts made by UNP is improving the quality of learning both in terms of material, process, and evaluation which is one of the main factors that must be done in realizing the quality of qualified graduates. Improving the quality of learning, also carried out at the Industrial Electrical Engineering Study Program (D4) which is one of the vocational education levels at the Electrical Engineering Department, Faculty of Engineering, UNP. Industrial Electrical Engineering Study Program (PSTEI) has a curriculum structure that is designed in accordance with the Vision, Mission, goals, objectives, and strategies that have been set. The PSTEI curriculum is based on suggestions and input from lecturers, industry and alumni in accordance with the demands of the needs of stakeholders in the field. The compiled curriculum must be able to develop the knowledge, skills and attitudes of graduates as expected by the world of work/ industry.

Industrial Electrical Installation is one of the courses in the curriculum structure of PSTEI FT UNP. Industrial Electrical Installation Courses Electrical Installation at PSTEI in the Department of Electrical Engineering FT-UNP is a course that leads students to get to know electrical installations that are developing in the industry and scientific insights related to the field of electrical installations in industry. The course emphasizes the activity of applying a theory in limited conditions and situations, such as laboratories, workshops, work spaces and so on. The application of this course is widely used in manufacturing industries. By mastering all the skills taught in each lecture material both theory and practice are expected to be a support for graduates truly prepared to work in industry. The Learning of Industrial Electrical Installation Practices is one of the courses in the curriculum structure of the Industrial Electrical Engineering Study Program (PSTEI) FT UNP. This course is a course that leads students to get to know electrical installations that are developing in the industry and scientific insights related to the field of electrical installations in the industry. The course emphasizes the activity of applying a theory in limited conditions and situations, such as laboratories, workshops, work spaces and so on. The application of this course is widely used in manufacturing industries. By mastering all the skills taught in each lecture material both theory and practice are expected to be a support for graduates truly prepared to work in industry.

In general, the description of this course is demanding that students be able to plan and install industrial motor installations, control systems, safety systems and the operation of motorcycles in industry. Various series of implementation of industrial motor control systems are practiced by students in repairs such as:

- Direct online operating motor control systems (DOL);
- Motor operating systems from many places;
- Sequential operating motor operating systems;
- Motor operating system turns;
- Forward-reverse motor operation system (reversing motor rotation);
- Motor operating system start - delta (Y-operation);
- Motor safety system and other loading.

Weaknesses of the teaching process that has been done, students are not able to imagine how the industrial electrical installation system in the field, namely in the business world/ industrial world. The practicum lecturing process carried out so far still prioritizes the achievement of competence in this course. So that students are more focused on doing series experiments and testing, this is why students have not yet thought to make the skills of arranging these sets of experiments to apply them in the field. Seeing the problems above, a media trainer is offered in the practice learning process, where the trainer is designed according to the needs of the workforce or the industrial world. The media trainers that use as a support for practicum are industrial electrical installation trainers with nuances within industry. This trainer is expected to be able to motivate students in learning so they feel a working atmosphere in the industry learning (Kustandi & Situmorang 2013, Bustami et al. 2018).

Vocational education has close links with the world of work or industry, so learning and practical training plays a key role in equipping graduates to be able to adapt to the workforce. As such, they must be formed through a series of exercises or learning and practical training that almost resembles the world of work (Wena 2009).

States that, to teach the practice of vocational skills it is necessary to use certain strategies so that students understand, both cognitively and motorically at the basic steps of a vocational skill. The majority of formal education institutions still do not apply the learning media needed by students in knowing new technologies that are developing in the industry (Wena 2009). Industrial Electrical Installation Trainer is one of the potentials that can be used by students to get to know more about the development of Installation technology in Industry. Industrial electrical installation trainers with the authority Within Industry Training are very suitable to be applied in introducing industry to students from an early age (Krismadinata et al. 2019).

Trainer in charge of Training Within Industry in the process of practical learning really guides students to learn to work gradually, sequentially and follow standard work procedures to master a work skill. In education there are two interrelated terms namely learning and the learning process. Learning is more focused on students, what they must do in accepting lessons. According to Sagala (2006), "Learning is a process in which an organism changes its behaviour as a result of experience". So, after following the learning process, students are expected to gain new experiences in interaction with the environment (Gagné 2003).

The important thing in learning and training in vocational practice is the mastery of practical skills, knowledge and behaviour that are directly related to these skills. One learning strategy for teaching basic vocational skills is the Training Within Industry learning strategy. The Within Industry Training Program consists of five main models: 1) work instructions, 2) work methods, 3) work relationships, 4) work safety, 5) program development. These programs are aimed at developing internal trainers and supervisors who can multiply their own hard work by teaching others. If each person who has been certified to provide training (main trainer) teaches a number of Supervisors (trainers), and each supervisor then trains 10 or more partners, then the success of the program will double. Training programs in Training Within Industry are known as J- Programs:

1.1 *Job instruction*

Designed to help supervisors "train" new employees who don't yet have expertise and are based on decades of practical experience. However, learning is used to help teachers train students who do not yet have expertise. Although the material underwent a slight revision over time, the basic premise remains the same as breaking down work into several elements, identifying key points, and carrying out operations to achieve success. The aim is to shorten the "training" period and improve work safety and quality through a better understanding of the work of vital elements of the workers.

1.2 *Job methods*

This TWI program provides techniques aimed at helping supervisors and workers to analyse all aspects of the work according to the method and to question every detail in the formation of needs, sequence, and responsibilities for each task. Efforts to question and evaluate will result in increased productivity and eliminating unnecessary steps and activities or waste.

1.3 *Job relations*

This program is intended to provide supervisors with a method for overcoming problems and for improving work relations. Many supervisors during wartime are inexperienced and have no knowledge of how to deal with problems and matters of concern to employees. Topics include giving feedback on employee performance, dealing with matters of concern to employees, rewarding great ideas or performance, communicating events or changes, and utilizing the abilities of each person.

1.4 *Job safety*

This program is intended so that each employee can work by prioritizing occupational safety and health (K3) so as to prevent accidents or unwanted events.

1.5 *Development program*

Program development is specifically aimed at people or individuals in each factory who want to identify specific needs or training, develop plans, get support from management, implement plans, train supervisors. Nolker & Schoenfeldt (in Wena 2009) states that "Within Industry Training Strategy consists of 5 stages of learning namely preparation, demonstration, imitation, practice, and evaluation". Description The stages of learning in the Within Industry Training strategy are as follows:

- Preparation. Broadly speaking the teacher's activities in this stage are preparing a worksheet (Jobsheet), explaining the learning and training objectives, explaining their importance, arousing student interest. Basically, the activities of the teacher in this stage are to plan, organize, and formulate the conditions of learning and training so that there are activities systematically with the strategy to be applied.
- Demonstration. In this stage the teacher or instructor has begun to enter the implementation phase. The teacher conveys material using learning media and practical training available. After that, students should be given time to give feedback on what they have seen. The next step is the teacher demonstrates in real time the work to be learned, explains how to work well in relation to the whole process, while taking positions so that students can follow the work process from the right point of view.
- Impersonation. After the demonstration phase is carried out carefully, then proceed with the imitation stage. In the imitation stage students do work activities imitating work activities that have been demonstrated by the teacher. In carrying out imitation activities, students must be organized and organized practical learning activities so that students are truly able to understand and carry out work activities in accordance with the learning objectives and practical training. In this stage the teacher must really pay attention to the stages of work done by students. The teacher must always monitor student work processes. If there are things that are not appropriate, the teacher must ask students to do repetitive work and help students until they can do work assignments correctly.
- Practice. After students are able to mimic how to work well, the next step is the implementation of practical activities. At this stage students repeat the work activity that has just been learned until the work skills learned are really fully mastered. The important thing that teachers need to do and pay attention to at this stage is the setting of management strategies and organizing learning and practical training, so that students are truly able to carry out practical learning activities optimally.
- Evaluation. In the learning strategy and practical training for the TWI model, evaluation activities are carried out at the practical stage. To evaluate learning and practical training, an instrument is used, namely Performance Assessment. The evaluation phase is the final stage that is important for every process of learning and training, especially in learning and training in vocational practices. By evaluating learning and practical training, students will know their abilities clearly so that students can improve and improve the quality of learning and training. Likewise, evaluation activities are very important for a teacher, because from the results of evaluations carried out it can be seen how far the set goals have been achieved.

Then imitation by students. At this stage students are divided into 5 heterogeneous groups and then students imitate the 5 trainers that are available until students are ready to practice and are able to mention the working principles correctly. After being able to imitate work continued with practice by students as well as evaluations by teachers. At this stage students practice 2 or 3 groups and students who have not yet practiced observing their peers who practice. In this stage students work from preparation for work preparing work equipment and practicing in groups that are determined to complete the work skilfully and correctly. During practical activities take place the teacher evaluates the achievement of student learning outcomes (Muskhir 2019).

From the evaluation results obtained results of the achievement of student practice values as a whole so that it becomes a material consideration for teachers for further practicum the application of TWI has its own advantages compared to the application of learning models that are generally used by teachers in learning. Learning that is generally used in practical learning is the method of lecture, question and answer and practice. While in the Training Within Industry strategy the learning process applies a combination of learning methods, namely: lecture, demonstration, question and answer methods, exercises or repetition of work by students, and practice in groups. Learning outcomes that can be achieved by students are in addition to students having skills in doing work and work attitudes in accordance with work standards, students also have a soul of leadership, confidence and activeness in learning (Elfizon et al. 2017). Based on these conditions, the authors see the potential of a trainer that can be used as a learning medium; therefore, researchers intend to develop an industrial electrical installation trainer that is expected to increase students' insights to work in industry so that the process so that the teaching and learning process can be done more optimally.

To carry out the development activities of the industrial electrical installation trainer development procedures are required. Development procedures are procedural steps that must be taken by the developer to reach the specified product. The trainer development procedure includes several stages, namely planning or preparing the trainer media design, production, testing and evaluation of the trainer. The industrial electrical installation trainer developed in this study belongs to the type of design media trainer because this trainer is not available in the Indonesian market, so design is needed to produce the media. Based on this, it is necessary to develop a learning type learning media, in broad outline the learning media development activities consist of three major steps that must be passed, namely planning, production and assessment activities (Bustami et al. 2018).

Industrial Electrical installation trainers are designed with nuances of Within Industry Training. Where students feel as if they are working in the industry. This Within Industry Training Strategy consists of five main models:

- Work instructions,
- Work methods,
- Work relationships,
- Work safety,
- Program development.

These programs are aimed at developing internal trainers and supervisors who can multiply their own hard work by teaching others. If each person who has been certified to provide training (main trainer) teaches a number of Supervisors (trainers), and each supervisor then trains 10 or more partners, then the success of the program will double (Candra et al. 2019).

2 RESEARCH METHODS

In writing this paper using the research and development method. The states Research and Development is a research method used to produce certain products, and test the effectiveness of these products. Development research in the field of education is a type of research that aims to produce products for learning purposes. In the implementation of R&D, there are several methods used, namely descriptive, evaluative and experimental methods. Descriptive research methods are used in initial research to collect data about existing conditions. Evaluative methods are used to evaluate the testing process of developing a product. The experimental method is used to test the efficacy of the product produced (Sugiyono 2008).

This type of research is an experimental research that is categorized into quasi-experimental type. The research was conducted at the Department of Electrical Engineering Faculty of Universitas Negeri Padang in Industrial Electrical Engineering Study Program in the Department of Electrical Engineering, Faculty of Engineering, UNP. As the subject of research is the 3rd semester students of Industrial electricity installations totalling 37 people, consisting of two classes 2TEA and 2TEB. Where 2TEA is an experimental class using implementation industrial electricity installation based on training within industry and 2TEB is a control class that uses conventional learning (Arikunto 1999). The determination of this class is done randomly from the existing class, this is done because the average score of student's GPA does not differ significantly. Thus, based on the t-test the two classes have the same initial capability. The research design used in this study is presented as following Table 1:

Table 1. Research design.

Class	Treatment	Result
Experiment	X1	O1
Control	X2	O2

Information:
X_1=Treatment with implementation industrial electricity installation based on training within industry
X_2 = Conventional learning
O_1 = The results of the experimental
O_2 = Results of a control class performance assessment

The type of instrument used in this study is the assessment of performance. According to Depdiknas (2006), "Performance assessment is an assessment done by observing the activities of learners in doing something". Validity in this research is content validity. Implementation Content validity is by arranging aspects to be assessed in the Industrial electricity installation courses according to the curriculum Industrial Electrical Engineering Study Program in the Department of Electrical Engineering FT UNP.

After data collected conducted analyses data. Prior to testing the research hypothesis, student learning outcomes must meet the requirements of normality test and homogeneity test (Riduwan 2007, Sudjana 1996).

3 RESULTS AND DISCUSSION

Based on data analysis, normality testing in the experimental class and control class obtained normal distribution data. Where X2 count <X2 table, i.e. for the experimental class X2 count is 4.61 and the control class X2 count is 0.63 while X2 table is 10.278. Homogeneity testing in both classes obtained an F count of 2.213 while the F table with a numerator = 32 and a mention = 30 was 3.46 at a significance level of 0.05. Thus, F arithmetic <F table means that both classes have homogeneous variance.

The results of testing the hypothesis with the t-test obtained tcount of 3.443 and for ttable 2.015, then tcount was compared with ttable with the test criteria if tcount> ttable then Ha was accepted, and the results obtained were 3.363> 2.015. Then the final conclusion Ha is accepted so that it can be concluded that there are differences in learning outcomes of 2TEIA students using implementation industrial electricity installation based on training within industry with conventional learning models in the Industrial Electrical Installation Practice Course.

Based on the results of the study, obtained an average student psychomotor learning outcome in the experimental class 82.5 while in the control class the average psychomotor student learning outcomes is 76.9. Thus, the learning outcomes of industrial electrical installation practicum using implementation industrial electricity installation based on training within industry are better than using conventional learning models.

Meanwhile, according to Moursund (in Wena 2009) some advantages of project-based learning include the following:

- Increased motivations;
- Increased problem-solving abilities;
- Improved library research skills;
- Increased collaboration;
- Increased resource-management skills.

So, it can be concluded that project-based learning is better than Drill and practice learning implementation industrial electricity installation based on training within industry can increase motivation and increase student independence in problem solving so that learning will be more meaningful, in contrast to Drill and practice learning which forms rigid habits in students (Elfizon et al. 2017).

This study concludes that there are differences in the average value of student learning outcomes taught using implementation industrial electricity installation based on training within industry and those taught by direct learning models. The average value in the experimental class with the project based learning model is 92,6 while the average value in the control class with the direct learning model is 81,2. Implementation industrial electricity installation based on training within industry has a positive influence by applying the project-based learning model to students becoming more active in learning in the laboratory of the students' experimental learning result is higher than the control class as can be seen in Table 2 bellow:

Table 2. Average and percentage completion of experiment and control class.

Class	Amount	Result $\overline{X}$
Experiment (2TEA)	18	92,6
Control (2TEB)	19	81,2

Based on Table 2, it appears that implementation industrial electricity installation based on training within industry is essentially a strategy that can facilitate students in the lecture so that students become skilled. By using the strategy Training model requires students to work in stages and structured, which includes: the preparation stage, demonstration, imitation, and practice (Candra et al. 2019).

Based on the description and analysis of data that has been done on student learning outcomes on electrical installation learning through learning model training in the experimental class and conventional learning on the electronics engineering education faculty of Universitas Negeri Padang, there are differences in learning outcomes between the experimental class and the control class. This difference can be seen from the highest value of the experimental of 97 class with an average of 92,6, while the control grade is at a high of 89 with an average of 81.2. Thus, it can be stated that the students 'learning outcomes in the experimental class is higher than the students' learning outcomes of the control class.

Strategy Training model implementation industrial electricity installation based on training within industry is essentially a strategy that can facilitate students in the lecture so that students become skilled. By using the strategy Training model requires students to work in stages and structured, which includes: the preparation stage, demonstration, imitation, and practice. The following is the normal curve of the experimental class and control class as follows (see Figure 1).

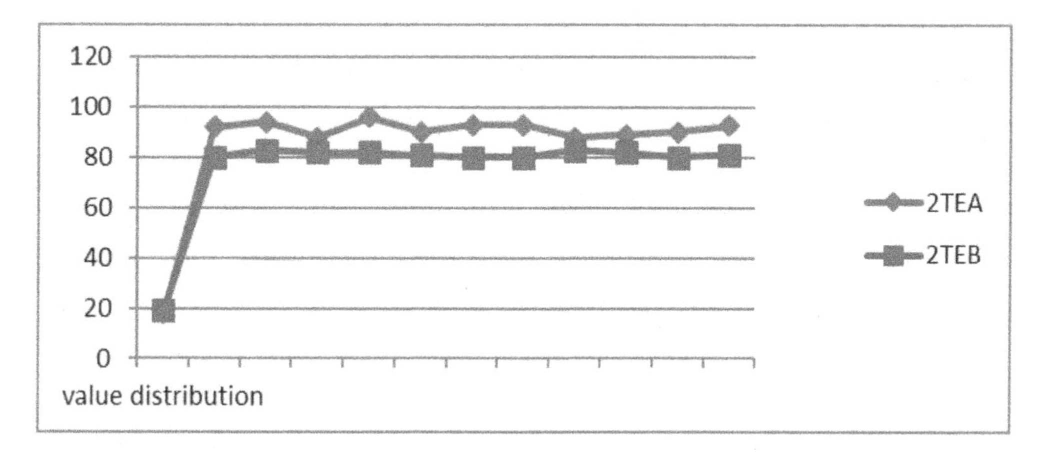

Figure 1. A figure Distribution of values and averages, experimental and control classes.

4 CONCLUSION

The research results show that the average value of performance is 82,3 % meanwhile the result of product assessment obtained is 88,75 %. The achievement of student learning outcomes with the implementation of this product can be classified into two categories, high category (46%) and very high category (54%). Therefore, it can be concluded that the implementation of industrial electrical installation trainers based on "training withing industry" is effective to improve the student learning outcomes in the Department of Electrical Engineering, Faculty of Engineering, UNP.

REFERENCES

Arikunto, S. 1999. *Dasar-Dasar Evaluasi Pendidikan*. Jakarta: Bumi Aksara.

Azhar, A. 2011. *Media Pembelajaran*. Jakarta: PT Raja Grafindo Persada.

Bustami, E., Candra, O. & Muskhir, M. 2018. Penerapan Strategi Training Within Industry Sebagai Upaya Meningkatan Motivasi Perkuliahan. *INVOTEK: Jurnal Inovasi Vokasional dan Teknologi* 18(2): 55–64.

Candra, O., Elfizon, E., Hendri, H., Aslimeri, A. & Aswardi, A. 2019. Peningkatan Keterampilan Bidang Pemasangan Instalasi Listrik Rumah Tangga Dan Bidang Service Peralatan Elektronik Bagi Pemuda Panti Budi Utama Di Lubuk Alung. *JTEV (Jurnal Teknik Elektro dan Vokasional)* 5(1): 31–36.

Candra, O., Pulungan, A.B. & Eliza, F. 2019. Development of Miniature Secondary Network of Electric Power Distribution System as a Learning Media for Electrical Engineering Students. *Journal of Physics: Conference Series* 1165(1).

Depdiknas 2006, *Panduan Penyusunan Kurikulum Tingkat Satuan Pendidikan*. Jakarta: BNSP Depdiknas,

Elfizon, E., Muskhir, M. & Candra, O. 2017. *Pengembangan Media Trainer Elektronika Dalam Pembelajaran Teknik Elektronika Pada Pendidikan Vokasi Teknik Elektro Fakultas Teknik Universitas Negeri Padang*. Padang: UNP.

Elfizon, E., Syamsuarnis, S. & Candra, O. 2017. The Effect of Strategy of Training Models in Learning Electrical Installation: 9–12.

Gagné, M. 2003. The Role of Autonomy Support and Autonomy Orientation in Prosocial Behavior Engagement. *Motivation and emotion* 27(3): 199–223.

Krismadinata, K., Elfizon, E., & Santika, T. 2019. Developing Interactive Learning Multimedia on Basic Electrical Measurement Course. *5th UPI International Conference on Technical and Vocational Education and Training (ICTVET 2018)*.

Kustandi, C. & Situmorang, R. 2013. Pengembangan Digital Library Sebagai Sumber Belajar. *Perspektif Ilmu Pendidikan* 27(1): 60–68.

Muskhir, M. 2019. Development of Industrial Electrical Installation Trainer Nuanced To Training within Industry for Students of Electrical Industrial Engineering Universitas Negeri Padang. *Journal of Physics: Conference Series* 1165(1).

Riduwan, M.B.A. 2007. *Skala Pengukuran Variabel-Variabel Penelitian*. Bandung: Alfabeta.

Sagala, S. 2006. *Konsep & Makna Pembelajaran*. Bandung: Alfabeta.

Sudjana, M.A. 1996. *Metoda Statistika* (6th ed). Bandung: Tarsito

Sugiyono 2008. *Metode Penelitian Pendidikan Pendekatan Kuantitatif, Kualitatif, dan R&D*. Bandung: Alfabeta.

Wena, M. 2009. *Strategi Pembelajaran Inovatif Kontemporer Suatu Tinjauan Konseptual Operasional*. Jakarta: Bumi Aksara.

Borderless Education as a Challenge in the 5.0 Society – Abdullah, Adriany & Abdullah (eds)
© 2021 Taylor & Francis Group, London, ISBN 978-0-367-61960-2

The achievement of four student competencies in domestic electrical installations using a project-based learning model

D.T.P. Yanto, E. Astrid & R. Hidayat
Universitas Negeri Padang, Padang, Indonesia

ABSTRACT: The research discusses the achievement of four competencies in electrical installation for vocational education students of electrical engineering: planning, preparing, installing, and testing competencies in the electrical installation fields that must be achieved by each student. The learning process is carried out with the help of project-based practice modules for each student. Validated performance assessment rubrics are used as data collection instruments to measure the level of competency achievement. The research subjects are 18 students who took part in the domestic electrical installation practice learning process. The aim of this research is to determine the level of student achievement in four competencies by using a project-based learning model. The results show that the total average of the achievement values for four competencies as a whole is 87.6, where separate scores for the competencies of planning, preparation, implementation, and testing are 84.8, 87.1, 91.3, and 87 respectively. It can be concluded that the implementation of project-based learning models is effective for students to achieve all four electrical installation competencies.

1 INTRODUCTION

Vocational education is a form of educational process oriented to specific expertise competencies that are tailored to needs in the work world. Thus, vocational education can produce competent graduates (Christidis 2019, Frattini & Meschi 2019, Yanto et al. 2017). In other words, vocational education can be defined as an educational process that creates graduates who are ready to work with certain competency skills.

The Electrical Engineering Study Program, Faculty of Engineering, at Universitas Negeri Padang is one of the vocational education programs in the field of electrical engineering diploma level 3 (D3), which aims to produce graduates who have competency in the field of electric power engineering. Based on the Indonesian Qualification Framework (KKNI), vocational education at D3 level is at level 5, where graduates are assigned to have at least four main competencies: planning, preparing, installing, testing/controlling in accordance with their field of expertise (Choi et al. 2019, Frattini & Meschi 2019, Yamashita & Yasueda, 2017). Hence, the implementation of the learning process in electrical engineering courses should be oriented to the four competencies (Arcidiacono et al. 2016, Efstratia 2014).

The domestic electrical installation practice is one of the learning processes that must be followed by every student in the electrical engineering education study program. The aim of this learning process is that students are able to plan, prepare, install, and test 1 phase and 3 phase electrical installations for residential use. There are many factors that can affect the achievement of a learning process, one of which is the application of learning models in accordance with the characteristics of the learning process. The learning model is a series of presentations of teaching material that covers all aspects – before and after learning – conducted by educators and all related facilities that are used directly or indirectly in the learning process (Candra et al. 2019, Sukardi et al. 2017). Learning models can also be interpreted as a way to organize the implementation of the learning process so as to achieve an optimal learning objective as well.

The currently learning model has many types, each of which has different characteristics. That is, each learning model has its own characteristics that need to be adjusted to the characteristics of the learning process to be carried out (Tascı 2015, Weichhart & Stary 2017, Yamashita & Yasueda 2017). Dealing with the objective of the domestic electrical installation practice, the learning model that corresponds to the characteristics and learning objectives of the learning process is project-based learning models, a learning process with the characteristics of honing student performance and competence.

The project-based learning model is one wherein implementation adopts the application of a project. By using this learning model, students gain the competencies required in the learning process and also gain experience in managing a project that is oriented to the world of work (Francese et al. 2015, Chen & Yang 2019). With these characteristics, the project-based learning model is considered capable of helping to achieve the optimal learning objectives from the domestic electrical installation practice. Thus, we conducted a study to determine the level of student achievement of the four competency skills of the learning process: planning, preparing, installing, and testing (Esteban & Arahal 2015, McGibbon &Van Belle 2015, Sykorova 2015), using a project-based learning model. There are many methods of measuring the level of achievement of learning objectives in previous research, such as the calculation of the gain score values for the pretest and posttest scores (Requies et al. 2018, Sykorova 2015), comparison of student score between the experimental class and the control class (Efstratia 2014, Copot et al. 2016), or percentage of increase in student score after carrying out the learning process (Ergül & Kargın 2014).

In this study, the level of competency by using a project-based learning model was assessed, then the value was converted into a percentage level for later interpretation in terms of five levels of achievement.

2 METHODS

The research design used in this study was a one-group subject design. The study was conducted on one class of students as research subjects, namely 18 higher education students in the Electrical Engineering Department, Faculty of Engineering, Universitas Negeri Padang. Students follow the learning process by using a project-based learning model assisted by learning media in the form of practice modules. During the learning the process was evaluated by observers using data collection instruments for each student (Francese et al. 2015, Weichhart & Stary 2017). The results of student performance are used as a reference in the analysis of the level of competency achieved by the students.

2.1 *Research instruments*

The data collection instrument used in this study is the student performance assessment rubric with five criteria. The instruments of this study were developed based on the competencies assigned as objectives of the learning process of the domestic electrical installation practice. The performance assessment rubric was developed based on student competency indicators in accordance with the competencies to be achieved, as presented in Table 1 (Christidis 2019, Copot et al. 2016).

Table 1. Indicators of research instrument.

Indicator	Total Item	Number
Planning	5	1 – 5
Preparation	5	6 – 10
Installation	6	11 – 16
Testing	4	17 – 20

Table 2. Interpretation of achievement levels.

No.	Final Score	Interpretation
1	86–100	Very High
2	71–85	High
3	56–70	Medium
4	41–55	Low
5	< 40	Very Low

The research instrument used is a research instrument that has been tested and for validity and reliability, in order to get the research instrument in the form of a work performance rubric that consists of 20 items in total.

2.2 *Techniques of data analysis*

Data obtained using research instruments in the form of student performance during the learning process, are subject to statistical analysis to get the final scores of each competency. The final grade will be used as a reference in measuring the level of achievement of the four student competencies. The final value (FS) for each student is obtained by dividing the total number of scores obtained for all items (ΣSc) by the number of items (TI) (Sart, 2014).

$$FS = \sum S_c / TI \tag{1}$$

Measurement of the level of competency is also carried out as a whole for all students where the grade taken is the classical average value of all students. Calculation of the classical average value of all students (FCS) for the four competencies achieved is done by dividing the total final grade of all students (S_{cs}) by the total number of students (TS) (Sart, 2014).

$$\mathrm{FCS} = \sum \mathrm{S_{cs}}/\mathrm{TS} \tag{2}$$

The final score for each competency is then interpreted as in Table 2 to get the level of achievement obtained (Blickle & Genau 2019, Sykorova 2015).

3 RESULT AND DISCUSSION

3.1 *Planning competency*

Planning competency is student competency in planning an electric power installation that is adjusted to the needs and conditions such as piping, single line, and maintenance. Planning competency is also an initial competency that can be measured in the learning process by using a project-based learning model. The results showed that the classical average of student performance on planning competencies in the learning process of the domestic electrical installation practice using a project-based learning model was 84.78. Thus, if interpreted in the achievement level table, the level of achievement of planning competence is at a high level.

From a total of 18 students, 13 students (72.2 %) got a performance score of ≥ 86, which means that the level of competency achievement was at a very high level. Five other people (27.8%) experienced a high level of competency achievement. So, it can be said that the project-based learning model can help the achievement of planning competencies in the learning process of the domestic electrical installation practice and the level of achievement that occurs is at a high level. A graph of the results of planning competency achievement for each student is presented in Figure 1.

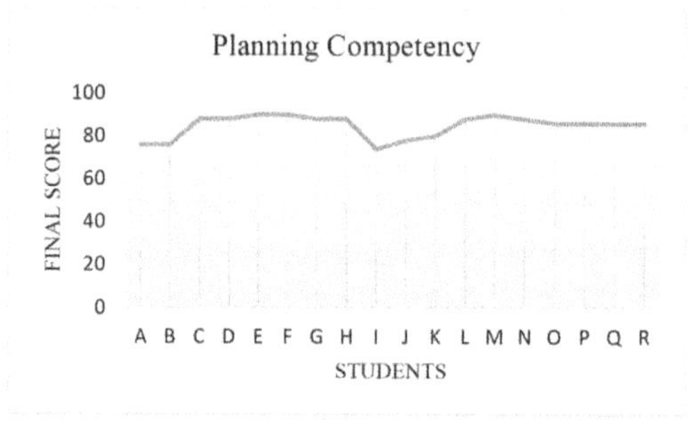

Figure 1. Graph of the results of planning competency achievement.

3.2 *Preparation competency*

Preparation competency is student competency in preparing everything needed during the implementation process, such as tools and materials. The preparation process is carried out by students based on the results of the planning they have carried out before. The results showed that the average value of the performance of all students in the learning process amounted to 87.11, which, if interpreted by the achievement level interpretation table, was at a very high level.

Of the 18 students, 15 (83.3%) received a performance score of ≥ 86, which means they are able to achieve this preparatory competency at a very high level of achievement. Meanwhile, three other people (16.7%) were able to obtain this preparatory competency at a high level of achievement. Hence, it can be concluded that a project-based learning model can help the achievement of competency preparation in the learning process of electrical installation practice with a very high level of achievement. The results of preparation competency achievement for each student is presented as a graph in Figure 2.

3.3 *Installation competency*

Installation competency is the core competency of the four, where students are able to carry out electrical installation activities in accordance with the planning and preparation that have been carried out previously. Competency in the use of tools and materials, as well as time efficiency is one of the sub-competencies that must be met in the installation competency. The results showed that the average value of performance for all students in the learning process was 91.33 or at a very high level of achievement.

Of the 18 students, 16 (88.9%) received performance scores on the installation competency ≥ 86 which means that their level of achievement was at a very high level. Meanwhile, two other

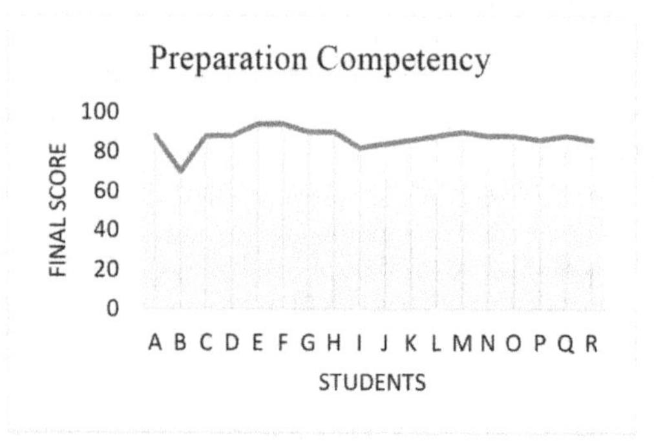

Figure 2. Graph of the results of preparation competency achievement.

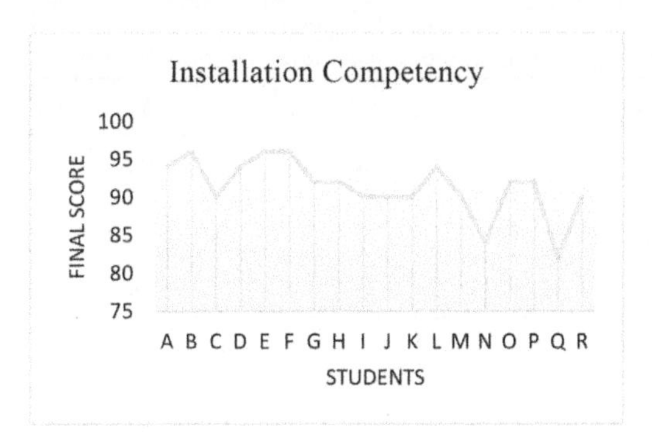

Figure 3. Graph of the results of installation competency achievement.

people (11.1%) experienced competency attainment at a high level. This indicates that the project-based learning model is able to facilitate the achievement of student competencies in the learning process in the domestic electrical installation practice. Achievements are at a very high level. The results of installation competency achievements are presented in the graph in Figure 3.

3.4 *Testing competency*

Testing competency is the competency of students in carrying out checking and testing activities on completed electric power installations. Tests are carried out to ensure that the electric power installation is in accordance with the plan and can operate properly. The results showed that the average value of performance for all students in the learning process of the domestic electric installation practice was 87 with very high levels of achievement.

Of the 18 students, 14 (77.8%) obtained a level of competency achievement of ≥ 86 which means that the level of achievement of competence was at a very high level. Meanwhile, four other students (22.2%) obtained the level of competency achievement at a high level. This indicates that a project-based learning model can help the achievement of testing competencies. Overall achievement for this testing competency is at a very high level. A graph of the results of testing competency achievement for each student is presented in Figure 4.

This study indicates that the implementation of the project-based learning model is able to facilitate achievement of the four main competencies of the learning process of the domestic electrical installation for students of the electrical engineering study program. Achievement of the four competencies as a whole is at a very high level (87.56). The descriptive analysis results

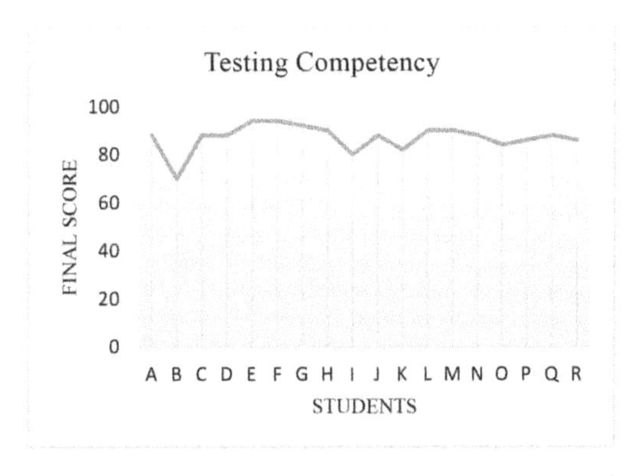

Figure 4. Graph of the results of testing competency achievement.

Table 3. Descriptive statistics of the achievement of four competencies.

Competencies	N	Minimum	Maximum	Mean
Planning	18	74	90	84.78
Preparation	18	70	94	87.11
Installation	18	82	96	91.33
Testing	18	70	94	87.00
Total	18	70	96	87.56

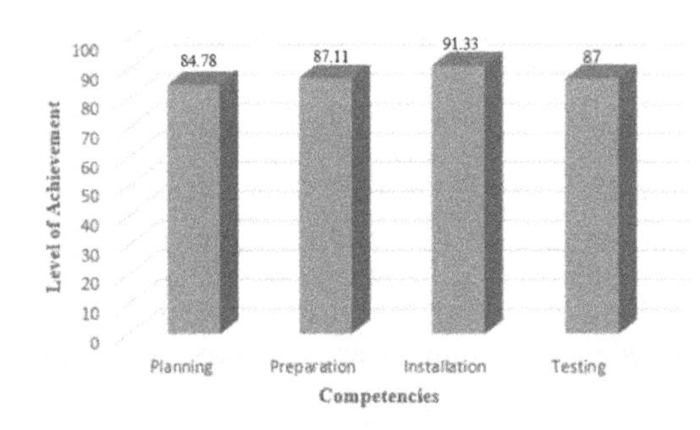

Figure 5. Chart of four competencies achievement levels.

for the data on the achievement of four competencies in the learning process is presented in Table 3. A graph of results of achieving the four main competencies of the learning process of domestic electric installation by applying a project-based learning model is presented in Figure 5.

4 CONCLUSION

The results of the analysis of the research data show that the four competencies relevant to the learning process of the Domestic Electricity Installation Practices overall can be optimally achieved by using a project-based learning model. Competency achievement is at a very high level. Thus, it can be concluded that the project-based learning model can assist students in achieving four competencies in the learning process of Electrical Installation Practices. Therefore, this project-based learning model can be used as an alternative choice applied in the field of Domestic Electrical Installation Practices so that the learning process can run optimally.

ACKNOWLEDGMENT

Thanks to the Rector of Universitas Negeri Padang (UNP) and Chairman of the Institute for Research and Society Dedi-cation (LP2M) UNP who have helped and facilitated the implementation of this study.

REFERENCES

Arcidiacono, G., Yang, K., Trewn, J., & Bucciarelli, L. 2016. Application of axiomatic design for project-based learning methodology. *Procedia CIRP* 53: 166–172.

Blickle, G., & Genau, H. A. 2019. The two faces of fearless dominance and their relations to vocational success. *Journal of Research in Personality* 81: 25–37.

Candra, O., Faradina, N., Islami, S., Yanto, D. T. P., & Dewi, C. 2019. Peningkatan Kompetensi Bidang Instalasi Listrik Domestik bagi Pemuda Panti Sosial Asuhan Anak Binaan Remaja (PSAABR) Budi Utama Lubuk Alung Kabupaten Padang Pariaman. *JTEV (Jurnal Teknik Elektro dan Vokasional)* 5(2): 22–27.

Chen, C. H., & Yang, Y. C. 2019. Revisiting the effects of project-based learning on students' academic achievement: A meta-analysis investigating moderators. *Educational Research Review* 26: 71–81.

Choi, S. J., Jeong, J. C., & Kim, S. N. 2019. Impact of vocational education and training on adult skills and employment: An applied multilevel analysis. *International Journal of Educational Development* 66: 129–138.

Christidis, M. 2019. Vocational knowing in subject integrated teaching: A case study in a Swedish upper secondary health and social care program. *Learning, Culture and Social Interaction* 21: 21–33.

Copot, C., Ionescu, C., & De Keyser, R. 2016. Interdisciplinary project-based learning at master level: control of robotic mechatronic systems. *IFAC-PapersOnLine* 49(6): 314–319.

Efstratia, D. 2014. Experiential education through project based learning. *Procedia-social and behavioral sciences* 152: 1256–1260.

Ergül, N. R., & Kargın, E. K. 2014. The effect of project based learning on students' science success. *Procedia-Social and Behavioral Sciences* 136: 537–541.

Esteban, S., & Arahal, M. R. 2015. Project based learning methodologies applied to large groups of students: Airplane design in a concurrent engineering context. *IFAC-PapersOnLine* 48(29): 194–199.

Francese, R., Gravino, C., Risi, M., Scanniello, G., & Tortora, G. 2015. Using project-based-learning in a mobile application development course: An experience report. *Journal of Visual Languages & Computing* 31: 196–205.

Frattini, T., & Meschi, E. 2019. The effect of immigrant peers in vocational schools. *European Economic Review* 113: 1–22.

McGibbon, C., & Van Belle, J.P. 2015. Integrating environmental sustainability issues into the curriculum through problem-based and project-based learning: a case study at the University of Cape Town. *Current Opinion in Environmental Sustainability* 16: 81–88.

Requies, J. M., Agirre, I., Barrio, V. L., & Graells, M. 2018. Evolution of Project-Based Learning in Small Groups in Environmental Engineering Courses. *Journal of Technology and Science Education* 8(1): 45–62.

Sart, G. 2014. The effects of the development of metacognition on project-based learning. *Procedia-Social and Behavioral Sciences* 152, 131–136.

Sukardi, S., Puyada, D., Wulansari, R. E., & Yanto, D. T. P. 2017. The validity of interactive instructional media on electrical circuits at vocational high school and technology. *The 2nd INCOTEPD*: 21–22.

Sykorova, J. 2015. Outputs of interactive exploration and project-based teaching at Mendel University in Brno, Czech Republic. *Procedia-Social and Behavioral Sciences* 174: 3224–3227.

Tascı, B. G. 2015. Project based learning from elementary school to college, tool: Architecture. *Procedia-Social and Behavioral Sciences* 186: 770–775.

Weichhart, G., & Stary, C. 2017. Project-based learning for complex adaptive enterprise systems. *IFAC-PapersOnLine* 50(1): 12991–12996.

Yamashita, K., & Yasueda, H. 2017. Project-based learning in out-of-class activities: Flipped learning based on communities created in real and virtual spaces. *Procedia computer science* 112: 1044–1053.

Yanto, D. T. P., Sukardi, D. P., & Puyada, D. 2017. Effectiveness of interactive instructional media on electrical circuits course: the effects on students cognitive abilities. *Proceedings of 4rd International Conference On Technical And Vocational Education And Training*: 75–80.

Borderless Education as a Challenge in the 5.0 Society – Abdullah, Adriany & Abdullah (eds)
© 2021 Taylor & Francis Group, London, ISBN 978-0-367-61960-2

Professional certification training model for alternative solutions for S-1 graduate teachers who do not achieve teacher professional education

O. Komar
Universitas Pendidikan Indonesia, Bandung, Indonesia

ABSTRACT: The problem that underlies this article is this phenomenon: thousands of strata S-1 graduates of teacher study programs currently do not get teaching job placement (unemployed), because the regulations of teacher placement require graduates' strata S-1 of teacher professional education. In addition, research findings on the relationship between academic competence and work performance of graduate's strata S-1 of teacher study programs show a mismatch (Dikti in 2016 and 2017). The unemployment of the teachers is suspected to have a dominant effect on social problems such as commotion in the community that disrupts the smooth development of government programs. An alternative solution is a professional certification training model in the field of tourism expertise and the field of hotel expertise for S-1 strata graduates who do not get professional teacher education. The alternative certification is expected to be a solution to the issue as the those who take it will have their professional certification written on their graduation diploma/certificate.

1 TEACHER POSITION

When education is the main sector in a country's development, the teacher position is an instrument of change in the national education system for empowerment and improvement of human quality. Teachers working in the front facilitate the implementation of educational policies and educational programs.

However, the community sees that there are four perspectives of the teacher's role (Supriadi 1999):

- The administration and management perspective shows teachers in four aspects: recruitment, appointment, placement, and coaching,
- The perspective of the teaching profession shows that teachers have difficulty of defining what, who, and how the profession is matched between criteria and the reality of teaching.
- The perspective of government bureaucracy shows the teacher as an "accomplice" to the bureaucracy in the school who must follow the demands of the bureaucracy.
- The perspective of the education system/school system shows that teachers are the focus of education implementation and agents of education reform.

Therefore, the role of multi-perspective teachers shows that teacher certification is not linear with teaching competence. So, the teacher's role shows no significant effect on social change and improvement in human quality, whereas the position of teachers and education support staff plays an important role in education reform. Teachers are agents of change to improve the quality of education. The development of teachers and education support staff is a crucial issue in education reform, so efforts must be made to improve the quality standards of the teaching profession.

It is urgent for the state to play a role in the rights and obligations of providing teacher needs and teacher quality. Preparing for professional teacher education must include (1) preparations ranging from recruitment of prospective teachers based on talent, interests, and attitudes to graduates of professional teacher education strata through teacher education programs designed based on teacher competency standards; and (2) the ability of the didactic, methodical, and pedagogical fields, ranging from the preparation of academic competency standards to the implementation strategy in the field, to be of sufficient time.

From the 1980s until now, the dynamic changes in professional teacher education have continued, in that change (name, role, model, and form) always becomes an academic/scientific discussion for education experts. There are developments in two models of teacher professional education. First, the concurrent model sees teacher professional education carried simultaneously between mastery of teacher competence and teaching training with mastery of subject matter, so as to obtain a S-1 teacher diploma and professional teacher certificate. Second, the executive model, sees teacher professional education carried out sequentially between obtaining an S-1 in mastering subject matter and mastering teacher competency, so as to obtain a professional teacher certificate.

Law No. 14 of 2005 concerning Teachers and Lecturers, in the Article 8 dictum "Teachers are required to have academic qualifications, competencies, educator certificates, be healthy physically and spiritually, and have the ability to realize national education goals." Then Article 9 states that "Academic qualifications as referred to in Article 8 are obtained through higher education undergraduate programs or four diploma programs." Furthermore, in Article 12: "Everyone who has obtained an educator's certificate has the same opportunity to be appointed as a teacher in a particular education unit." The article indicates that the teaching profession is for S1graduate from LPTK and from Non-LPTK.

Law No. 14 of 2005 concerning Teachers and Lecturers tends to be professional teacher education with consecutive models. Consecutive models are teacher professional education models conducted by recruiting S1 graduates from mastery of subject matter and from S1 graduates of teachers who win the recruitment of professional teacher education that will be taken for 6–12 months held at LPTK, thus, obtaining professional teacher certification (Law No. 14 of 2005).

2 RECOGNITION OF TEACHER CERTIFICATION

Teacher professionalism is characterized by ownership of teacher certification mandated by Law No. 14 of 2005 concerning Teachers and Lecturers. The implementation of teacher certification refers to Law No. 19 of 2005 concerning National Education Standards. Article 28, Paragraph 1 teacher standards have academic qualifications and teacher competencies. Educators' academic qualification standards have a minimum academic level of S-1 or D4 diploma. Article 10 paragraph (1) reads: "Teacher competencies ... including pedagogical competencies, personality competencies, social competencies, and professional competencies obtained from professional teacher education" (Government Regulation No. 19 of 2005).

The implementation of teacher certification has two organizing patterns. First in 2006, for teachers in positions, there were three implementation schemes: (1) held through a portfolio pattern (prior learning), by showing evidence of teacher performance, (2) held through face-to-face patterns in class, with Teacher's Professional Education and Training, and (3) organized through the pattern of Teacher Education and Professional Training that is refined, with participants first being monitored for their needs assessment online, then by Teacher Professional Education and Training. Second, in the academic year 2018/19 teacher certification for preservice teachers was through implementation of the specified study program.

However, implementation of teacher certification in the LPTK is incomplete, both in the implementation of concurrent model teacher certification, and in the implementation of consecutive model teacher certification. In concurrent model teacher certification, the process is not complete in the S-1 graduates, and teachers only get a diploma and not get a professional teacher certificate.

In consecutive model teacher certification, the process is not completed in S-1 subject-matter-only graduates who obtain a diploma and cannot automatically continue to professional teacher education to obtain a professional teacher certificate. The implementation of teacher certification in teacher professional education, varies for participants who graduated as S-1 teachers (concurrent model) and as S1 subject matter (consecutive model).

The consequences of implementing teacher certification in the LPTK are incomplete, indicating that thousands of S-1 teachers have not succeeded in achieving teacher professional education. The reasons include: (a) strict recruitment requirements on teacher professional education, (b) limited teacher development, and (c) increased public interest in teacher professional education. Solutions are needed, including a model of professional certification training in tourism expertise and hotel expertise for graduate S1 teachers who do not achieve teacher professional education.

We describe the solution through the general question: "How does the implementation of professional certificate training in the field of tourism and hotel professional certification work for graduate S1 teachers who do not achieve teacher professional education?" General questions are elaborated into specific questions as follows.

- What is the plan for professional certificate training in tourism and hotel professional certification for graduate S1 teachers who do not achieve teacher professional education?
- What is the training recruitment rating level?
- What are the results of the analysis of the training implementation (design, curriculum, instructor, implementation, and evaluation)?
- What is the effectiveness of the training (learning, output, and outcome)?
- What is the opinion of the workplace stakeholders on training graduates?

3 RESULTS AND RECOMMENDATIONS

The training consists of three stages: pretraining (preparation before training), ongoing training (implementation of training) and posttraining (after training). The pretraining phase with training planning activities, consists of needs assessment through regular meetings discussing how the training will be carried out, recruitment of participants, future outcomes after training, training materials, facilities, and infrastructure.

Implementation of the training program in the form of delivery of material follows the learning plan and learning strategy. Supporting factors for implementation include: professional leadership, qualification of instructors, harmonious cooperation from all stakeholders, as well as cool and noise-free environmental conditions. The inhibiting factors for the implementation of training are library materials, equipment, and time.

Effectiveness of the training is measured by evaluating four levels: level of reaction, level of learning, level of behaviour, and level of results. Evaluation of reaction levels is done by measuring the reaction of trainees to the training conducted; level of learning by measuring the ability of trainees to understand the material provided; level of behaviour by measuring the change in behaviour that occurs in trainees; and level of results by measuring the absorption of training results.

The opinion of the training graduates' workplace stakeholders on the performance of training graduates, is that employees feel part of the company; have information in doing work; understand the company's vision, mission, culture, and values; and help employees socialize in the company.

Recommendations are in place for trainees to be disciplined during training activities.

REFERENCES

Supriadi, D. 1999. Mengangkat Citra dan Martabat Guru. Yogyakarta: Adicita Karya Nusa.
Government Regulation No. 19 of 2005, concerning National Education Standards.
Law No. 14 of 2005, concerning Teachers and Lecturers.

Borderless Education as a Challenge in the 5.0 Society – Abdullah, Adriany & Abdullah (eds)
© 2021 Taylor & Francis Group, London, ISBN 978-0-367-61960-2

The effectiveness of job sheets in studying road and bridge construction in grade XI of Building Information and Modelling Design (BIMD) in State Vocational High School 5 Bandung

D. Purwanto & E. Susanto
Universitas Pendidikan Indonesia

ABSTRACT: Job sheet learning media is a type of media that is used to mediate the learning process and the dissemination of information for students to obtain better cognitive, affective, and psychomotor aspects. In studying Construction of Roads and Bridges Grade XI BIMD of State Vocational School 5 Bandung, to facilitate the delivery of material in both theory and practice, it was necessary that there be a medium that can help the learning process for teachers and students in the practicum process. Job sheets were adopted as a learning medium. The objective of this study was to examine the effectiveness of utilizing job sheets in studying construction of roads and bridges in the school in question. The research method used was descriptive qualitative and the sample was 34 students of Grade XI BIMD V State Vocational School 5 Bandung. On the basis of the learning completion, the results indicate that job sheets were utilized effectively with a percentage of 79.4%.

1 INTRODUCTION

Vocational schools are secondary educational institutions whose objective is to prepare students to become a competent and independent workforce by honing their abilities and skills in particular fields so that they can later easily adapt themselves to the world of employment.

State Vocational High School (SMKN) 5 Bandung is a state-owned vocational secondary educational institution that offers the competency of Building Information and Modelling Design (BIMD) encompassing sciences of building construction drawings, building construction, land measurements, construction drawings using 2D and 3D computer applications, interior and exterior designs, construction of roads and bridges, cost calculation, construction reports, etc. The primary goal of this program is to cultivate students who are expected to professionally function as a drafter/designer in planning and execution of a construction.

One of the things the students learn in this program is road and bridge construction, whose goal is to enable students to design roads and bridges. To simplify their work of designing, job sheets were created in which students can follow the instructions contained in the job sheet and can consult with and receive assistance from the teacher.

Slamet (2005) remarks that job sheets are a printed type of teaching aid that supports teachers in teaching skills especially in workshops. The aid contains a set of visual and verbal directions on how to make or complete a job. Job sheets are used as a guide for students in studying and mastering competencies taught by educators (Sukardi 2010).

The components of job sheets include (1) the core material of the practice, which consists of steps of activities/processes that students must perform, practical guides on how to properly use the equipment, and a pretest before the practice; (2) a tool of evaluation used; and (3) work safety (Sarbiran 2009). According Sukardi (2010), there are two types of job sheets that are used in teaching practice namely production job and combination job (combining exercises and production jobs).

This research was conducted to gauge the level of effectiveness of job sheess in studying Road and Bridge Construction Grade XI BIMD at State Vocational School 5 Bandung viewed from the mastery of learning completion.

2 RESEARCH METHODS

The method used in this research was a descriptive research method using a qualitative approach, whose goal was to examine the effectiveness of the use of job sheets in Road and Bridge Construction Grade XI BIMD in State Vocational School 5 Bandung seen from student learning completion. The sample was 34 students of Grade XI BIMD V State Vocational School 5 Bandung.

2.1 *Data analysis technique*

2.1.1 *Testing effectiveness*

To discover the effectiveness level of the use of job sheets as seen from student learning completion, the following formula was used:

Notes:

P = Percentage of learning completion

n = Number of students who completed

N = Number of overall students

Means were identified from the analysed data and effectiveness levels were assessed as in Table 1:

Table 1. Table of effectiveness categories.

Percentages (%)	Categories
0 – 20	Ineffective
21 – 40	Less Effective
40 – 60	Somewhat Effective
61 – 80	Effective
81 – 100	Very Effective

The student learning completion can be stated as follows.

- Individual absorption, a student said to be complete when s/he obtains the results ≥ 75 out of a maximum score of 100.
- Classical absorption, a class is said to be complete if at least 75% of the students have obtained ≥ 75 (adjusted with the minimum completion criteria (MCC) of the school under examination).

3 RESEARCH RESULTS

This section presents the level of effectiveness of the utilization of job sheets seen from the student learning completion. The percentage of student learning completion in the cognitive aspect is as in Table 2:

From this result, it can be said that on the cognitive aspect, student learning completion reached 79.4%. Based on the criteria of effectiveness, the effectiveness level on this aspect can be categorized as effective. However, in terms of the criteria of student learning completion, the students cannot be said to be complete because some students have not fulfilled the MCC, but via the classical lens, student performance can be said to be complete because more than 75% of students have obtained ≥ 75 (MCC).

The percentage of student learning completion on the psychomotor is as follows.

$$P = \frac{n}{N} \; x \; 100\% = \frac{34}{34} \; x \; 100\% = 100\%$$

Table 2. Student scores.

No.	Name	Score	No.	Name	Score	No.	Name	Score
1	Respondent 01	80	13	Respondent 13	85	25	Respondent 25	70
2	Respondent 02	85	14	Respondent 14	90	26	Respondent 26	70
3	Respondent 03	75	15	Respondent 15	80	27	Respondent 27	90
4	Respondent 04	70	16	Respondent 16	75	28	Respondent 28	80
5	Respondent 05	80	17	Respondent 17	75	29	Respondent 29	80
6	Respondent 06	85	18	Respondent 18	70	30	Respondent 30	85
7	Respondent 07	70	19	Respondent 19	80	31	Respondent 31	85
8	Respondent 08	85	20	Respondent 20	80	32	Respondent 32	70
9	Respondent 09	90	21	Respondent 21	85	33	Respondent 33	80
10	Respondent 10	75	22	Respondent 22	70	34	Respondent 34	85
11	Respondent 11	80	23	Respondent 23	80			
12	Respondent 12	85	24	Respondent 24	85			

From these results, it can be said that on the psychomotor aspect, student learning completion has reached 100%. On the basis of the criteria of effectiveness, the results of this study on the psychomotor are in the category of very effective. Meanwhile, in terms of the criteria of learning completion, the student results can be considered complete both via the individual and classical absorption lens especially because on the psychomotor, the all students have already met the minimum requirement of MMC.

4 DISCUSSION

Job sheets are a learning medium that is suitable for use in the process of learning practice especially in vocational schools. From several previous research results, including the use of job sheets in the field of Automotive Engineering and field of Measure Soil Class X Engineering Drawing Buildings in Vocational High Schools, job sheets have produced satisfactory value and all students finish working on the task in time. In the job sheets there are steps that students should work on so that students can understand what they will do.

5 CONCLUSION

Based on the above results, the following conclusion can be drawn. The average scores of student performance on the cognitive aspect have fulfilled the minimum completion criteria. Although some students have not passed, on the whole, average student scores are above the MMC set by the school. Likewise, on the psychomotoric aspect, the student scores have met the MMC. The effectiveness level of job sheets in roads and bridge construction on the basis of the mean scores of student performance on the cognitive and psychomotoric aspects has met minimum completion criteria, that is, above the MMC. Thus, it can be said that the use of job sheets is considered effective in providing student knowledge and skills.

REFERENCES

Sarbiran. 2009. *Handout Untuk Mata Kuliah Praktik*. Yogyakarta: UII Yogyakarta.

Slamet, T. 2005. Teknik Pembuatan Job sheet. In Seminar dan Lokakarya Program Hibah Kompetensi A-1. *Jurusan Pendidikan Teknik Otomotif Fakultas Teknik Universitas Negeri Makasar*: 28–30.

Sukardi, T. 2010. Penerapan Work Preparation dan Intensitas Pendampingan pada Capaian Prestasi Praktik Pemesinan Mahasiswa Jurusan Mesin Fakultas Teknik Universitas Negeri Yogyakarta. *Jurnal Cakrawala Pendidikan* 2(2).

Borderless Education as a Challenge in the 5.0 Society – Abdullah, Adriany & Abdullah (eds)
© 2021 Taylor & Francis Group, London, ISBN 978-0-367-61960-2

Author Index